Editors' addresses:

Dr. B. Olivier
Dr. J. Mos
CNS-Pharmacology
Duphar B.V.
P.O. Box 900
1380 DA Weesp
The Netherlands

Prof. Dr. J. L. Slangen
Department of Psychopharmacology
Faculty of Pharmacy
University of Utrecht
Sorbonnelaan 16
3584 CA Utrecht
The Netherlands

Deutsche Bibliothek Cataloging-in-Publication Data

Animal models in psychopharmacology / ed. by B. Olivier . . . –
Basel ; Boston ; Berlin : Birkhäuser, 1991
(Advances in pharmacological sciences)
ISBN 3-7643-2503-8 (Basel . . .) Gb.
ISBN 0-8176-2503-8 (Boston) Gb.
NE: Olivier, Berend [Hrsg.]

Printed from the editors' camera-ready manuscripts on acid-free paper in Germany

ISBN 3-7643-2503-8
ISBN 0-8176-2503-8

Preface

As part of the XIth International Congress of Pharmacology, a post-congress Satellite Meeting entitled ANIMAL MODELS IN PSYCHOPHARMACOLOGY was held on July 7 and 8, 1990 in Amsterdam. The organizers (B.Olivier, J.Mos and J.L.Slangen) thought it a very convenient time to deal with various new developments in the field of drug treatment and pharmacological prevention of abnormal human behaviour.

The emergence of better classifications of psychiatric disorders, e.g. in anxiety, has created realistic opportunities for developing new psychoactive agents for separate diagnostic categories such as panic disorders or obsessive-compulsive disorder. It appeared that drugs with new mechanisms of action were active in a therapeutic area, like in anxiety, where the existing and effective treatment was dominated by the use of benzodiazepines. Such findings led to an explosion of preclinical research aiming for detecting new drugs, with new mechanisms of action. A fascinating example is the serotonergic field, where the serendipitous finding of the anxiolytic activities of buspirone steered research into the role of 5-HT_{1A} agonists in anxiety, which has been very successful.

However, these new drugs appeared not easily detectable with the classical animal models in use for the particular area. In the case of the anxiolytics, animal models of anxiety appeared biased towards detecting benzodiazepines. Consequently, new animal models had to be developed which are sensitive towards other mechanisms involved in anxiety, like serotonin or NMDA.

The central aim of the present volume is the description and explanation of the developments in the various fields exploiting animal models for psychiatric disorders. The emphasis of this research focusses on three methodological disciplines, viz. pharmacological, psychological and ethological models. Because of the broadness of the fields, we restricted ourselves to animal models of anxiety, depression, psychosis, aggression and sexual behaviour. Moreover, drug-discrimination was chosen as a topic of interest because this technique appears of value for every therapeutic area.

We invited for every discipline leading experts in their respective fields. For anxiety, Sanger (pharmacological models), Barrett (psychological models) and Insel (ethological models) cover the new developments in this field. These main contributions are complemented with several interesting papers from invited poster contributions. For depression Porsolt (pharmacological models), Willner (psychological models) and Panksepp (ethological models) set the scene, complemented by interesting poster contributions and one invited commentary (Archer). For psychosis Costall (pharmacological models) and Ellenbroek (ethological models) present intriguing thoughts on animal models of psychosis. Aggression (Mos and Olivier) and sexual behaviour (Everitt) are covered by single chapters, primarily because the clinical fields are not yet so well exploited and understood as the preceding ones. Preclinically, however, recent developments promise exciting avenues for treatment of human disorders (e.g. the serenics). Drug-discrimination has emerged during the last decade as a very promising and helpful technique in the search for the in vivo mechanism of action of psychoactive drugs. The technique can be used for every discipline and directly complements in vitro studies on the mechanism of action of psychotropic drugs. In the miscellaneous section a very interesting chapter was contributed by Miczek on ultrasonic vocalizations and the possible application in animal models for various therapeutic areas.

We think that the broadness and scope of this volume describing animal models in psychopharmacology, will be of help for students of behaviour and pharmacology as well as for clinicians involved in drug studies in humans. It is hoped that the enormous gaps between the preclinical and clinical understanding of psychiatric diseases and animal analogues of those, can be narrowed by this contribution.

The editors wish to thank all the scientists for their excellent contributions. Many people have contributed to the success of this symposium and the present volume. We want to thank Duphar B.V. for their generous support, especially Dr.H.Roseboom and Dr.A. de Jonge.

This book would not have appeared without the excellent technical assistance of Marijke Mulder, who assisted in all the "invisible" work behind the production of this volume.

Berend Olivier
Jan Mos
CNS-pharmacology
Department of Pharmacology
Duphar B.V.
Weesp.

Jef Slangen
Laboratory of Psychopharmacology
Faculty of Pharmacy
State University of Utrecht
Utrecht

CONTRIBUTORS

Alberto, A.
Department of Physiology, Pharmacology and Biophysics, Institute of Biological Sciences, Federal University of Rio Grande do Sul, Porto Alegre, RS, 90.049, Brazil

Archer, T.
Department of Psychology, University of Göteborg, P.O.Box 14158, S-400 20 Göteborg, Sweden

Awouters, F.H.L.
Janssen Research Foundation, B-2340 Beerse, Belgium

Barrett, J.E.
Department of Psychiatry, Uniformed Services University of the Health Sciences, 4301 Jones Bridge Road, Bethesda, Maryland 20889, U.S.A.

Berendsen, H.H.G.
CNS-Pharmacology Department, Organon, P.O.Box 20, Oss, The Netherlands

Blanchard, D.C.
Békesy Laboratory of Neurobiology and John A. Burns School of Medicine, University of Hawaii, Honolulu, HI 96822, U.S.A.

Blanchard, R.J.
Department of Psychology, University of Hawaii, Honolulu, HI 96822, U.S.A.

Bohus, B.
Department of Animal Physiology, University of Groningen, Biological Centre, 9750 AA Haren, The Netherlands

Bouws, G.A.
Department of Animal Physiology, University of Groningen, Biological Centre, 9750 AA Haren, The Netherlands

Brand, T.
Department of Endocrinology and Reproduction, Faculty of Medicine and Health Sciences, Erasmus University, P.O.Box 1738, 3000 DR Rotterdam, The Netherlands

Brandao, M.L.
Department of Physiology, University of Wales College of Cardiff, Cardiff, CF1 1SS, United Kingdom

Broekkamp, C.L.
CNS-Pharmacology Department, Organon, P.O.Box 20, Oss, The Netherlands

Broersen, L.M.
Department of Psychopharmacology, Faculty of Pharmacy, University of Utrecht, Sorbonnelaan 16, 3584 CA Utrecht, The Netherlands

Bruinvels, J.
Department of Pharmacology, Workgroup Biological Psychiatry, Erasmus University Rotterdam, P.O.Box 1738, 3000 DR Rotterdam, The Netherlands

Casas, J.
Department of Neurosciences, School of Medicine, Cádiz, Spain

Christensen, A.V.
St. Hans Hospital Department P, DK-4000 Roskilde, Denmark

Colin, C.
Laboratoire de Biologie du Comportement, URA C.N.R.S. 1293, B.P. 239, 54506 Vandoeuvre les Nancy CEDEX, France

Commissaris, R.L.
Department of Pharmaceutical Sciences, College of Pharmacy, Wayne State University, Detroit, Michigan 48202, U.S.A.

Costall, B.
Postgraduate Studies in Pharmacology, The School of Pharmacy, University of Bradford, Bradford, BD7 1DP, United Kingdom

de Almeida, R.M.M.
Department of Physiology, Pharmacology and Biophysics, Institute of Biological Sciences, Federal University of Rio Grande do Sul, Porto Alegre, RS, 90.049, Brazil

de Boer, S.F.
Department of Psychopharmacology, Faculty of Pharmacy, University of Utrecht, Sorbonnelaan 16, 3584 CA Utrecht, The Netherlands

de Haes, P.L.A.J.
Janssen Research Foundation, B-2340 Beerse, Belgium

Desor, D.
Laboratoire de Biologie du Comportement, URA C.N.R.S. 1293, B.P. 239, 54506 Vandoeuvre les Nancy CEDEX, France

Domeney, A.M.
Postgraduate Studies in Pharmacology, The School of Pharmacy, University of Bradford, Bradford, BD7 1DP, United Kingdom

Dortmans, C.
CNS-Pharmacology Department, Organon, P.O.Box 20, Oss, The Netherlands

Ellenbroek, B.A.
Psychoneuropharmacological Research Unit, Department of Pharmacology, University of Nijmegen, P.O. Box 9101, 6500 HB Nijmegen, The Netherlands

Elorza, J.
Department of Neurosciences, School of Medicine, Cádiz, Spain

Enginar, N.
Department of Pharmacology, Istanbul Medical Faculty, University of Istanbul, Istanbul, Turkey

Eroglu, L.
Department of Pharmacology, Istanbul Medical Faculty, University of Istanbul, Istanbul, Turkey

Everitt, B.J.
Department of Anatomy, University of Cambridge, Downing Street, Cambridge CB2 3DY, United Kingdom

Filho, R.
Department of Physiology, Pharmacology and Biophysics, Institute of Biological Sciences, Federal University of Rio Grande do Sul, Porto Alegre, RS, 90.049, Brazil

Fontana, D.J.
Department of Pharmaceutical Sciences, College of Pharmacy, Wayne State University, Detroit, Michigan 48202, U.S.A.

Gamble, E.E.
Department of Psychiatry, Uniformed Services University of the Health Sciences, 4301 Jones Bridge Road, Bethesda, Maryland 20889-4799, U.S.A.

Geoffroy, M.
St. Hans Hospital Department P, DK-4000 Roskilde, Denmark

Gibert-Rahola, J.
Departments of Neurosciences and Toxicology, Neuropsycho-pharmacology Unit, University of Cádiz, Spain

Gleeson, S.
Department of Psychiatry, Uniformed Services University of the Health Sciences, 4301 Jones Bridge Road, Bethesda, Maryland 20889-4799, U.S.A.

Gómez-Cama, M.C.
Department of Neurosciences, School of Medicine, Cádiz, Spain

Ikemoto, S.
Department of Psychology, Bowling Green State University, Bowling Green, Ohio 43403, U.S.A.

Insel, T.R.
Laboratory of Clinical Science, National Institute of Mental Health, P.O. Box 289, Poolesville, MD 20837, U.S.A.

Jenck, F.
CNS-Pharmacology Department, Organon, P.O.Box 20, Oss, The Netherlands

Joly, D.
Synthélabo Recherche (L.E.R.S), 31 ave P.V. Couturier, 92220 - Bagneux, France

Kelly, M.E.
Postgraduate Studies in Pharmacology, The School of Pharmacy, University of Bradford, Bradford, BD7 1DP, United Kingdom

Ketelaars, C.E.J.
Polikliniek Kinder- en Jeugdpsychiatrie A.Z.G., P.O.Box 30001, 9700 RB Groningen, The Netherlands

Koene, P.
Department of Comparative and Physiological Psychology, Psychological Laboratory, University of Nijmegen, 6525 HR Nijmegen, The Netherlands

Koolhaas, J.M.
Department of Animal Physiology, University of Groningen, Biological Centre, 9750 AA Haren, The Netherlands

Korte, S.M.
Department of Animal Physiology, University of Groningen, Biological Centre, 9750 AA Haren, The Netherlands

Krafft, B.
Laboratoire de Biologie du Comportement, URA C.N.R.S. 1293, B.P. 239, 54506 Vandoeuvre les Nancy CEDEX, France

Kroonen, J.
Department of Endocrinology and Reproduction, Faculty of Medicine and Health Sciences, Erasmus University, P.O.Box 1738, 3000 DR Rotterdam, The Netherlands

Krijzer, F.
CNS-Pharmacology, Duphar B.V., P.O. Box 900, 1380 DA Weesp, The Netherlands

Lenègre, A.
I.T.E.M.-Labo, 93 Avenue de Fontainebleau, 94276 Le Kremlin-Bicêtre Cedex, France

Leonsegui, I.
Departments of Neurosciences and Toxicology, Neuropsycho-pharmacology Unit, University of Cádiz, Spain

Lopez-Garcia, J.A.
Department of Physiology, University of Wales College of Cardiff, Cardiff, CF1 1SS, United Kingdom

Lucion, A.B.
Department of Physiology, Pharmacology and Biophysics, Institute of Biological Sciences, Federal University of Rio Grande do Sul, Porto Alegre, RS, 90.049, Brazil

McArthur, R.A.
I.T.E.M.-Labo, 93 Avenue de Fontainebleau, 94276 Le Kremlin-Bicêtre Cedex, France

Meert, T.F.
Janssen Research Foundation, B-2340 Beerse, Belgium

Micó, J.A.
Departments of Neurosciences and Toxicology, Neuropsychopharmacology Unit, University of Cádiz, Spain

Miczek, K.A.
Department of Psychology, Research Building, Tufts University, 490 Boston Avenue, Medford MA 02155, U.S.A.

Molewijk, H.E.
CNS-Pharmacology, Duphar B.V., P.O.Box 900, 1380 DA Weesp, The Netherlands

Morêl, E.
Synthélabo Recherche (L.E.R.S), 31 ave P.V. Couturier, 92220 - Bagneux, France

Mos, J.
CNS-Pharmacology, Duphar B.V., P.O.Box 900, 1380 DA Weesp, The Netherlands

Muscat, R.
Department of Psychology, City of London Polytechnic, Old Castle St, London E1 7NT, United Kingdom

Nader, M.A.
Department of Psychiatry, Uniformed Services University of the Health Sciences, 4301 Jones Bridge Road, Bethesda, Maryland 20889-4799, U.S.A.

Naylor, R.J.
Postgraduate Studies in Pharmacology, The School of Pharmacy, University of Bradford, Bradford, BD7 1DP, United Kingdom

Nelson, E.
Department of Psychology, Bowling Green State University, Bowling Green, Ohio 43403, U.S.A.

Olivier, B.
CNS-Pharmacology, Duphar B.V., P.O.Box 900, 1380 DA Weesp, The Netherlands

Panksepp, J.
Department of Psychology, Bowling Green State University, Bowling Green, Ohio 43403, U.S.A.

Perrault, G.
Synthélabo Recherche (L.E.R.S), 31 ave P.V. Couturier, 92220 - Bagneux, France

Porsolt, R.D.
I.T.E.M.-Labo, 93 Avenue de Fontainebleau, 94276 Le Kremlin-Bicêtre Cedex, France

Roberts, M.H.T.
Department of Physiology, University of Wales College of Cardiff, Cardiff, CF1 1SS, United Kingdom

Rodgers, R.J.
Department of Psychology, University of Leeds, Leeds LS2 9JT, United Kingdom

Rijnders, H.J.
Department of Psychopharmacology, Faculty of Pharmacy, University of Utrecht, Sorbonnelaan 16, 3584 CA, Utrecht, The Netherlands

Sanger, D.J.
Synthélabo Recherche (L.E.R.S), 31 ave P.V. Couturier, 92220 - Bagneux, France

Schmidt, W.J.
University of Tübingen, Department of Neuropharmacology, Mohlstrasse 54/1, D-7400 Tübingen, Germany

Sheehan, M.J.
Neuropharmacology Department, Glaxo Group Research Ltd, Ware, Herts, SG12 0DP, United Kingdom

Slangen, J.L.
Department of Psychopharmacology, Faculty of Pharmacy, University of Utrecht, Sorbonnelaan 16, 3584 CA Utrecht, The Netherlands

Slob, A.K.
Department of Endocrinology and Reproduction, Faculty of Medicine and Health Sciences, Erasmus University, P.O.Box 1738, 3000 DR Rotterdam, The Netherlands

Smit, J.
Department of Animal Physiology, University of Groningen, Biological Centre, 9750 AA Haren, The Netherlands

Tejedor del Real, P.
Departments of Neurosciences and Toxicology, Neuropsycho-pharmacology Unit, University of Cádiz, Spain

Thuillier, F.
Laboratoire de Biologie du Comportement, URA C.N.R.S. 1293, B.P. 239, 54506 Vandoeuvre les Nancy CEDEX, France

Tilders, F.J.H.
Department of Pharmacology, Medical Faculty, Free University, Van der Boechorststraat 7, 1081 BT Amsterdam, The Netherlands

Tolboom, J.
Department of Statistics, Duphar B.V., P.O.Box 900, 1380 DA Weesp, The Netherlands

Tornatzky, W.
Department of Psychology, Research Building, Tufts University, 490 Boston Avenue, Medford MA 02155, U.S.A.

van den Broek, I.
Department of Psychopharmacology, Faculty of Pharmacy, University of Utrecht, Sorbonnelaan 16, Utrecht

van der Gugten, J.
Department of Psychopharmacology, Faculty of Pharmacy, University of Utrecht, Sorbonnelaan 16, 3584 CA Utrecht, The Netherlands

van der Heyden, J.A.M.
CNS-Pharmacology, Duphar B.V., P.O.Box 900, 1380 DA Weesp, The Netherlands

van der Molen, R.
CNS-Pharmacology, Duphar B.V., P.O. Box 900, 1380 DA Weesp, The Netherlands

van der Poel, A.M.
CNS-Pharmacology, Duphar B.V., P.O.Box 900, 1380 DA Weesp, The Netherlands

van Dijken, H.H.
Department of Pharmacology, Medical Faculty, Free University, Van der Boechorststraat 7, 1081 BT Amsterdam, The Netherlands

van Hest, A.
CNS-Pharmacology, Duphar B.V., P.O.Box 900, 1380 DA Weesp, The Netherlands

Vermote, P.C.M.
Janssen Research Foundation, B-2340 Beerse, Belgium

Vivian, J.
Department of Psychology, Research Building, Tufts University, 490 Boston Avenue, Medford MA 02155, U.S.A.

Vollmer, F.
CNS-Pharmacology, Duphar B.V., P.O. Box 900, 1380 DA Weesp, The Netherlands

Vossen, J.H.M.
Department of Comparative and Physiological Psychology, Psychological Laboratory, University of Nijmegen, 6525 HR Nijmegen, The Netherlands

Willner, P.
Department of Psychology, City of London Polytechnic, Old Castle St, London E1 7NT, United Kingdom

Winslow, J.T.
Laboratory of Clinical Science, National Institute of Mental Health, P.O. Box 289, Poolesville, MD 20837, U.S.A.

Woudenberg, F.
Department of Psychopharmacology, Faculty of Pharmacy, University of Utrecht, Sorbonnelaan 16, 3584 CA Utrecht, The Netherlands

Yates, G.
Department of Psychology, Bowling Green State University, Bowling Green, Ohio 43403, U.S.A.

Ybema, C.E.
Department of Psychopharmacology, Faculty of Pharmacy, University of Utrecht, Sorbonnelaan 16, Utrecht

Zethof, T.J.J.
CNS-Pharmacology, Duphar B.V., P.O.Box 900, 1380 DA Weesp, The Netherlands

Zivkovic, B.
Synthélabo Recherche (L.E.R.S), 31 ave P.V. Couturier, 92220 - Bagneux, France

CONTENTS

PART 6: **MISCELLANEOUS**

part 1:

ANXIETY

Animal Models in Psychopharmacology
Advances in Pharmacological Sciences

ANIMAL MODELS AND RECENT DEVELOPMENTS IN THE SEARCH FOR NOVEL ANXIOLYTICS

D.J. Sanger, G. Perrault, E. Morel, D. Joly and B. Zivkovic

Synthélabo Recherche (L.E.R.S), 31 ave P.V. Couturier, 92220 - Bagneux, France

SUMMARY: There have been a number of important recent developments in the search for novel drugs for treating anxiety disorders, which have implications for the development and validation of animal models. It has been proposed that anxiolytic activity may be shown by compounds acting at serotonin receptors ($5HT_{1A}$, $5HT_2$, $5HT_3$) and at excitatory amino acid receptors, as well as by ω (BZ) receptor ligands which differ structurally from benzodiazepines. The behavioural effects of such compounds, however, often show differences from those of the benzodiazepines and barbiturates. Future psychopharmacological research will need to develop and refine animal models in the light of new information from the clinic.

For over 20 years benzodiazepines have been the most widely used drugs for the treatment of anxiety. There can be little doubt that these agents have been very effective and this fact, together with their relative lack of important side effects, has led to the enormous success of benzodiazepines and the almost complete exclusion from clinical use of other anti-anxiety agents. However, there has been increasing concern about the disadvantages associated with benzodiazepine therapy. These include the occurrence of physical dependence under some circumstances (for review see Woods et al., 1987) and the fact that benzodiazepines have a complex pharmacological profile which, in addition to anxiolytic activity, includes anticonvulsant, muscle-relaxant, amnestic and sedative effects. Although all these activities have clinical utility they are often considered both unnecessary and undesirable in the treatment of pathological anxiety. There has thus been considerable recent research activity aimed at the development of novel, more selectively anxiolytic drugs. In particular, there is hope that anti-anxiety agents might be discovered without, or with greatly reduced, sedative, muscle-relaxant and amnestic activities.

The psychopharmacology of benzodiazepines and other drugs with similar profiles (barbiturates, meprobamate) has been well researched. These drugs have many behavioural effects in experimental animals and a variety of procedures have been proposed as animal models of anxiety, useful in screening and developing new anxiolytics (Sanger, 1990a). Relatively few of these models, however, have been subject to rigorous psychological and pharmacological validation (Lister, 1990; Treit, 1985).

The most widely used procedures have involved the suppression of ongoing conditioned or unconditioned behaviour by contingent aversive stimuli (punishment or conflict tests) and the analysis of reactions to novel stimuli or environments. In the latter case it is often assumed that exploratory behaviour is reduced by anxiety provoked by novelty and consequently facilitation of exploration would be expected when anxiety is reduced. Such procedures have seemed to have some face validity and certainly have produced tests which predict clinical anxiolytic activity with benzodiazepines (Lister, 1990 ; Pollard and Howard, 1990). However, as described below, they often prove to be insensitive to putative anxiolytics acting through other mechanisms.

NOVEL AGENTS ACTING AT ω (BZ) RECEPTORS

Since the reports that benzodiazepines bind to high affinity sites within the central nervous system (Mohler and Okada, 1977; Squires and Braestrup, 1977) much research has been aimed at the discovery of structurally novel compounds active at the same sites. Table 1 shows a list of several classes of chemical agents with affinity for these sites (now called ω receptors - Langer and Arbilla, 1988) with example compounds described in the literature as having anxiolytic or potentially anxiolytic activity. More thorough review of this category of compounds has been provided by Gardner (1988).

Among this list of potential anxiolytics, which is certainly not exhaustive, bretazenil, abecarnil, alpidem and suriclone are perhaps most extensively investigated in preclinical and clinical studies (Martin et al., 1988; Stephens et al., 1990; Zivkovic et al., 1990; Julou et al., 1985). As might be expected, these compounds have many effects in common with more traditional benzodiazepines. All have anticonvulsant activities, for example, and, as shown in Figure 1, they all increase drinking suppressed by electric shock punishment in rats. However, these compounds also differ in a number of ways from benzodiazepines. Even in other tests of punished responding they are not consistently active. Thus, while bretazenil and suriclone have been reported to increase rates of operant responding suppressed by punishment, alpidem and abecarnil are not active in such procedures. Similarly, although bretazenil appears to be devoid of behavioural depressant effects abecarnil, alpidem and suriclone do reduce both conditioned and unconditioned behaviour at high doses although there may be very wide separations between anxiolytic and depressant doses. Furthermore, in studies of the discriminative stimulus properties of these drugs both alpidem and abecarnil show differences from benzodiazepines and other ω-receptor ligands (Andrews and Stephens, 1989; Zivkovic et al., 1990).

Table 1 List of Novel Compounds With Affinity for ω Receptors and Potential Anxiolytic Activity

Chemical Series	Example Compound
Benzodiazepines derivatives	Bretazenil (Ro 16-6028)
β-Carbolines	Abecarnil (ZK 112-119)
Imidazopyridines	Alpidem
Cyclopyrrolones	Suriclone
Triazolopyridazines	CL 218,872
Pyrazoloquinolinones	ICI 190,622
Isoindolinones	DN-2327
Imidazoquinazolines	RU 32514

Such differences between the behavioural profiles of different ω receptor ligands give rise to two questions: which laboratory tests provide the best predictors of clinical anxiolytic activity, and thus the best animal models of anxiety for this type of compound and what differences in mechanisms of action produce these different profiles? No answer can yet be given to the first question as there are relatively few studies which have systematically investigated a range of ω receptor ligands in a battery of behavioural tests and, in any event, further information about the clinical activity of the drugs will be necessary to validate the animal models. To deal with the second question, mechanistic studies have put forward two hypotheses to explain such novel pharmacological profiles; these compounds may have partial agonist activity at ω-receptors or may have selective affinity for ω receptor subtypes.

In the case of bretanezil there is some clear evidence for a partial agonist profile. Thus, in addition to showing only some of the pharmacological effects associated with benzodiazepines, bretazenil also has some antagonist actions and requires a relatively high level of receptor occupancy to produce pharmacological effects (Martin et al., 1988; Potier et al., 1988; Sanger, 1987). The profiles of alpidem and abecarnil seem more complex. Stephens and colleagues (1990) provide little evidence for a partial agonist or agonist/antagonist profile for abecarnil although this compound did appear to produce some antagonism of the loss of righting reflex produced by diazepam. Similarly, alpidem has not been shown to exert antagonist activity although it has been argued that it has a relatively low intrinsic activity at ω receptors (Zivkovic et al., 1990). Alpidem has, however, been reported to be highly selective for the ω_1 and ω_3 receptor subtypes (Langer et al., 1990). These authors have proposed that this particular profile may be responsible for the unusual anxiolytic profile of alpidem, a hypothesis which certainly deserves further investigation.

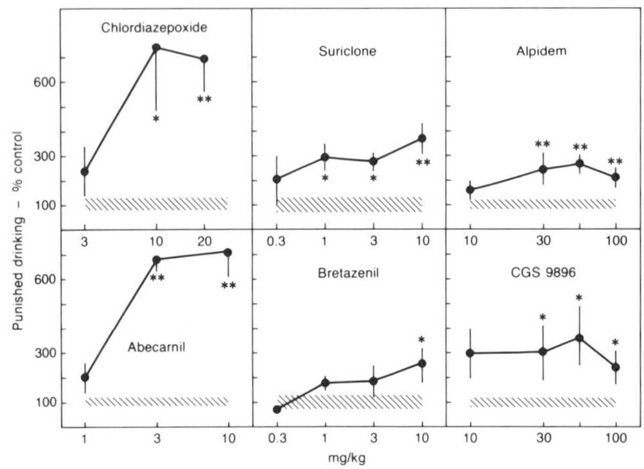

Fig. 1 Effects of several ω receptor ligands on punished drinking in rats. For further details of the method used see Sanger et al. (1985). $*p<0.05$; $**p<0.01$ difference from controls.

SEROTONIN (5-HT) RECEPTORS

The hypothesis that brain serotonergic mechanisms are important in anxiety and in the mechanisms of action of anxiolytic drugs including benzodiazepines (Gardner, 1986; Iversen, 1984) has been given emphasis by a number of recent findings. There is now considerable evidence for the existence of a variety of serotonin receptors ($5\text{-}HT_{1A}$, $5\text{-}HT_{1B}$, $5\text{-}HT_{1C}$, $5\text{-}HT_{1D}$, $5\text{-}HT_2$, $5\text{-}HT_3$, $5\text{-}HT_4$) and compounds with selective affinity for these different sites show different pharmacological profiles (Glennon, 1990). There are suggestions that drugs acting at 5-HT_{1A}, 5-HT_2 and 5-HT_3 receptor sites may exert anxiolytic or anxiolytic-like activity.

5-HT_{1A} receptors: The view that agonist or partial agonist activity at 5-HT_{1A} sites produces anxiolytic effects arose from the findings that the novel anxiolytic drug, buspirone, has high affinity for these sites (Glaser and Traber, 1983). Buspirone also affects other receptors, in particular dopamine receptors, but a number of buspirone analogues have been described, including gepirone and ipsapirone, which have greater selectivity for 5-HT_{1A} sites. These compounds have similar (though not identical) pharmacological profiles to that of buspirone and have also been reported to have anxiolytic activity in clinical trials. It is possible, therefore, that activity at the 5-HT_{1A} receptor is responsible for anxiolytic effects although this has not been firmly established.

Buspirone and related compounds have pharmacological effects in the laboratory very different from those associated with benzodiazepines. In animal models involving punishment or novelty there have been reports of benzodiazepine-like effects although most researchers find little or no activity. For example, most systematic studies of punished operant responding in rodents or monkeys find only very small increases in response rates after buspirone, ipsapirone and gepirone (Howard and Pollard, 1990; Witkin and Perez, 1990; Weissman et al., 1984). In marked contrast, these compounds produce large, dose-related increases in punished responding in pigeons (Brocco et al., 1990; Gleeson et al., 1989) an effect which appears to involve serotonin (Witkin et al., 1987). In tests of exploratory behaviour in rats, such as the staircase test or elevated plus maze, buspirone is also generally inactive (Moser, 1989; Sanger, 1990a). A large number of apparently contradictory findings have been reported in this area, however.

It is clearly a matter for concern if buspirone and benzodiazepines show different behavioural effects in the laboratory as a number of clinical trials have reported that buspirone has efficacy similar to that of anxiolytic benzodiazepines such as diazepam in the treatment of generalised anxiety disorder (Goa and Ward, 1986). However, there are a small number of behavioural procedures in which these compounds have similar activity and which may, therefore, provide a more satisfactory prediction of clinical anxiolytic activity. These procedures include the potentiated startle response (Kehne et al., 1988) and a conditioned emotional response (CER-Sanger, 1990b).

In the latter study it was found that the operant responding of rats suppressed by a stimulus which preceded an unavoidable footshock could be increased in rate by chlordiazepoxide, clorazepate, buspirone and ipsapirone but not by a variety of other drugs. As shown in Figure 2 the results were particularly impressive when considered in terms of suppression ratios. Although this result suggests that the CER technique may provide a more sensitive method for evaluating buspirone-like anxiolytics, it is not at all clear which aspect of the procedure gives rise to this apparent sensitivity. In the same study buspirone produced no statistically significant increase in rates of operant responding suppressed by contingent footshock (i.e. punishment). The critical differences between the CER and punishment procedures are presently unknown.

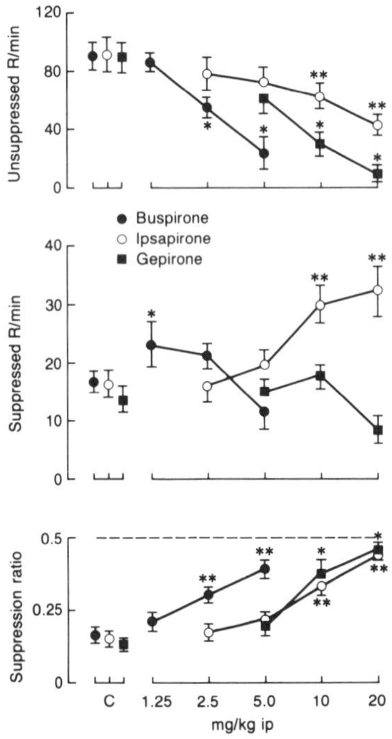

Fig. 2 Effects of buspirone, ipsapirone and gepirone on the responding of rats maintained by a CER procedure. The figure shows response rates during a pre-shock stimulus, response rates when the stimulus was not present and a suppression ratio between these two rates. This ratio could take values between 0, indicating complete response suppression during the stimulus, and 0.5, indicating that response rates were not changed during the stimulus $*p<0.05$; $**p<0.01$ difference from control values. From Sanger (1990).

5-HT$_2$ receptors: A number of early experiments reported that compounds known to have 5-HT antagonist effects could increase rates of punished operant responding in rats or pigeons (Graeff and Schoenfeld, 1970; Graeff, 1974). Although most of the compounds used in such studies are now known to have affinity for more than one 5-HT receptor subtype, similar effects have been reported for agents known to be relatively selective antagonists at 5-HT$_2$ receptors. Ritanserin is of particular interest as its pharmacology has been investigated in detail (Awouters et al., 1988) and it has been reported to have anxiolytic activity in patients suffering from generalised anxiety disorder (Ceulemans et al., 1985). Ritanserin increases rates of punished responding in pigeons although its effect seems smaller and less consistent that that produced by benzodiazepines (Brocco et al., 1990; Gleeson et al., 1989). In rats ritanserin and other

5-HT_2 antagonists can increase rates of punished drinking (Table 2) although they do not invariably do so and ritanserin has not been found to increase rates of operant responding suppressed by punishment in this species (Brocco et al., 1990; Kilts et al., 1982; Meert and Janssen, 1989). Other animal models in rats, such as the elevated plus maze, have detected activity with ritanserin although effects tend to be small and variable (Colpaert et al., 1985 ; Meert and Janssen, 1989).

Table 2 Effects of several 5-HT_2 Antagonists and Diazepam on Punished Drinking in Rats.

Treatment	Dose mg/kg, ip	N° of Shocks Received in 3 min test
Control	-	4.3 ± 0.9
Diazepam	3.0	28.2 ± 5.6**
Ritanserin	0.3	7.4 ± 2.5
Ritanserin	1.0	13.4 ± 5.7
Ritanserin	3.0	22.0 ± 6.2*
Pirenperone	0.03	6.0 ± 1.8
Pirenperone	0.1	14.2 ± 4.2*
Pirenperone	0.3	18.0 ± 2.7**
Cyproheptadine	1.0	15.1 ± 4.3
Cyproheptadine	3.0	35.1 ± 6.7**
Cyproheptadine	10.0	31.1 ± 3.6**

Values are the means ± S.E.M. for groups of at least 8 rats. *$p<0.05$; **$p<0.01$ difference for controls. For further methodological details see Sanger et al. (1985).

5-HT_3 receptors: Great interest has been caused recently by the description of a number of compounds with high affinity and selectively for 5-HT_3 receptors. Compounds such as ondansetron and zacopride have demonstrated anti-emetic activity and, it has been claimed, have a variety of potentially important psychopharmacological actions. Both of these drugs are currently being investigated in clinical trials for anxiolytic activity.

5-HT_3 antagonists have been reported to show no activity in tests of punished drinking in rats and punished operant responding in pigeons (Jones et al., 1988 ; Gleeson et al., 1989). However, they apparently do show activity over surprising wide dose-ranges in a test of exploration in a two compartment box in mice and in the social interaction model in rats. They also exert a calming effect on marmosets and cynomolgus monkeys without inducing sedation and muscular incoordination (Costall et al., 1988; Jones et al., 1988).

As these compounds have been available for only a relatively short period of time it has not yet been possible for such results to be systematically replicated and extended in many laboratories although problems of replicability have already become apparent (File and Johnston, 1989). However, because of the potential clinical importance of these compounds much work is currently underway and it is likely that the pharmacological profiles of 5-HT_3 antagonists will soon be defined in detail.

EXCITATORY AMINO ACID RECEPTORS

Excitatory amino acid receptors exist in a variety of subtypes and have been subject to very extensive investigation in recent years. A number of compounds have been described with relatively selective activity as antagonists at these receptors, and particularly at the NMDA (N-methyl-d-aspartate) subtype. Most pharmacological research with these compounds has been directed at their clinical potential in the treatment of epilepsy or of cerebrovascular or neurodegenerative disorders, but it has also been suggested that NMDA anatagonists may have anxiolytic activity (Meldrum, 1985; Olney, 1989).

In 1982, Clineschmidt and co-workers reported that MK-801, a compound with anticonvulsant activity, produced anxiolytic-like effects in tests of punished drinking and CER in rats although it did not produce an antipunishment effect in squirrel monkeys. It was subsequently reported that MK-801 is a very potent and selective non-competitive NMDA antagonist which acts by blocking the ion channel associated with the receptor complex (Wong et al., 1986). Other experiments have confirmed that MK-801 and other non-competitive NMDA antagonists with similar mechanisms of action, such as phencyclidine and ketamine, can increase punished responding in rats although effects are often small and restricted to very limited dose ranges (Porter et al., 1989; Sanger and Jackson, 1989). For example, Figure 3 shows that MK-801 produced small and variable increases in punished responding in rats whereas phencyclidine did not have a statistically significant effect. There was also no effect of ifenprodil, an anti-ischaemic drug which also acts as an NMDA antagonist probably through the polyamine binding site (Carter et al., 1988, 1989).

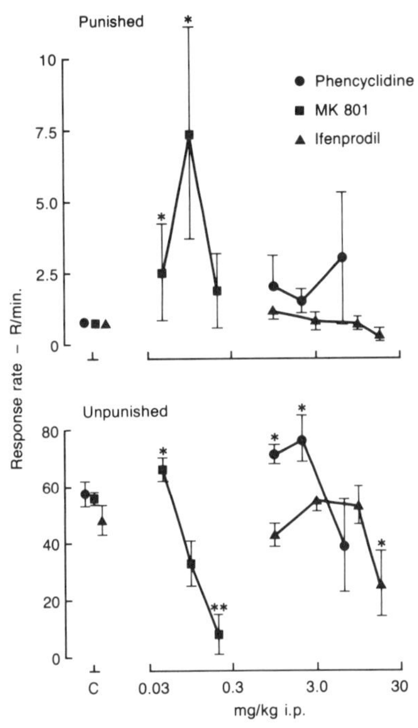

Fig. 3 Effects of MK-801, phencyclidine and ifenprodil on rates of punished and unpunished operant responding in rats. $*p<0.05$; $**p<0.01$ difference from control values. From Sanger and Jackson (1989).

It seems unlikely that compounds such as phencyclidine would have clinical utility as anxiolytics as phencyclidine is known to produce anaesthesia and psychotomimetic effects in humans. However, it is possible that other NMDA antagonists with a different site of action might be anxiolytic without having these undesirable actions. As noted above, ifenprodil was not active in increasing punished operant responding in rats, suggesting that activity at the polyamine site is probably not associated with anxiolytic effects. However, several publication have reported that competitive NMDA antagonists including CPP and AP7 show some activity in tests of punished behaviour and exploration (Bennett and Amrick, 1986 ; Dunn et al., 1989; Stephens et al., 1986). A recent report also describes activity in the elevated plus-maze test for compounds with affinity for the glycine site associated with the NMDA receptor (Trullas et al., 1989).

CONCLUSIONS

This brief review indicates that several avenues are currently being explored in the search for novel anxiolytic drugs. A variety of interesting new compounds have been described recently with a number of mechanisms of action. In most cases the claims for the anxiolytic potential of these compounds have been based on results obtained in behavioural tests in the animal laboratory. However, it is becoming clear that when the overall psychopharmacological profiles of compounds acting at 5-HT_{1A}, 5-HT_2, 5-HT_3, NMDA and even ω receptors are compared with those of traditional anxiolytics many differences become apparent. It remains to be established whether models such as punished responding and exploratory behaviour will provide the best predictors of clinical anxiolytic activity or whether these tests will need to be refined or new ones developed to deal with this new generation of anti-anxiety drugs. It will certainly be necessary for pre-clinical researchers to pay careful attention to the results of clinical research and to develop their experimental tools in the light of new information.

REFERENCES

Andrews, J.S., Stephens, D.H. (1989) Soc. Neurosci. Abs. 15, 633.
Awouters, F., Niemegeers, C.J.E., Megens, A.A.H.P., Meert, T.F., Janssen, P.A.J. (1988) Drug Develop. Res. 15, 61-73.
Bennett, D.A., Amrick, C.L. (1986) Life Sci. 39, 2455-2461.
Brocco, M.J., Koek, W., Degryse, A.D., Colpaert, F.C. (1990) Behav. Pharmacol. 1, 403-418.
Carter, C.J., Benavides, J., Legendre, P., Vincent, J.D., Noel, F., Thuret, F., Lloyd, K.G., Arbilla, S., Zivkovic, B., Mackenzie, E.T., Scatton, B., Langer, S.Z. (1988) J. Pharmacol. Exp. Ther. 247: 1222 - 1232.
Carter, C.J., Rivy, J.P., Scatton, B. (1989) Eur. J. Pharmacol. 146, 611-612.
Ceulemans, D.L.S., Hoppenbrouwers, J.A., Gelders, Y.G., Reyntjens, A.J.M. (1985) Pharmacopsychiat. 18, 303-305.
Clineschmidt, B.V., Williams, M., Witoslawski, J.J., Bunting, P.R., Risley, E.A., Totaro, J.A. (1982) Drug Dev. Res. 2, 147-163.
Colpaert, F.C., Meert, T.F., Niemegeers, C.J.E., Janssen, P.A.J. (1985) Psychopharmacology 86, 45-54.
Costall, B., Domeney, A.M., Gerrard, P.A., Kelly, M.E., Naylor, R.J. (1988) J. Pharm. Pharmacol. 40, 302-305.
Dunn, R.W., Corbett, R., Fielding, S. (1989) Europ. J. Pharmacol. 169, 1-10.
File, S.E., Johnston, A.L. (1989) Psychopharmacology 99, 248-251.
Gardner, C.R. (1986) Pharmac. Biochem. Behav. 24, 1479-1485.
Gardner, C.R. (1988) Drug Dev. Res. 12, 1-28.
Glaser, T., Traber, J. (1983) Eur. J. Pharmacol. 88, 137-138.
Gleeson, S., Ahlers, S.T., Mansbach, R.S., Foust, J.M., Barrett, J.E. (1989) J. Pharmacol. Exp. Ther. 250, 809-817.
Glennon, R.A. (1990) Neurosci. Biobehav. Rev. 14, 35-47.

Goa, K.L., Ward, A. (1986) Drugs 32, 114-129.
Graeff, F.G. (1974) J. Pharmacol. Exp. 189, 344-350.
Graeff, F.G., Schoenfeld, R.I. (1970) J. Pharmacol. Exp. Ther. 173, 277-283.
Howard, J.L., Pollard, G.T. (1990) Drug Dev. Res. 19, 37-49.
Iversen, S.D. (1984) Neuropharmacology 23: 1553-1560.
Jones, B.J., Costall, B., Domeney, A.M., Kelly, M.E., Naylor, R.J., Oakley, N.R., Tyers, M.B. (1988) Brit. J. Pharmacol. 93, 985-993.
Julou, L., Blanchard, J.C., Dreyfus, J.F. (1985) Pharmacol. Biochem. Behav. 23, 653-659.
Kehne, J.H., Cassella, J.V., Davis, M. (1988) Psychopharmacology 94, 8-13.
Kilts, C.D., Commissaris, R.L., Cordon, J.J., Rech, R.H. (1982) 78, 156-164.
Langer, S.Z., Arbilla, S. (1988) Fund. Clin. Pharmacol. 2, 159-170.
Langer, S.Z., Arbilla, S., Tan, S., Lloyd, K.G., George, P. Allen, J., Wick, A.E. (1990) Pharmacopsychiat. 23, 103-107. Suppl.
Lister, R.G. (1990) Pharmac. Ther. 46, 321-340.
Martin, J.R., Pieri, L., Bonetti, E.P., Schaffner, R.,Burkard, W.P., Cumin, R., Haefely, W.E. (1988) Pharmacopsychiat. 21, 360-362.
Meert, T.F., Janssen, P.A.J. (1989) Drug Dev. Res. 18, 119-144.
Meldrum, B. (1985) Clin. Sci. 68, 113-123.
Mohler, H., Okada, T. (1977) Science 198, 849-851.
Moser, P.C. (1989) Psychopharmacology 99, 48-53.
Olney, J.W. (1989) Biol. Psychiatry 26, 505-525.
Pollard, G.T., Howard, J.L. (1990) Pharmac. Ther. 45, 403-424.
Porter, J.H., Wiley, J.L., Balster, R.L. (1989) J. Pharmacol. Exp. Ther. 248, 997-1002.
Potier, M.C., Prado de Carvalho, L., Venault, P., Chapouthier, G., and Rossier, J. (1988) Europ. J. Pharmacol. 156, 169-172.
Sanger, D.J. (1987) Psychopharmacology 93, 365-368.
Sanger, D.J. (1990a) In: Neuromethods, vol. 20 (A.A.Boulton, G.B.Baker and M.T.Martin-Iverson, Eds), Humana/Clifton N.J. (In press).
Sanger, D.J. (1990b) J. Pharmacol. Exp. Ther. 254, 420-426.
Sanger, D.J., Jackson, A. (1989) J. Pharmacol. Exp. Ther. 248: 1215-1221.
Sanger, D.J., Joly, D., and Zivkovic, B. (1985) J. Pharmacol. Exp. Ther. 232, 831-837.
Squires, R.F and Braestrup, C. (1977) Nature 266, 732-734.
Stephens, D.N., Meldrum, B.S., Weidmann, R., Schneider, C., Grutzner, M. (1986) Psychopharmacology 90, 166-169.
Stephens, D.N., Schneider, H.H., Kehr, W., Andrews, J.S., Rettig, K.J., Turski, L., Schmiechen, R., Turner, J.D., Jensen, L.H., Petersen, E.N., Honore, T., Bondon Hansen, J. (1990) J. Pharmacol. Exp. Ther. 253, 334-343.
Treit, D. (1985) Neurosci. Biobehav. Rev. 9, 203-222.
Trullas, R., Jackson, B., Skolnick, P. (1989) Pharmacol. Biochem. Behav. 34, 313-316.
Weissman, B.A, Barrett, J.E., Brady, L.S., Witkin, J.M., Mendelson, W.B., Paul, S.M., Skolnick, P. (1984) Drug Devel. Res. 4, 83-93.
Witkin, J.M., Perez, L.A. (1990) Behav. Pharmacol. 1, 247-254.
Witkin, J.M., Mansbach, R.S., Barrett, J.E., Bolger, G.T., Skolnick, P., Weissman, B. (1987) J. Pharmacol. Exp. Ther. 243, 970-977.
Wong, E.H.F., Kemp, J.A., Priestley, T., Knight, A.R., Woodruff, G.N., Iversen, L.L. (1986). Proc. Natl. Acad. Sci. U.S.A. 83, 7104-7108.
Woods, J.H., Katz, J.L., Winger, G. (1987) Pharmacol. Rev. 39, 251-419.
Zivkovic, B., Morel, E., Joly, D., Perrault, G., Sanger, D.J., Lloyd, K.G. (1990) Pharmacopsychiat. 23, 108-113, Suppl.

Animal Models in Psychopharmacology
Advances in Pharmacological Sciences

RAT PUP ULTRASONIC VOCALIZATIONS: AN ETHOLOGICALLY RELEVANT BEHAVIOUR RESPONSIVE TO ANXIOLYTICS

Thomas R. Insel, M.D., James T. Winslow, Ph.D.

Laboratory of Clinical Science, National Institute of Mental Health, P.O.Box 289, Poolesville, MD 20837, U.S.A.

SUMMARY: Rat pups emit ultrasonic calls during brief periods of social isolation. These calls are reduced by most clinically anxiolytic compounds and increased by several different classes of anxiogenics. In addition to detecting anxiolytic efficacy, this behaviour may be a useful tool for studying selective aspects of the neurobiology of anxiety. It should be noted however that pharmacologic results with rat pup studies depend on both the ontogeny of the neurochemical system involved and the maturity of various factors affecting pharmacokinetics and pharmacodynamics.

The search for animal models of psychopathology may be, by its very nature, ill-fated. Although virtually every other branch of medicine can point to convincing and even essential animal preparations relevant to the pathophysiology of a human ailment, the psychiatrist is at a distinct disadvantage. Mental illnesses are without demonstrable neuropathology, include psychological symptoms often without behavioural signs, and cannot be entirely defined by their response to a specific treatment. The situation is somewhat more hopeful if one seeks to model a specific affect rather than a psychiatric syndrome. Anger, contentment, sadness, and fear may arguably have their counterparts in other mammals. In addition, anxiety, which has an intimate relation to fear, may arise in other mammals under conditions which reliably induce the corresponding human affect.

The search for naturalistic or ethologically-relevant behavioural paradigms of anxiety has included studies in rodents and non-human primates. Others have reported on the study of defensive burying (Treit, 1985), intruder responses (Miczek and Winslow, 1987; Poshivalov, 1987), social interaction (File, 1980), and exploratory behaviour

(Crawley, 1985) as useful models for the study of anxiolytic drugs. Each of these paradigms has specific advantages for pharmacologic testing and each yields somewhat different profiles of anxiolytic efficacy. In the current review, we describe a relatively new paradigm for the study of anxiolytics: the rat pup ultrasonic vocalization (USV). This simple test has been shown to detect a wide range of anxiolytics and may also provide a useful model for investigating the ontogeny and the physiology of anxiety.

RAT PUPS AND THEIR ULTRASONIC VOCALIZATIONS

The USVs of infant rats are monotonic, whistle-like sounds with a frequency range between 35 and 45 kHz. As these calls are associated with social isolation, they have been variously described as "distress vocalizations" or "isolation calls," placing them in the context of the high frequency "cries" which are emitted by infants across the mammalian order. Like these other "cries," rat pup USVs appear to be laryngeal in origin (Roberts, 1975) and are extremely potent stimuli for maternal retrieval (Allin and Banks, 1972; Elwood, 1979; Smotherman et al., 1974) and maternal prolactin release (Terkel et al., 1979; Voloschin, 1984; although also see Stern et al., 1984). Their high frequency conforms to a region of high auditory sensitivity for adult rats (Noirot, 1966). In addition, the biophysical properties of these sounds may make them difficult to localize, possibly providing protection from a predator while alerting the mother when she is near the nest.

Although pups begin to emit ultrasonic vocalizations almost immediately after removal from the nest, the precise stimuli that elicit these calls remain unclear. Calls are evident within the first day of postnatal life, increase in number and intensity in the second week and decrease abruptly around day 14, about the time of eye opening (Fig. 1a). Within these first fourteen days of postnatal life, the physiologic responses to environmental conditions change rapidly in the pup. For instance, pups are poikilothermic right after birth and do not begin to develop homeothermy until day 5 or 6. Possibly for this reason, a decrease in ambient temperature has been noted to stimulate ultrasonic calls most noticeably near the end of the first week (Allin and Banks, 1972; Okon, 1971). The power of this effect can be seen in Fig. 1b, where data from a group of 6-day-old pups recorded at various temperatures demonstrate a near absence of calls at 36° C, the temperature within a huddled litter. One interpretation of this finding is that "isolation" for a 6-day-old rat pup, with immature visual and auditory systems, may be perceived primarily by the change in temperature. Consistent with this idea, the rate of USV decreases across the first 10 minutes of isolation (Fig. 1c). Between 10 and 14 days of age, thermal regulation has matured considerably and yet ultrasonic calls can still be predictably elicited by isolating pups. Hofer and Shair (1987) have demonstrated that the isolation calls from pups at this stage are reduced by exposure to littermates as well as by exposure to the mother. Analytic studies of the sensory inputs from social contact indicate that texture, contour, warmth, and odor have additive effects for reducing isolation calls; no single component appears to be prepotent (Hofer and Shair, 1987). Furthermore, previous social experience appears to be unnecessary, as isolated pups raised without littermates respond to the addition of pups with a robust decrease in calling rate (Hofer and Shair, 1987). In addition, rat pups raised with artificial mothers, as well as being isolated from other pups, emit normal appearing ultrasonic vocalizations.

Although the importance of hunger as a stimulus for pup ultrasonic calls has never been documented, recent studies have investigated a role for gustatory input. Blass and co-workers (1987) have recently demonstrated that intraoral infusions of 7.5% sucrose induced a rapid suppression of calls. As this effect was not apparent with intragastric infusions of sucrose or with intraoral infusions of water, it appeared that the tasting of sucrose was essential. Moreover, the opiate antagonist naltrexone prevented this sucrose effect on vocalization, suggesting that an opiate mechanism might mediate the neural and behavioural response to this particular gustatory stimulus.

DETECTION TECHNIQUES

There are three features of USV that can be readily measured. Most investigators count the number of calls using a high frequency microphone connected to a simple ultrasonic signal detection device, such as a bat detector, which transduces the sounds into the audible range. This provides a record of the occurrence, but says nothing about the two other features of these calls: the amplitude and the frequency. For these latter features, a sonographic analysis is essential. Calls need to be recorded and then analyzed individually. Several computer-based systems have been developed for analyzing rat pup USV (Burkholder et al., 1982; Mos and Olivier, 1989; van der Poel et al., 1989). These additional measures have proven useful for certain pharmacologic studies (Insel et al., 1986) as well as for detecting temporal changes in the character of USV across a burst of calling (van der Poel et al., 1989).

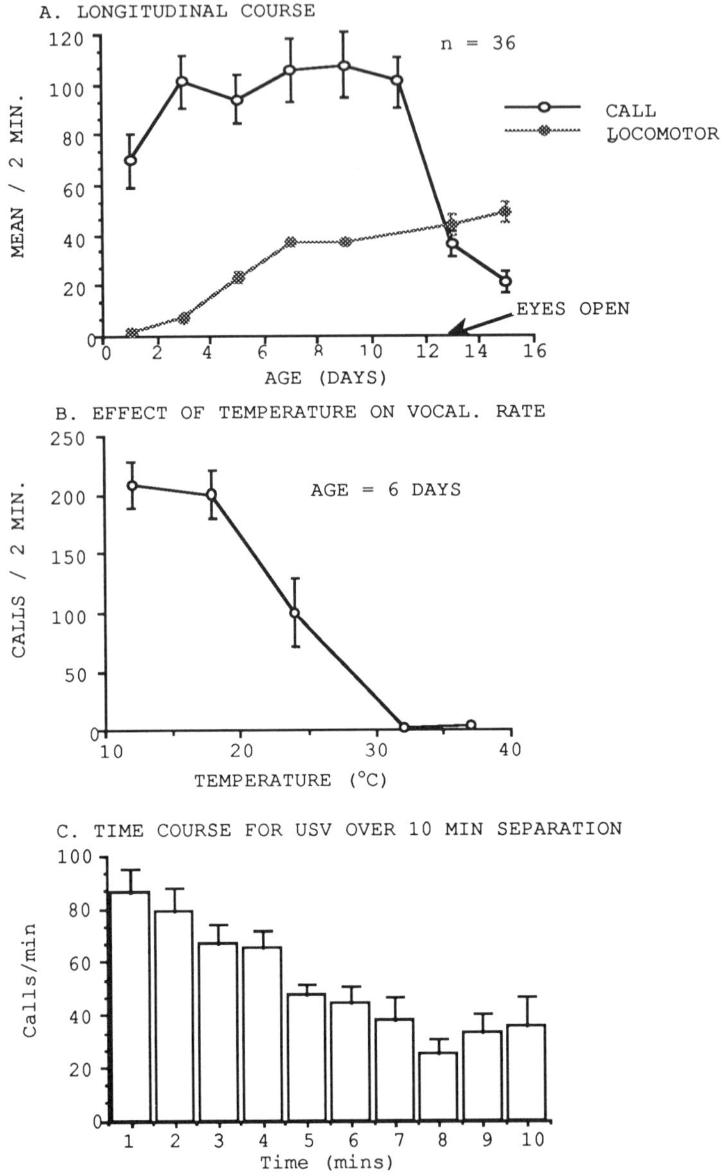

Fig. 1 Various influences on rat pup USV include (A) age when tested (locomotor refers to number of blocks crossed on floor grid of test chamber), (B) ambient temperature (measured in test chamber), and (C) duration of isolation (from littermates).

NON-PHARMACOLOGIC FACTORS AFFECTING USV

In the rat pup, the production of USV is an attractive dependent measure for behavioural studies not only for its ease of assessment, but because it is such a reliable and robust phenomenon. In most laboratories, rat pups from 6 to 12 days of age will regularly emit at least 50 calls per minute. There does not appear to be a gender difference in the rate of USV across the first 14 days (Insel et al., 1988). Previous experience with isolation also does not appear to alter the number of USV, suggesting that pups can be used repeatedly. Indeed, pups separated on a daily basis continue to produce USV along the same developmental course as those tested only once.

There is increasing recognition however, that other non-vocal variables need to be monitored concurrently to determine if pharmacologic effects are specific for vocal behaviour. For instance, changes in temperature (both ambient and body temperature), locomotion, and coordination (geotaxis) are all potential confounds in studies of USV. One additional confounding variable, respiratory rate, is generally overlooked in studies of USV. Rat pups emit USV during the expiratory phase, so that alterations in respiratory rate should predictably affect the number if not the character of USV. Future studies will need to address this factor in determining if a change in USV rate is truly specific.

In addition to these physiologic factors affecting USV production, most investigators have noted a striking variability in the number of USVs across litters. The source of this variability is not clear, but genetic differences may be responsible because robust differences have been noted between inbred strains of rodents. For instance, as shown in Fig. 2, pups from Maudsley reactive rats, a strain which has been isolated for "emotional reactivity" (Broadhurst, 1975), emit five times as many USV as pups from the Maudsley non-reactive congener (Insel and Hill, 1987). Similarly, pups of the prairie vole, Microtus ochrogaster, emit 50-100 USV/min. whereas pups of the closely related montane vole, Microtus montanus (Shapiro and Insel, in press) give few if any calls and do not show a glucocorticoid response to social isolation. As these two species of vole are dichotomous for their affiliative behaviour in adulthood (prairie voles are monogomous, montane voles are relatively solitary), the difference in USV may provide a developmental index of species-typical patterns of adult social attachment (Shapiro et al., 1986).

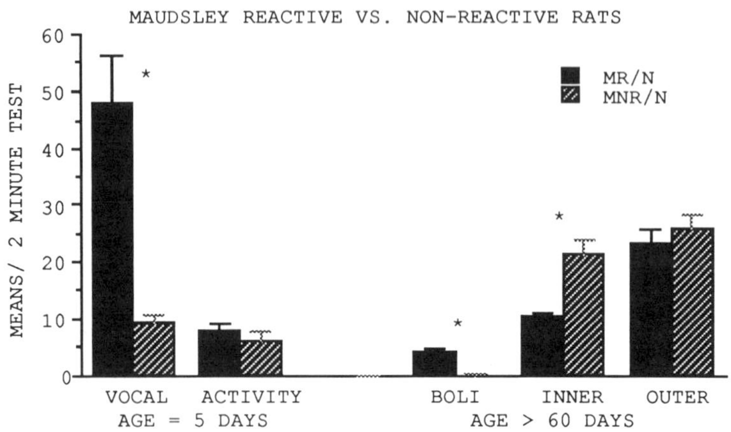

Fig. 2 Adult Maudsley reactive rats show increased defecation and decreased exploration in an open field test (* refers to $p<0.05$). At age 5 days the strains cannot be differentiated by differences in activity, but reactive pups emit approximately 5 fold more USV. Modified from Insel and Hill, 1987.

PHARMACOLOGIC STUDIES OF RAT PUP USV

Attempts to alter rat pup USV pharmacologically can be classified into three general groups: drugs affecting channel-linked, monoaminergic, and peptidergic systems. These studies have been undertaken for various reasons, from developing an animal model of clinical separation anxiety to demonstrating functional efficacy of a system in development, but as a group they provide a useful profile of the pharmacologic sensitivity of this specific behaviour.

CHANNEL-LINKED SYSTEMS

Benzodiazepines: The benzodiazepines are the gold standard for describing anxiolytic efficacy. Although these compounds are not entirely selective for clinical anxiety (their behavioural profile includes sedative, anticonvulsant, and muscle relaxant effects), they are remarkably useful for pharmacologic study because of the array of compounds that bind to the benzodiazepine site and the various tools now available to assess the physiologic state of this receptor and the GABA-A receptor and the Cl^-

channel to which it is coupled. Gardner (1985) was the first to observe that two prototype benzodiazepine agonists, chlordiazepoxide and diazepam, decreased the number of USV in isolated rat pups. These effects were specific in that effective doses for reducing vocalization did not alter locomotor activity. In a subsequent study, shown in Fig. 3, we corroborated the rate reducing effects of diazepam, demonstrating that doses as low as 0.5 mg/kg nearly eliminated USV without decreasing locomotor activity (Insel et al., 1986). Diazepam not only decreased the number of calls, it decreased both the peak frequency (from 38.8± 0.6 to 36.7± 0.5 kHz) and the mean power of calls. These effects appeared to be mediated via the benzodiazepine receptor as a 5.0 mg/kg dose of the benzodiazepine antagonist Ro 15-1788 entirely reversed the effects of diazepam (0.5 mg/kg). As further evidence for the relevance of USV to the pharmacologic study of anxiety, the anxiogenic compounds FG-7142 (25 mg/kg) and pentylenetetrazol (20 and 40 mg/kg) increased the number of calls (Insel et al., 1986).

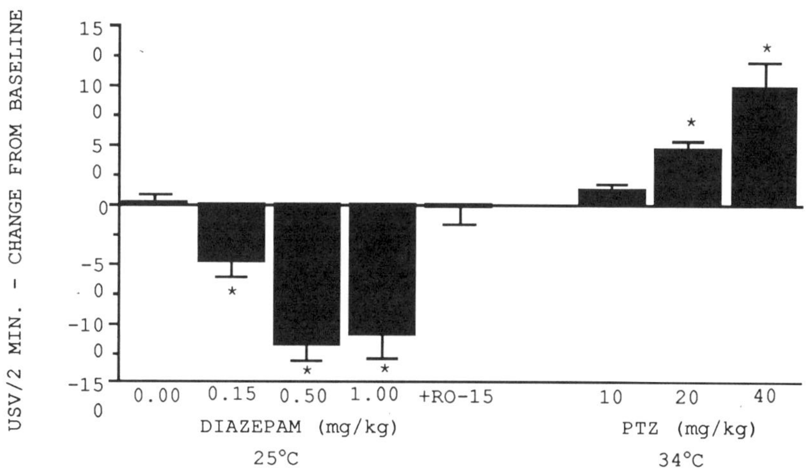

Fig. 3 Diazepam decreases and pentylenetetrazol increases the number of USV in 10 day old rat pups during the first 2 minutes of social isolation (* refers to $p<0.05$). Data shown represent means for individual differences from baseline tests 30 minutes earlier. Ro 15-1788 (5 mg/kg), the benzodiazepine receptor antagonist, reverses the effects of diazepam (0.5 mg/kg). Modified from Insel et al., 1986.

One intriguing aspect of the benzodiazepine studies of USV has been the suggestion that the benzodiazepine-GABA-Cl⁻channel might be involved in the physiologic mediation of these calls. The first such evidence was the observation that Ro 15-1788, at doses that generally lacked intrinsic activity, reduced the rate of USV (Insel et al.,

1986). Although Ro 15-1788 may function as a partial agonist under several circumstances, one interpretation of its effects on rat pup USV was that an endogenous ligand released during social isolation was prevented from acting in the presence of Ro 15-1788, hence the apparent reduction in calls. Further evidence along these lines came from subsequent experiments using ^{3}H-Ro 15-1788 for in vivo receptor binding studies in socially isolated pups (Insel et al., 1989). In these studies, 25 minutes of social isolation was associated with a regionally specific reduction in receptor binding when 3H-Ro 15-1788 was given in vivo but not when the receptor binding experiment was performed in vitro (Fig. 4). That is, there was not a change in receptor number following social isolation, but an apparent change in receptor availability during the interval of separation. This change in availability, representing about a 40% decrease in binding in vivo, was restricted to neocortex, hippocampus, and the superior and inferior colliculi. Concurrent behavioural studies demonstrated that the pups given tracer amounts of ^{3}H-Ro 15-1788 were indeed vocalizing during the period of separation.

These binding data along with the pharmacologic responses noted above might implicate an endogenous ligand for the benzodiazepine receptor in the physiologic response to social isolation. Administration of ^{14}C-2-deoxyglucose instead of ^{3}H-Ro 15-1788 to investigate regional cerebral metabolic changes during an identical social separation paradigm also found about a 40% decrease in tracer accumulation in the neocortex and hippocampus (Fig. 4). While the concordance in metabolic and receptor binding data would appear to confirm the importance of the benzodiazepine receptor complex in the mediation of this behaviour, several problems remain in the interpretation of these results. First, we have been unable to preclude a change in regional blood flow as a factor in the in vivo receptor and ^{14}C-2-deoxyglucose studies. As partial evidence against blood flow as a factor, in vivo binding of another label, ^{3}H-diprenorphine, does not change in cortex and hippocampus during social isolation. But two other concerns remain. First, ^{35}S-TBPS binding, a probe for the functional state of the Cl^- channel, does not increase during social isolation as one would predict if the benzodiazepine-GABA receptor complex were activated (author's unpublished data). This probe is a sensitive indicator of stress in adulthood, but, for unclear reasons, appears insensitive to a variety of stressors in rat pups prior to weaning. In addition, lesions of neocortex using either neurotoxins or electrolytic techniques, have failed to alter the production of USV with social isolation (Cagiano et al., 1987; author's unpublished data). It is not clear, therefore, that the binding changes with benzodiazepine receptor ligands are of functional importance. Whatever the physiologic

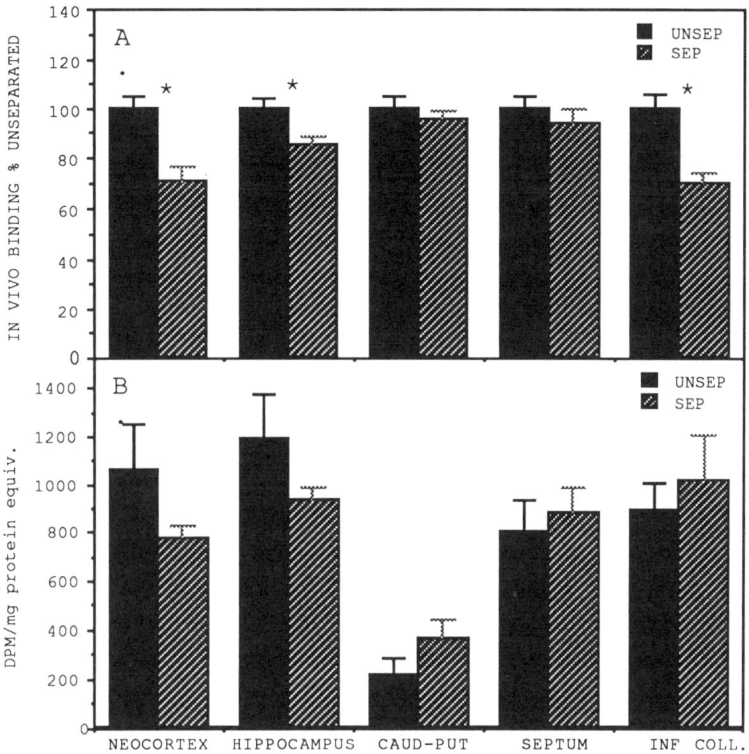

Fig. 4 During 25 minutes of social isolation, 10 day old rat pups show (A) reduced benzodiazepine receptor availability as manifested by decreased concentrations of ^{3}H-Ro 15-1788 binding to the cortex, hippocampus, and inferior colliculus using in vivo receptor autoradiography. Following the same behavioural protocol, (B) decreases in regional cerebral metabolism were noted in neocortex and hippocampus using 2-deoxyglucose with semi-quantitative imaging.

role of this system, the pharmacologic data demonstrate that the rat pup USV is exquisitely sensitive to manipulation with anxiolytic and anxiogenic drugs.

Excitatory amino acids: The N-methyl-D-aspartate (NMDA) subtype of the glutamate receptor is one of several receptors that mediate excitatory amino acid synaptic transmission (Cotman and Iversen, 1987). The NMDA receptor is thought to be one constituent of a hetero-oligomer consisting of several interdependent domains that include a transmitter receptor, a cation (Ca^{++}) channel, and a strychnine insensitive glycine modulatory site (Foster and Fagg, 1987; Monaghan et al., 1989). Activation (opening of the NMDA-gated cation channel) appears to depend on the binding of both an appropriate excitatory amino acid (for example glutamate or NMDA) and glycine to their respective recognition sites on this supramolecular complex (Huettner, 1989; Kleckner and Dingledine, 1989).

Both competitive NMDA antagonists (Bennett and Amrick, 1986; Leibman and Bennett, 1988; Stephens and Andrews, 1988; Stephens et al., 1986) and use dependent channel blockers (Clineschmidt et al., 1982; Leibman and Bennett, 1988) exhibit anticonflict actions in animals models which are predictive of anxiolytic effects in man. In addition, a competitive antagonist (7-chlorokynurenic acid) (Kemp et al., 1988) and a partial agonist (1-aminocyclopropanecarboxylic acid, ACPC) (Marvizon et al., 1989) at strychnine-insensitive glycine receptors have been reported to exhibit anxiolytic actions in an elevated plus-maze (Trullas et al., 1989). These findings are consistent with the hypothesis that reducing activity at NMDA-coupled cation channels via either the NMDA receptor or the glycine modulatory site could result in an anxiolytic action (Leibman and Bennett, 1988; Stephens and Andrews, 1988). In general however, these compounds appear significantly less efficacious than the benzodiazepines (Trullas et al., 1989).

In contrast to the results with other models of anxiety, NMDA functional antagonists appear as effective as the benzodiazepines in reducing USV (Winslow et al., submitted). For instance, ACPC produced a dose-dependent reduction in isolation-induced distress calls at doses that did not affect locomotor activity (Table 1). This effect was long-lasting, with significant reductions in USV evident 4 hours post drug administration. Moreover, this effect was antagonized by pre-treatment with a dose of glycine which did not alter USV, but was sufficient to elevate hippocampal glycine levels by 85%. In addition, the glycine receptor antagonist 7-chlorokynurenic acid (Kemp et al., 1988) reduced the number of USV (Table 1). Receptor binding studies have demonstrated that ACPC is a high affinity ligand at strychnine-insensitive glycine receptors which, in terms of ^{3}H-MK-801 binding, acts as a partial agonist (Marvizon et al., 1989). Our behavioural results, demonstrating a reversal of ACPC effects with glycine and a reduction of USV with 7-chlorokynurenic acid strongly suggest that, in functional terms, ACPC reduces cation channel activation via the glycine receptor.

Like ACPC, the competitive NMDA antagonist AP-7 produced a dose-dependent reduction in USV without affecting either motor activity or geotaxis (Table 1). A similar reduction in USV was observed with CPP (Table 1). The effects of AP-7 on vocal behaviour were antagonized by a dose of NMDA which by itself did not alter USV, consistent with the hypothesis of NMDA receptor involvement. In addition, doses of glycine sufficient to antagonize the effect of ACPC did not block the effects of AP-7

(Table 1), indicating the NMDA and glycine receptors while functionally coupled, are independent binding sites in the pup brain.

The non-competitive NMDA antagonists ifenprodil and MK-801 both significantly reduced USV, but these effects were associated with a pronounced disturbance of both motor activity and geotaxis (Table 1). Since Anly one dose of each agent was employed, it is possible that lower doses of these compounds might distinguish a decrease in USV from changes in motor parameters. While ifenprodil has been proposed to act at a polyamine site in the NMDA receptor complex (Carter et al., 1990; Shoemaker et al., 1990) it appears also to affect sigma receptors. The inability to clearly resolve the putative anxiolytic properties of MK-801 from its effects on various aspects of motor performance has been previously reported. This confound may account for differences in the anxiolytic efficacy of MK-801 and related channel blockers across testing paradigms and species (Clineschmidt et al., 1982; Stephens et al., 1986). NMDA like the anxiogenic benzodiazepine receptor inverse agonists FG-7142 and ß-CCE increases the rate of ultrasonic calling by isolated rat pups at subconvulsant doses (Table 1). A similar parallelism between NMDA and benzodiazepine receptor inverse agonists has been observed in social interaction and plus-maze tests (Dunn et al., 1989) as well as drug discrimination paradigms (Woods et al., 1988). These findings, coupled with the ability of NMDA antagonists to effect behavioural changes similar to those of clinically effective anxiolytics indicates that glutamatergic pathways may be involved in the development and/or expression of anxiety.

In previous studies in adult rats, the low potencies of CPP, AP-7, and ACPC in vivo relative to their potencies in vitro has been attributed to the limited penetration of these charged molecules across the blood brain barrier (Loscher et al., 1988; Trullas et al., 1989). These compounds were more than one order of magnitude more potent in reducing rat pup vocalization than in effecting anticonvulsant or anticonflict actions in adult animals (Skolnick et al., 1989; Trullas et al., 1989) which may in part be attributable to the immature blood-brain barrier in the neonatal rodent (see for example Vorbrodt, 1990). Differences in pharmacokinetic factors between immature and adult animals may also contribute to the relatively long duration of action of ACPC in pups which has been reported to be rapidly cleared from the adult CNS (Rao et al., 1990) and possesses a short duration of action as an anticonvulsant in adult rats (Skolnick et al., 1989). However, developmental differences in potency are probably not due simply to increased penetration of the blood-brain barrier, as a similar increased sensitivity has been reported for the neurotoxic effects of centrally administered NMDA in infant rats

Table 1 Compounds affecting the cation channel modulate rat pup USV

			Vocalizations	Activity
			% Pre-treatment ±s.e.m	
Glycine Receptor Ligands	Dose	n=	Call/ 2min	Cells/ 2 min
vehicle	0.0	6	108.0 ± 7.4	96.8 ± 4.5
ACPC	12.5	6	75.6 ± 15.8*	90.8 ± 10.5
	50.0	9	40.6 ± 4.3*	113.9 ± 9.1
	100.0	5	40.7 ± 7.4*	107.7 ± 12.5
7-Cl-Kynurenic acid	25.0	5	52.6 ± 9.3*	57.8 ± 6.4*
Glycine	200.0	5	119.3 ± 17.8	106.7 ± 18.5
ACPC (50)+Glycine (200)		6	83.4 ± 11.0	127.5 ± 9.0
ACPC (50)+NMDA (1.0)		4	30.6 ± 9.6*	118.8 ± 3.0
NMDA Receptor Ligands				
vehicle	0.0	6	104.8 ± 9.4	101.0 ± 8.9
AP-7	7.0	5	38.4 ± 5.9*	
	10.0	6	34.1 ± 4.5*	74.9 ± 19.0
	30.0	4	23.0 ± 7.8*	65.8 ± 18.8
±CPP	2.5	7	36.4 ± 6.3*	71.2 ± 4.4
NMDA	1.0	9	83.9 ± 7.6	103.3 ± 11.8
	2.5	9	148.4 ± 11.2*	124.9 ± 15.4*
	5.0	9	127.9 ± 11.6*	124.8 ± 8.6*
AP-7 (10)+Gly (200)		4	32.7 ± 7.7*	106.3 ± 11.7
AP-7 (10)+NMDA (1.0)		7	70.9 ± 4.4	108.5 ± 9.2
Other sites coupled to the cation channel				
MK801	0.05	6	40.3 ± 9.8*	137.2 ± 17.0*
Ifenprodil	5.0	6	17.3 ± 3.4*	20.3 ± 6.5*

* Represents significant Dunnett's t comparison of drug effects with vehicle control ($p < 0.05$) in the presence of significant ANOVA.

(McDonald et al., 1988). Moreover, the rat brain is exquisitely sensitive to hypobaric ischemia, probably as a result of the increased sensitivity of the NMDA-glycine complex (Olney et al., 1989). The apparent hypersensitivity of immature rats to the effects of NMDA (and antagonists) may be attributed in part to developmental idiosyncrasies of glutamate and glycine receptors. It may be relevant that during the first two postnatal weeks, glutamate potently stimulates phosphoinositol hydrolysis in olfactory bulb, cerebellar and hippocampal slices- an effect virtually absent in the mature rat brain (Nicolletti et al., 1986; Palmer et al., 1990). In addition, binding to NMDA and glycine receptors in the pup is increased over adult levels and shows a unique anatomic distribution (Insel et al., 1990; McDonald and Johnston, 1990).

MONOAMINES

Serotonin: The most consistent clinical evidence for anxiolytic effects with monoaminergic drugs may come from studies with serotonergic compounds. Several 5-$HT1_A$ agonists have been demonstrated to be anxiolytic while drugs such as m-CPP, which in the rat may be 5-$HT1_B$ agonists, may be anxiogenic. Serotonergic modulation of USV's has been demonstrated in several pharmacological studies. Consistent with the clinical observations, administration of the $5HT_{1A}$ agonist drugs 8-OH-DPAT, buspirone, and ipsapirone selectively reduced vocal behaviour at doses which did not affect motor actvity or coordination. Administration of the $5HT_{1B}$ agonist TFMPP produced a biphasic dose-related change in the rate of USV: low doses (0.1, 0.3 mg/kg, sc) increased calling while a higher dose (3.0 mg/kg, sc) decreased vocalizations (Winslow and Insel, 1990) (Fig. 5). Although the $5HT_2$ receptor agonist DOI also reduced vocalizations this effect could not be dissociated from profound effects on motor activity at any dose tested (Mos and Olivier, 1989; Winslow and Insel, 1990).

In addition, serotonin reuptake inhibitors such as clomipramine, fluvoxamine, citalopram, paroxetine, and zimeldine selectively reduce the rate of ultrasonic calling (Mos and Olivier, 1989; Winslow and Insel, 1989). These effects were not reduced by co-administration of opioid or benzodiazepine receptor inhibitors. Serotonin reuptake inhibitors have been shown to selectively reduce the symptoms of a clinical anxiety syndrome, obsessive-compulsive disorder (Insel and Winslow, 1990).

Serotonin may also have a role in the physiology of the pup's vocalizations. Selective destruction of serotonergic systems with chemical neurotoxins significantly reduces calling by rat pups. Hard and coworkers (1982) produced a significant decrease in the rate of calling by pups treated with 5,7 DHT (25 μg/kg, ICV, postnatal day 1). Calling was measured repeatedly and remained low up to 15 days after injection, at which time ultrasonic vocal behaviour normally disappears. Motor development in 5,7-DHT treated pups was not significantly affected.

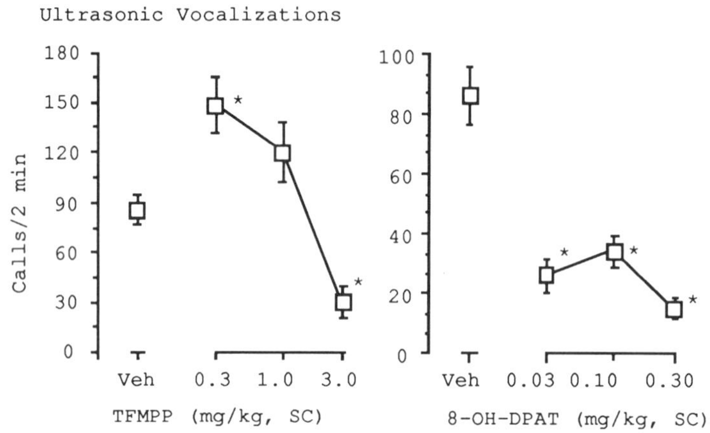

Fig. 5 $5\text{-}HT_1$ agonists have opposite effects on ultrasonic calling at low doses depending on the receptor subtype. TFMPP, the putative $5\text{-}HT_{1B}$ receptor agonist, increased vocalizations while the $5\text{-}HT_{1A}$ agonist 8-OH-DPAT decreased calling. Both drugs decreased calling at the highest dose tested. * refers to $p<0.05$ relative to littermate controls injected with saline.

Recently, we investigated the effects of the serotonin neurotoxin MDMA (also known as "Ecstacy"). Repeated administration of 10 mg/kg MDMA produced a long-lasting decrease in the rate of calling that was evident as late as 12 days after the last injection of MDMA (Table 2). The cortical density of serotonergic uptake sites and concentrations of 5-HT and 5-HIAA were also significantly reduced in these pups at 17-18 days after the last injection. The concentrations of other monoamines were unaffected by MDMA.

Administration of the drug during prenatal development (days 13-16 of gestation) did not induce the loss of serotonergic innervation nor did it confer any change in USV production. In a subsequent study we also evaluated the effects of MDMA induced depletion of serotonin on the changes in vocal behaviour produced by TFMPP and 8-OH-DPAT (Winslow and Insel, 1990). MDMA blocked the rate increasing effects of TFMPP consistent with the hypothesis that some subset of $5\text{-}HT_{1B}$ receptors are located presynaptically on the terminals. The USV rate decreasing effects of 8-OH-DPAT were not affected by MDMA, evidence that post-synaptic 5-HT_{1A} receptors were spared by MDMA.

Table 2 Long-term effects of MDMA

Drug (dose mg/kg)	Ultrasonic Vocal./ 2 min	Cell Crossing	Temp (°C)	Weight
Saline	112.12 ± 9.26	70.50 ± 4.11	34.39 ± 0.17	20.42 ± 0.54
MDMA (10)	68.33 ± 14.26*	74.22 ± 8.33	34.17 ± 0.33	18.94 ± 1.47*
MDMA (20)	23.77 ± 6.25*	66.18 ± 6.52	33.94 ± 0.34	17.68 ± 0.65*

* Represents significant Dunnett's t comparison of drug effects with vehicle control ($p<0.05$) in the presence of significant ANOVA. Drug or saline given daily on postnatal days 1-4. Testing on day 9.

Noradrenaline: The role of brain catecholamines in the mediation of clinical anxiety has been based largely on data from challenges of α_2-agonists, such as clonidine, and α_2-antagonists, such as yohimbine with non-human primates, as well as from brain-stimulation and lesion studies (Redmond, 1979). In addition, the beta receptor antagonist, propranolol, has been useful for the treatment of many of the somatic aspects of anxiety.

In general, rat pup responses to noradrenergic compounds are not correlated with clinical anxiolytic effects. One of the most potent compounds for increasing the number of USV is clonidine (Hård et al., 1988; Insel et al., 1988; Kehoe, 1988) which clinically appears sedating and anxiolytic. The mechanism for clonidine's effects on USV are not clear. These effects are antagonized by yohimbine and prazosin, suggesting mediation via the α_2-adrenergic receptor, but receptor binding studies suggest that the full complement of α_2-receptors does not mature until after 14 days of age. Evidently, even a few such receptors are sufficient for a behavioural response. The direction of the response remains paradoxical however, as clonidine in the immature rat increases locomotor behaviour even at doses in the low μg/kg range, whereas the same dosage of clonidine is sedating in the adult rat. This paradox is not explained by differential ontogeny of pre-synaptic α_2-receptors: neither 6-OHDA (which lesions the presynaptic sites), nor reserpine combined with AMPT (which deplete CA stores) alter the paradoxical effects of clonidine (Hård and Engel, 1988; Winslow and Insel, 1989). The explanation for clonidine's stimulatory effects may reside in the late development of a second messenger pathway or in regional differences in receptor development.

In contrast to the effects of serotonin uptake blockers noted above, acute administration of catecholamine uptake blockers, such as desipramine, nortriptyline and mazindol selectively increase the rate of USV. These effects on vocalization are not associated with consistent changes in locomotor activity or body temperature (Winslow and Insel, 1989). Furthermore, tolerance to the rate increasing effects of desipramine was not evident even after 12 days of twice daily administration of 5 mg/kg desipramine.

A role for NA in the physiologic mediation of USV is only partly supported by recent studies with 6-OHDA and DSP4. Lesioning of noradrenergic pathways on postnatal day 1 alters the rate of USV although this disruption is not immediately evident, and may depend on depletions of 80% or more. For example, NA depletions of 50-60% associated with injections of 6-OHDA (100μg/kg, ICV) or DSP4 (50 μg/kg, SC) had no significant effect on USV (Hård et al., 1982; Winslow and Insel, 1989). An 82%, selective depletion of NA associated with 6-OHDA (50-100 mg/kg/day, 4 days, SC) did produce a large decrease in the rate of calling by rat pups at 8 but not 2 days after the last injection (Winslow and Insel, 1989).

Dopamine: There are few studies of the role of dopamine in the mediation of ultrasonic vocalization. Repeated administration of the non-selective dopamine receptor antagonist haloperidol significantly reduced the rate of calling and increased the average call duration throughout a 14 day treatment period (Cagiano et al., 1986). The more selective D_1 antagonist SCH 23390, did not affect the rate of calling, but did increase the duration of calls and modified the average peak frequency (kHz) of calls. Sulpiride, a selective D_2 receptor antagonist did not significantly affect calling (Cuomo et al., 1987). Although these data suggest an excitatory role for dopamine on vocal behaviour, apomorphine, a dopamine receptor agonist has also been reported to reduce vocal behaviour (Hård et al., 1982). Thus, both stimulation and antagonism of dopamine appear to reduce vocal behaviour. Repeated prenatal administration of haloperidol also significantly disrupted the subsequent postnatal development of rat pup ultrasonic vocalizations resulting in a decreased rate of calling on day 4 (Cagiano et al., 1987, 1988). However it is not clear from these studies if this disruption may be accounted for by changes in the rate of development.

NEUROPEPTIDES

Neuropeptides have become an increasing focus for developmental studies because their receptors are generally present in prodigious quantities early in development and they can be readily administered intraventricularly to pups prior to day 7 when the skull is largely calcified. Interest in the role of the neuropeptides in the control of behaviour has focused primarily on the opioid peptides, however recent experiments have also demonstrated potential involvement of corticotropin-releasing hormone and oxytocin.

Opioids: Although opiates have been implicated in the mediation of the stress response, there is less evidence that opiates are clinically useful anxiolytics or that opiate antagonists increase anxiety. Opiates such as morphine decrease rat pup USV, but in our own studies this effect has only been observed at doses that also decrease locomotor activity (Winslow and Insel, in press). In addition, opiate antagonist such as naloxone or naltrexone have not increased the rate of calling, although others have described a blockade in the normal decrement in calling over a 5 minute separation test following naltrexone administration (Kehoe and Blass, 1986). Reducing the number of μ receptors by 40% with the alkylating agent β-funaltrexamine does not alter the number of calls with separation, but this "receptor lesion" does decrease the sensitivity to exogenous opiates (Winslow and Insel, in press). Finally, in vivo binding of the μ receptor antagonist ^{3}H-diprenorphine does not change during 25 minutes of social isolation associated with high rates of USV (Winslow and Insel, in press). Thus far, the available data do not implicate the μ receptor in the mediation of rat pup USV even though distress calls in other species such as chicks and guinea pigs appear more closely involved with brain opioids (Panksepp et al., 1985).

Corticotropin Releasing Factor: Corticotropin releasing factor (CRF) is a 41 amino acid neuropeptide with physiologic and behavioural effects related to stress. The physiologic effects of CRF include the release of corticotropin (ACTH) and β-endorphin from the pituitary and the stimulation of various autonomic centers within the CNS. Central administration of CRF to adult rats has been associated with behavioural activation and enhanced response to novelty (Britton et al., 1986; Dunn and File, 1987), and decreased appetitive behaviours (Parrott, 1990). Given this profile of physiologic and behavioural effects, one might predict that CRF would resemble the anxiogenic compounds in rat pups. Surprisingly, when injected into the lateral ventricle of 6 day old rat pups, CRF decreased rather than increased the number of USV (Insel and Harbaugh, 1989). This effect was evident at doses as low as 0.01 μg, lasted for at least 90 minutes, and was reversed by co-administration of the CRF receptor antagonist, alpha-helical CRF (1.0 μg). Indeed, the CRF antagonist which has not generally been associated with behavioural effects, actually decreased the number of USV (Insel and

Harbaugh, 1989). CRF, when injected into the lateral ventricle, did not alter body temperature or plasma corticosterone. By contrast, peripheral administration failed to affect vocal behaviour, but increased plasma corticosterone markedly.

The explanation for the increase in USV following central administration of CRF is not obvious. Extensive studies of the ontogeny of rat CRF receptors demonstrate early development of both receptor binding and second messenger responses in the brain (Insel et al., 1988). Whether the pathways subserved by CRF function differently in the immature brain remains unclear. We have considered one functional explanation for the paradoxical reduction of USV by CRF (Insel and Harbaugh, 1989). Following prolonged separations, pups normally decrease their rate of calling, in effect decreasing the possibility of discovery by a predator. It is possible, although certainly unproven, that a peptide associated with stress rather than anxiety, might shift pups into this quiet state prematurely.

SUMMARY

In summary, most of the pharmacologic data support the notion that rat pup USV may be a relatively simple and useful model for detecting anxiolytic efficacy. As can be seen in Table 3, across a wide range of neurochemical systems, anxiolytics decrease while anxiogenics increase the number of USV. In general, the results of the pup isolation test are at least as good as (and in the case of serotonergic compounds somewhat better than) social interaction, plus maze, and conflict paradigms for predicting clinical efficacy of anxiolytics. The major exception remains the α_2-noradrenergic receptor ligands which have paradoxical effects on USV. In addition, CRF, which one might presume would increase the number of calls, decreases the rate of calling at doses approaching the physiologic range. Both of these exceptions may reflect ontogenetic differences in the underlying neurochemical system. Why clonidine is such a potent stimulant until day 18 when it becomes a potent depressant remains an intriguing pharmacologic question. This observation also reminds us of one of the major constraints of the rat pup isolation test- we are working with a developing system in which pharmacokinetic and pharmacodynamic variables may differ markedly from the mature animal. The studies reviewed herein demonstrate that these developmental differences can work in favor of the investigator (as with excitatory amino acids) as well as leading to "paradoxical" results as with clonidine. At the very least, the rat pup isolation test appears useful within specific classes of compounds for distinguishing

Table 3 Anxiolytic and anxiogenic effects assessed on five different measures

Drug	USV	Conflict	Plus Maze	Social Interaction	Defensive Burying	Clinical
Diazepam	↓	↓	↓	↓	↓	↓
PTZ	↑	↑	↑	↔	↔	↑
FG-7142	↑	↑	↑	↑	↔	↑
ACPC	↓	↔	↓	↔	?	?
AP-7	↓	↓	↓	↓	?	?
NMDA	↑	↔	↑	↑	?	?
8-OH-DPAT/Busp.	↓	↓↔	↔↑	↓↑	↓	↓
TFMPP/mCPP	↑	↑	↔	↑	?	↑
5-HT uptake inh.	↓	↑	↑	↔	↑	↓
Clonidine	↑	↓↑	↔	?	↓	↓
Yohimbine	↓	↓↑	↑	↑	↑	↑
NE uptake inh.	↑	↔	↔	?	↔	↓↔
Gardner 86 Hard 88 Insel 86,88 Kehoe 88 Mos 89 Winslow 90abc	Broekkamp 89 Engel 84 Kruse 81 Soderpalm 88 Soderpalm 89 Vogel 71	Broekkamp 89 Critchley 87 Dunn 86 Johnston 89 Moser 89 Pellow 86 Pellow 85 Trullas 89	Beck 86 Broekkamp 89 Dunn 89 File 82 Kennett 89 Pellow 85	Beardsley 90 Craft 88 Fernandez Gusti 89 Treit 87	Charney 83 Cooper 88 Rickels 88 Wamboldt 88	

Anxiolytic ↓, Anxiogenic ↑, No Change ↔, Undetermined ?

anxiolytics and anxiogenics. In addition, it may provide an important model for studying the ontogeny and neurobiology of affect, although the relationship of rat pup USV to any specific human affect (e.g. separation distress, anxiety, fear) remains speculative.

REFERENCES

Allin, J., Banks, E. (1972) Anim. Beh. 20, 175.
Beardslee, S.L., Papadakis, E., Fontana, D.J., Commissaris, R.L. (1990) Pharmacol. Biochem. Behav. 35, 451.
Beck, C.H., Cooper, S.J. (1986) Psychopharmacology (Berlin) 89, 203.
Bennett, D., Amrick, C. (1986) Life Sci. 39, 2455.
Blass, E., Fitzgerald, E., Kehoe, P. (1987) Biochem. Beh. 26, 483.
Britton, K.T., Lee, G., Dana, R., Risch, S.C., Koob, G.F. (1986) Life Sci. 39, 1281.
Broadhurst, P. (1975) Behav. Genet. 5, 229.
Broekkamp, C., Jenck, F. (1989) In: Behavioural Pharmacology of 5-HT. (P.Bevan, A.Cools, and T.Archer, Eds), Lawrence Erlbaum Assoc., Hillsdale, New Jersey, p 321.
Burkholder, J., Hill, J., Vaughn, W., Cascio, H. (1982) Behav. Res. Meth. Instrument 14, 511.
Cagiano, R., Barfield, R., White, N.R., Pleim, E., Weinstein, M., Cuomo, V. (1988) Eur. J. Pharmacol. 157, 45.
Cagiano, R., De Salvia, M., Renna, G., Racagni, G., Cuomo, V. (1987) Arch. Toxicol. 11, 148.
Cagiano, R., Sales, G., Renna, G., Racagni, G., Cuomo, V. (1986) Life Sci. 38, 1417.
Carter, C., Rivy, J., Scatton, B. (1990) Eur. J. Pharm. 164, 611.
Charney, D., Redmond, D., Jr. (1983) Neuropharmacol. 22, 1531.
Clineschmidt, B., Williams, M., Witoslawski, J., Bunting, P., Risley, A., Totaro, J. (1982) Drug Dev. Res. 2, 147.
Cooper, S., Abbott, A. (1988) Tr. Pharmacol. Sci. 9, 269.
Cotman, C., Iversen, L. (1987) TINS, 10, 263.
Craft, R.M., Howard, J.L., Pollard, G.T. (1988) Pharmacol. Biochem. Behav. 30, 775.
Crawley, J.N. (1985) Biobehav. Rev. 9, 37.
Critchley, M.A., Handley, S.L. (1987) Psychopharmacology (Berlin), 93, 502.
Cuomo, V., Cagiano, R., Renna, G., De Salvia, M., Racagni, G. (1987) Neuropharmacol. 26, 701.
Dunn, A.J., File, S.E. (1987) Horm. Behav. 21, 193.
Dunn, R.W., Corbett, R., Fielding, S. (1989) Eur. J. Pharmacol. 169, 1.
Elwood, R. (1979) Dev. Psychobiol. 12, 281.
Engel, J., Hjorth, S., Svensson, K., Carlsson, A., Liljequist, S. (1984) Eur. J. Pharm. 105, 365.
Fernandez-Guasti, A., Hong, E. (1989) In: Behavioural pharmacology of 5-HT (P.Bevan, A.Cools, and T.Archer, Eds), Lawrence Erlbaum Assoc., Hillsdale, NJ, p 377.
File, S.E. (1980) J. Neurosci. Meth. 2, 219.
File, S.E., Lister, R.G., Nutt, D.J. (1982) Neuropharmacology, 21, 1033.
Foster, A., Fagg, G. (1987) Nature, 329, 395.
Gardner, C. (1985) Drug Dev. Res. 5, 185.
Gardner, C. (1986) Pharmacol. Biochem. Behav. 24, 1479.
Hård, E., Engel, J. (1988) Neuropharmacology, 27, 981.

Hård, E., Engel, J., Lindh, A.-S. (1988) Journal of Neural Transm. 73, 217.
Hård, E., Engel, J., Musi, B. (1982) Scand. J. Psychol. Suppl. 1, 90.
Hofer, M., Shair, H. (1987) Devl. Psychobiol. 20, 465.
Huettner, J. E. (1989) Science 243, 1611.
Insel, T., Winslow, J. (1990) J. Clin. Psychiatry 51, 1.
Insel, T., Miller, L., Gelhard, R. (1990) Neurosci. (In press).
Insel, T., Miller, L., Gelhard, R., Hill, J. (1988) In: The Physiological Control of Mammalian Vocalization (J.D.Newman, Ed), Plenum Publishing Corporation. New York, p 331.
Insel, T.R., Gelhard, R.E., Miller, L.P. (1989) Devl. Psychobiol. 22, 509.
Insel, T.R., Harbaugh, C.R. (1989) Pharmacol. Biochem. Behav. 32, 197.
Insel, T.R., Hill, J.L. (1987) Biol. Psychiatry 22, 783.
Insel, T.R., Hill, J.L., Mayor, R.B. (1986) Pharmacol. Biochem. Behav. 24, 1263.
Johnston, A.L., File, S.E. (1989) Pharmacol. Biochem. Behav. 32, 151.
Kehoe, P. (1988) In: The physiological control of mammalian vocalization (J.Newman, Ed), Plenum Press. New York, p 301.
Kehoe, P., Blass, E.M. (1986) Devl. Psychobiol. 19, 385.
Kemp, J., Foster, A., Leeson, P., Priestley, T., Tridgett, R., Iversen, L., Woodruff, G. (1988) Proc. Natl. Acad. Sci. 85, 6547.
Kennett, G., Whitton, P., Shah, K., Curzon, G. (1989) Eur. J. Pharmacol. 164, 445.
Kleckner, N.W., Dingledine, R. (1989) Mol. Pharmacol. 36, 430.
Kruse, H., Dunn, R., Theuer, K., Novick, W., Shearman, G. (1981) Drug Devl. Res. 1, 137.
Leibman, J., Bennett, D. (1988) In: Frontiers in excitatory amino acid research (E.Cavalhieiro, J.Lehmann, and L.Turski, Eds), Alan R Liss, New York, p 301.
Loscher, W., Nolting, B., Honack, D. (1988) Eur. J. Pharmacol. 152, 9.
Marvizon, J., Lewin, A., Skolnick, P. (1989) J. Neurochem. 52, 992.
McDonald, J., Johnston, M. (1990) Brain Res. Rev. 15, 41.
McDonald, J., Silverstein, F., Johnston, M. (1988) Brain Res. 459, 200.
Miczek, K.A., Winslow, J.T. (1990) In: Experimental Psychopharmacology (A.J.Greenshaw and C.T.Dourish, Eds), Humana Press, Clifton, N.J. p 37.
Monaghan, D., Bridges, R., Cotman, C. (1989) Ann. Rev. Pharmacol. Toxicol. 29, 365.
Mos, J., Olivier, B. (1989) In: Behavioural Pharmacology of 5-HT (T.Archer, P.Bevan and C.A.Archer, Eds), Lawrence Earlbaum, New Jersey, p 361.
Moser, P. (1989) Psychopharmacology 99, 48.
Nicolletti, F., Iadarola, M., Wroblewski, J., Costa, E. (1986) Proc. Natl. Acad. Sci. 83, 1931.
Noirot, E. (1966) Anim. Behav. 14, 459.
Okon, E. (1971) J. Zool. Lond. 164, 122.
Olney, J., Ikonomidou, C., Mosinger, J., Shahid Salles, K., Frierdich, G. (1989) Neurosci. Abstr. 15, 417.
Palmer, E., Nangeltaylor, K., Krause, J., Roxas, A., Cotman, C. (1990) Devl. Brain Res. 51, 132.
Panksepp, J., Siviy, S., Normansell, L. (1985) In: The Psychobiology of Attachment and Separation (M.Reite and T.Field, Eds), Academic Press, New York, p 3.
Parrott, R.F. (1990) Physiol. & Behav. 47, 519.
Pellow, S., Chopin, P., File, S.E. (1985a) Neurosci. Lett. 55, 5.
Pellow, S., Chopin, P., File, S.E., Briley, M. (1985b) J. Neurosci. Meth. 14, 149.
Pellow, S., File, S.E. (1986) Pharmacol. Biochem. Behav. 24, 525.
Poshivalov, V. (1987) In: Ethopharmacology of agonistic behaviour in animals and humans (B.Olivier, J.Mos and P.Brain, Eds), Martinus Nijhoff, Dordrecht, The Netherlands, p 122.
Rao, T., Cler, J., Compton, R., Emmett, M., Mick, S., Sun, E., Iyengar, S., Wood, P. (1990) Neuropharmacology 29, 305.

Redmond, D. (1979) In: Phenomenology and treatment of anxiety (W.Fann, I.Karacan, A.Pokorny and R.Williams, Eds), Spectrum, New York, p 153.
Rickels, K., Schweitzer, E., Csanalosi, I., Case, W., Chung, H. (1988) Arch. Gen. Psychiatry 45, 444.
Roberts, L. (1975) Journal of Zoology London 175, 243.
Söderpalm, B., Engel, J.A. (1988) Pharmacol. Biochem. Behav. 30, 471.
Soderpalm, B., Engel, J.A. (1989) J. Neural. Transm. 76, 191.
Shapiro, J.P., Austin, D., Ward, S., Dewsbury, D.A. (1986) Anim. Behav. 34, 90.
Shapiro, L., Insel, T. (1990) Devl. Psychobiol. (In press).
Shoemaker, H., Allen, T., Langer, S. (1990) Eur. J. Pharmacol. 176, 249.
Skolnick, P., Marvizon, J., Jackson, B., Monn, J., Rice, K., Lewin, A. (1989) Life Sci. 45, 1647.
Smotherman, W., Bell, R., Starzec, I., Ellias, J., Zachman, T. (1974) Behav. Biol. 12, 55.
Stephens, D., Andrews, J. (1988) In: Frontiers in excitiatory amino acid research (E.Cavalheiro, J.Lehmann and L.Turski, Eds), Alan R Liss, New York, p 309.
Stephens, D., Meldrum, B., Weidmann, R., Schneider, C., Grützner, M., (1986) Psychopharmacology 90, 166.
Stern, J., Thomas, D., Rabii, J., Barfield, R. (1984) Horm. Behav. 18, 86.
Terkel, J., Damassa, D., Sawyer, C, (1979) Horm. Behav. 12, 95.
Treit, D. (1985) Neurosci. Biobehav. Rev. 9, 203.
Treit, D. (1987) Brain Res. Bull. 19, 401.
Treit, D., Pinel, J.P., Fibiger, H.C. (1981) Pharmacol Biochem. Behav. 15, 619.
Trullas, R., Jackson, B., Skolnick, P. (1989) Pharmacol. Biochem. Behav. 34, 313.
Van der Poel, A., Noach, E., Miczek, K. (1989) Psychopharmacology 97, 147.
Vogel, J., Beer, B., Clody, D. (1971) Psychopharm. 21, 1.
Voloschin, L., Tramezzani, J. (1984) Endocrinology 114 (2), 618.
Vorbrodt, A., Lossinsky, A., Dobrogowska, D., Wisniewski, H. (1990) Devl. Brain Res. 52, 31.
Wamboldt, M., Insel, T. (1988) In: Handbook of Anxiety Disorders (C.Last and M.Hersen, Eds), Pergamon Press, New York.
Winslow, J., Insel, T. (1989) Soc. Neurosci. Abstr. 15, 1018.
Winslow, J., Insel, T. (1990) J Pharm Exp Therap. (In press).
Winslow, J., Insel, T. (1990) Behav. Neurosci. (In press).
Winslow, J., Insel, T.R., Trullas, R., Skolnick, P. (1990) Eur. J. Pharm. (In press).
Winslow, J.T., Insel, T.R. (1990) Neuropsychopharm. 3, 51.
Woods, J., France, C., Hartman, J., Baron, S., Cook, J. (1988) In: Frontiers in excitatory amino acid research (E.Cavalheiro, J.Lehmann and L.Turski, Eds), Alan R. Liss, New York, p 301.

Animal Models in Psychopharmacology
Advances in Pharmacological Sciences

ANIMAL BEHAVIOR MODELS IN THE ANALYSIS AND UNDERSTANDING OF ANXIOLYTIC DRUGS ACTING AT SEROTONIN RECEPTORS

James E. Barrett

Department of Psychiatry, Uniformed Services University of the Health Sciences, 4301 Jones Bridge Road, Bethesda, Maryland 20889, U.S.A.

SUMMARY: Behavioral models have been used extensively in the development and evaluation of anxiolytic drugs. The introduction of clinically-effective anxiolytic drugs that appear to function through mechanisms other than those related to benzodiazepine (BZ) compounds has resulted in a critical re-evaluation of traditional models and concepts of anxiety, as well as the introduction of new animal models. This chapter provides an overview and evaluation of current animal behavior models used in the analysis of anxiolytic drugs with a focus on novel anxiolytics acting through serotonin (5-HT) receptor systems.

INTRODUCTION

The use of animal behavior to study psychotherapeutic drugs is based on the critical need for sensitive, selective and reliable predictors of drug activity in humans. Behavioral models are particularly appropriate because they represent the integrated activity of the intact organism and permit the simultaneous expression of drug effects on several different systems. Together with approaches that focus on more isolated preparations (e.g., receptor binding), they serve as a valuable component of drug development research. Animal behavior models can help to provide an empirical basis for evaluating the potential clinical application of drugs and, as a consequence, should also be helpful in clarifying the neurobiological basis of various disorders. Animal models for evaluating anxiolytic drugs were established early in psychopharmacological research and have been valid and useful models for preclinical drug evaluation. Recently, however, anxiolytic drugs with novel neurochemical mechanisms of action, such as buspirone, which appear to exert their effects predominantly through 5-HT rather than through the GABA/BZ system have not been detected reliably in standard behavioral procedures. The inconsistency in results has raised important questions concerning animal models of anxiety and has provided an impetus for the development of newer models.

This chapter reviews briefly the basic animal behavior models used currently in evaluating anxiolytic drugs. Since all models are sensitive to BZs (reviews by Barrett and Witkin, 1990; Broekkamp et al., 1989; Gardner, 1988; Treit, 1985), the main focus will be on drugs acting at selected 5-HT receptor subtypes.

There are basically two types of animal behavior models used in the study of anxiolytic drugs. Models based on conditioned behavior involve responses developed and/or controlled by operant or respondent conditioning procedures. Models based on unconditioned behavior consist of some form of exploratory behavior and require no specific training. Because these models rely on species-typical behavior (e.g., social interactions), they have often been designated "ethologically-based models" (Lister, 1990). Table 1 summarizes those models currently used most frequently in the behavioral analysis of anxiolytic drugs.

Table 1 Animal behavior models of anxiolytic drug effects

Conditioned Behavior	Unconditioned Behavior
Punishment or Conflict	Exploratory Behavior
Potentiated Startle	Elevated Plus Maze
Periaqueductal Gray Stimulation	Defensive Burying
Drug Discrimination (PTZ)	Ultrasonic Vocalization
Conditioned Taste Aversion	Other Behaviors

CONDITIONED BEHAVIOR MODELS

Punishment or conflict: In one version of this procedure a food-deprived animal engages in a particular response such as pressing a lever that results in the presentation of food. At certain times throughout the session, usually indicated by a distinctive stimulus such as a light, lever pressing produces both food and shock; subsequently, responding is suppressed during the punishment-correlated stimulus (Geller and Seifter, 1960). Behavior suppressed by punishment has been a very reliable indicator of the anxiolytic actions of BZs, as well as of a variety of other drugs such as pentobarbital and ethanol. These compounds produce reliable and robust increases in punished behavior of a variety of species including fish, rodents, pigeons, non-human primates and humans (review by Pollard and Howard, 1990). A variation of this procedure uses water-deprived rats whose licking is suppressed by shock (Vogel et al., 1971). The latter procedure requires less training and, although sensitive to a number of anxiolytics, may also show positive results with certain non-anxiolytic compounds (Barrett and Witkin, 1990; Pollard and Howard, 1990). The effects of numerous drugs on behavior under these procedures have been studied extensively and, as a result, many BZ and non-BZ compounds are well characterized.

1. **Buspirone and 5-HT$_{1A}$ drugs:** Despite the consistency in effects of BZs on behavior under the conflict procedure, the effects of buspirone and other 5-HT$_{1A}$ compounds on punished responding are quite inconsistent in rats and monkeys (Barrett and Witkin, 1990). Table 2 summarizes the effects of buspirone and the selective 5-HT$_{1A}$ agonist 8-OH-DPAT using punished lever pressing or drinking in the rat. Results are mostly negative with punished lever pressing. Even when extensive efforts have been made to study systematically a wide range of variables, such as route of administration, strain of rat, variations of the schedule of punishment, time course and drug history, buspirone still failed to increase punished responding (Howard and Pollard, 1990). Effects using punishment of drinking appear more consistent with buspirone and the majority of studies report positive effects. It should be noted, however, that in almost every experiment in which a BZ was compared with buspirone, the effects of buspirone were much smaller. The somewhat marginal effects of buspirone on suppressed drinking can be increased by administering buspirone chronically following test sessions for an 8-week period (Schefke et al., 1989). These results parallel clinical data suggesting that the onset of buspirone's clinical effects may require an initial period of approximately two weeks (Goa and Ward, 1986). Further studies that examine the effects of chronic buspirone administration are necessary since the effects of buspirone do not always change following a period of chronic administration (Wettstein, 1988).

Effects with 8-OH-DPAT are inconsistent regardless of the type of procedure. It should be noted, however, that when either 8-OH-DPAT or buspirone have been studied using punished drinking, increases have been obtained when injections were made into the median raphe (Carli and Samanin, 1988; Carli et al., 1989).

Effects of buspirone and 8-OH-DPAT in monkeys are also most often negative. Buspirone has been shown to produce increases in punished lever pressing in some cases (e.g., Geller and Hartmann, 1982; Weissman et al., 1984), but not in others

(Gleeson and Barrett, 1990; Sepinwall, 1985; Wettstein, 1988). When increases in punished responding have occurred with monkeys, they were often much smaller in comparison to those obtained with BZs and may have been related to very low control response rates. The most robust and long-lasting increases with buspirone occurred in cynomolgus monkeys (Geller and Hartmann, 1982), whereas all other experiments were performed using squirrel monkeys.

In contrast to the inconsistent effects of buspirone and 8-OH-DPAT in mammals, reliable, robust and dose-dependent increases in punished responding occur with buspirone, 8-OH-DPAT and a wide range of other 5-HT_{1A} drugs such as flesinoxan, SM-3997 and WY 47,846 in pigeons (Barrett et al., 1989; Brocco et al., 1990; Gleeson et al., 1989; Mansbach et al., 1988; Witkin and Barrett, 1986; unpublished data). Further, the magnitude of the increases in punished responding with these drugs has been comparable to that obtained with BZs (Barrett et al., 1986; Brocco et al., 1990; Howard et al., 1990; Mansbach et al., 1988). Thus, the pigeon appears to be a very good species with which to evaluate the effects of potential anxiolytics acting at the 5-HT_{1A} site.

2. **Other 5-HT Compounds**: Compounds with activity at other 5-HT sites have also been studied using the conflict procedure. RU 24969, a mixed 5-$HT_{1A/B}$ agonist, did not increase punished licking behavior of rats (Carli et al., 1988; Gardner, 1985) or food-maintained lever pressing of monkeys (Gleeson and Barrett, 1990); small increases in punished food-maintained responding were obtained with RU 24969 in rats (Gardner, 1986). In contrast, RU 24969, as well as eltoprazine, a compound similar to RU 24969 in terms of activity at the 5-$HT_{1A/B}$ receptor, produced large increases in punished key-pecking of pigeons (Gleeson et al., 1989; unpublished data). Other compounds with actions at 5-HT_1 subtypes, e.g., the 5-$HT_{1B/C}$ agonists TFMPP and mCPP, have not shown consistent increases in punished responding of any species studied thus far (Gardner, 1988; Gleeson et al., 1989).

Ritanserin, a 5-HT_2 antagonist, increases punished responding of pigeons (Brocco et al., 1990; Gleeson et al., 1989) but does not appear to produce consistent, dose-related increases in punished lever pressing of rats (Colpaert et al., 1985; Gardner, 1986; Meert and Janssen, 1989). The 5-HT_2 antagonist ketanserin increased punished responding of squirrel monkeys (Brady and Barrett, 1985) but few other studies with non-human primates have been conducted with these drugs.

Recently, considerable attention has been directed towards the potential role of 5-HT_3 receptors in anxiety. A number of 5-HT_3 antagonists have been evaluated using conflict procedures but, regardless of the species and procedure, these drugs have shown no anxiolytic-like activity (Gleeson et al., 1989; Jones et al., 1988). ICS205930, GR 38032F and diazepam, however, decreased the latency to move from an elevated platform to a grid floor that, on the previous day, resulted in the delivery of electric shock (Papp and Przegalinski, 1989).

Table 2 Effects of buspirone and 8-OH-DPAT on punished behavior (rat)

Drug	Species-Procedure	Effect	Reference
Buspirone	Rat - Lever Pressing	+	Geller & Hartmann, 1982
		-	Goldberg et al., 1983
		-	Gardner, 1986
		+	Young et al., 1987
		-	Mason et al., 1987
		-	Howard & Pollard, 1990
		-	Witkin & Perez, 1990
		-	Brocco et al., 1990
	Rat - Drinking	+	Oakley & Jones, 1983
		+	Weissman et al., 1984
		-	Sanger et al., 1985
		+	File, 1985
		-	Gardner, 1986
		+	Merlo Pich & Samanin, 1986
		+	McCloskey et al., 1987
		+	Shimizu et al., 1987
		+	Gower & Trickelbank, 1988
		+	Przegalinski et al., 1989
		+	Moser et al., 1990
8-OH-DPAT	Rat - Lever Pressing	+	Hodges et al., 1987
		-	Witkin and Perez, 1990
	Rat - Drinking	+	Engel et al., 1984
		-	Carli and Samanin, 1988

Data taken from published articles where more than a single dose was administered. (+) = increase; (-) = decrease or no effect.

In summary, data from the conflict procedure with compounds acting at the 5-HT_{1A} and 5-HT_2 receptor subtypes are inconsistent in mammals. However, consistent, reliable and dose-dependent increases in punished responding occur with pigeons. Evaluation of these agents as potential anxiolytics, therefore, would appear best conducted using the pigeon conflict procedure.

Potentiated Startle: The potentiated startle procedure has been proposed as a measure of conditioned fear and is sensitive to a number of anxiolytic drugs (review by

Davis, 1988). In brief, the startle response to a loud sound is potentiated or enhanced in the presence of a stimulus previously paired with shock when compared to the amplitude of the startle response in the absence of the conditioned stimulus. BZs, as well as buspirone and gepirone, block the potentiated startle response at doses that do not decrease unconditioned startle to the sound alone; however, 8-OH-DPAT did not do so and ipsapirone, another 5-HT_{1A} ligand, did so only at very high doses (40 mg/kg). Davis et al. (1988) found that p-chloroamphetamine did not alter the potentiated startle response, nor did electrolytic raphe lesions modify buspirone's effects, leading to the conclusion that buspirone's anxiolytic effects are not mediated by actions at 5-HT_{1A} receptors. This result cannot presently be reconciled with what appears to be a large amount of evidence implicating 5-HT in the anxiolytic effects of compounds such as buspirone and with data reported by Mansbach and Geyer (1988) showing a blockade of potentiated startle with 8-OH-DPAT and ipsapirone. It has been shown recently that the 5-HT_3 antagonist GR 38032F reduces potentiated startle (Glenn and Green, 1989) leaving the involvement of 5-HT mediation in this procedure still plausible. Perhaps further work with this method will clarify some of the variables accounting for the differences and resolve the role of 5-HT in mediating this effect.

Periaqueductal Gray Stimulation: Electrical stimulation of the dorsal portion of the periaqueductal gray (PAG) produces a number of responses that have been characterized as aversive and animals will respond to terminate stimulation to this area (Graeff, 1987). Anxiolytic compounds such as BZs elevate the threshold intensity required to produce the escape response, resulting in the suggestion that this procedure may provide a model for the study of anxiety (Jenck et al., 1989a). Drugs with activity at 5-HT_{1A} sites such as 8-OH-DPAT, however, lower the threshold and facilitate the escape response; 5-HT_2 antagonists (ketanserin, trazodone and spiperone), as well as those with agonist actions at the 5-HT_{1C} site (mCPP, TFMPP and DOI), and 5-HT uptake inhibitors, increase the escape threshold (Jenck et al., 1989b, 1990b). Although ritanserin, a 5-HT_2 antagonist, increased the threshold, this compound is also a 5-HT_{1C} antagonist, which may account for its different effects from other 5-HT_2 antagonists (Jenck et al., 1990a). The 5-HT_3 antagonists, ICS205930, MDL72222 and GR38032F were without effect under this procedure (Broekkamp et al., 1989; Jenck et al., 1989a). Based on these studies, together with experiments with other compounds, it would appear that 5-HT_{1C} receptors play a prominent role in the aversion induced by PAG stimulation, leading to the suggestion that this receptor may be involved in anxiety. Full evaluation of this possibility must await the development of selective antagonists for this receptor. Whether this model is suitable for evaluating potential anxiolytic drugs, or whether it is more appropriate for the evaluation of potential antidepressants, however, must await further research. Since compounds such as 8-OH-DPAT which, like buspirone, act at the 5-HT_{1A} receptor produce effects opposite those of the BZs in this system, it may be that the PAG model more appropriately detects actions of drugs other than their anxiolytic effects or, perhaps best detects compounds with mixed anxiolytic/antidepressant potential.

Drug Discrimination: Procedures using drug discrimination methods as a model of anxiety establish an anxiogenic compound such as pentylenetetrazol (PTZ) as a discriminative stimulus by reinforcing one response in the presence of that drug and a different response when saline is administered (Lal and Emmett-Oglesby, 1983). Animals trained with PTZ will respond as though PTZ has been administered when given other drugs with purported anxiogenic effects such as ß-carboline carboxylic acid (ß-CCM) and FG 7142 (Lal and Emmett-Oglesby, 1983; Stephens et al., 1984; Vellucci et al., 1988). The PTZ stimulus is antagonized by the administration of diazepam and chlordiazepoxide. Baboons trained with PTZ generalize not only to ß-CCE but to buspirone (Ator et al., 1989). Winter (1988) found that yohimbine, also reported to be anxiogenic in humans, occasioned drug-appropriate responding in rats trained to discriminate either 8-OH-DPAT or ipsapirone from saline. Winter and Rabin (1989) found comparable effects in rats trained with ipsapirone or 8-OH-DPAT; yohimbine, buspirone and gepirone all resulted in drug-appropriate responding. In rats trained with yohimbine, however, neither 8-OH-DPAT nor buspirone fully substituted, both yielding less than 80% responding on the drug-appropriate lever. Although it is clear that yohimbine, ipsapirone, 8-OH-DPAT and buspirone can occasion similar discriminative stimulus effects under some conditions, this common mediation most likely is not through the 5-HT_{1A} receptor since Winter and Rabin (1989) showed that yohimbine had a very low affinity for the site labeled by [^{3}H]8-OH-DPAT. Further work with PTZ and yohimbine is necessary to better understand the pharmacological and behavioral significance of these findings and their relevance to the clinical presentation and management of anxiety.

Conditioned Taste Aversion: The intake of distinctively-flavored solutions can be suppressed when the consumption of those solutions is paired with the administration of a drug such as lithium chloride (LiCl). The suppression of fluid intake can be attenuated by chlordiazepoxide (Cappell and LeBlanc, 1973). This procedure has been proposed as a potential model for the study of anxiolytic drugs and has been evaluated using a variety of compounds (Ervin et al., 1987; Ervin and Cooper, 1988). In these experiments, conditioned taste aversions to saccharin were developed by LiCl or 1-5-

hydroxytryptophan (l-5-HTP) and were antagonized by several BZs, as well as by buspirone, phenobarbital and meprobamate. Several other drugs, including imipramine, morphine, chlorpromazine and atropine did not attenuate suppressed drinking, nor did the administration of the otherwise "dipsogenic" agents isoproterenol or histamine. Ervin and Cooper (1988) also manipulated the dose of the aversion-inducing drugs and found that higher doses of either LiCl or l-5-HTP resulted in greater suppression of drinking and less antagonism by chlordiazepoxide. These authors have characterized this procedure as a "conflict" between drinking induced by a period of deprivation and drinking that is suppressed by the conditioned aversion to the saccharin solution. The evidence for specificity appears to be quite good and this procedure warrants further exploration.

UNCONDITIONED BEHAVIOR MODELS

Exploratory Behavior: The use of exploratory behavior in an open field as a measure of anxiolytic drug effects is problematic because non-anxiolytic compounds such as amphetamine produce effects similar to BZs and because of the difficulty in separating exploratory and locomotor behavior. Lister (1990) has, in fact, concluded that these features make the measurement of locomotor activity in an open field "an unsuitable index of anxiety in psychopharmacological research." Presumably, this also implies that exploratory activity is not particularly useful for evaluating anxiolytic drugs. Similar concerns exist with regard to the use of exploratory behavior measured by head dipping in a holeboard.

Another technique that has been used quite extensively is that in which transitions of mice between a large brightly lit (often painted white) and smaller dimly illuminated area (painted black) are measured (sometimes called the two-compartment test). Mice spend more time in the area with a low illumination level but transitions into the brightly-illuminated area are increased by BZs at doses that do not increase locomotor activity measured in a separate apparatus (review by Crawley, 1985). Although amphetamine also produces effects similar to those of the BZs, unlike the BZs, amphetamine also increases general locomotor activity at doses that increase transitions; it is suggested that such "false positives" can be eliminated if they produce increases in both locomotor activity and light-dark transitions at comparable doses. Merlo Pich and Samanin (1989) have recently demonstrated effects with BZs in the rat that are similar to those obtained with mice.

Buspirone, administered i.p. as well as into the dorsal raphe nucleus of the mouse, increased the amount of time spent in the white compartment and decreased rearing as well as line crossings within the black compartment (Costall et al., 1988b). Although transitions between the two compartments were not increased, the overall effects were interpreted as reflecting an anxiolytic action of buspirone (cf. Costall et al., 1989). The 5-HT_2 antagonist ritanserin, as well as methysergide, metergoline and cyproheptadine did not affect behavior using this procedure. Effects similar to those obtained with buspirone were reported for diazepam, the 5-HT_3 antagonists GR38032F, MDL72222, ICS205930, BRL43694 and zacopride, as well as for sulpiride and other substituted benzamides (Costall et al., 1988a; Jones et al., 1988)). Mos et al. (1989), however, found no indication of anxiolytic-like actions in this procedure for GR38032F. Onaivi and Martin (1989) have performed a neuropharmacological and physiological validation using a computer-controlled black-white test apparatus and have also reported that many compounds, including diazepam, buspirone, ICS205930 and MDL72222 did not increase transitions between the light and dark compartments but did increase activity in the white chamber. In contrast to these reports, however, Carli et al. (1989) have shown increases in light-dark transitions after s.c. and median raphe injections of buspirone and 8-OH-DPAT in the rat; Fernández-Guasti and López-Rubalcava (1990) found that the 5-HT_{1A} agonists ipsapirone and indorenate increased transitions of mice. These differences may depend on the strain of mice or rats, dimensions or configuration of the test apparatus, or other details of the procedure which warrant further study to determine whether the different effects reflect different pharmacological actions. In general, though, the pattern of results with this procedure is somewhat mixed.

Elevated-Plus Maze: This procedure has gained increasing use for assessing anxiolytic drug effects. The description, use and validation of this procedure has been summarized by Pellow et al. (1985) and Pellow (1986). The apparatus consists of an elevated maze with intersecting arms, two of which are open and two closed. The primary measures are the proportion of entries into the open arms and the time spent on the open arms expressed as a percentage of the total time spent on both open and closed arms. Under nondrug conditions rodents spend more time in the closed arms of the maze. An anxiolytic effect is suggested when the drug increases open arm entries without altering the total number of arm entries; an anxiogenic effect is suggested when open arm entries are decreased without altering the total number of entries.

Several studies have shown that BZs increase the percentage of entries and time spent on the open arms while reducing the total number of arm entries; putative anxiogenic drugs such as FG7142, PTZ and yohimbine produced opposite effects, whereas other compounds, such as haloperidol and imipramine showed no significant effects in this procedure (reviews by Pellow, 1986; Pellow et al., 1985). In some studies buspirone and 8-OH-DPAT were without effects, whereas quipazine, ritanserin and RU24969 showed anxiogenic-like activity; ipsapirone reduced both measures of activity (review by Pellow, 1986; Pellow and File, 1986). In other studies with the elevated plus maze, however, 8-OH-DPAT produced anxiogenic-like actions, whereas effects with ritanserin were characteristic of anxiolytic activity (Critchley and Handley, 1987). Intra-hippocampal injections of buspirone have been reported to increase the time spent on the open arms of the maze (Kostowski et al., 1989). Several other reports with buspirone are either negative or suggest a possible anxiogenic-like action, leading to the conclusion that this test may either be insensitive to certain types of anxiolytics or may be measuring something unrelated to human anxiety (Moser, 1989). Although Söderpalm et al. (1989) and Dunn et al. (1989) have reported anxiolytic-like effects with 8-OH-DPAT, buspirone and gepirone using this procedure, it is clear that the findings differ and that the relevant variables are as yet unknown (cf. Moser et al., 1990).

Compounds active at $5\text{-HT}_{1B/C}$ sites have recently been shown to produce anxiogenic-like actions in the elevated plus maze, findings that parallel reports of their ability to induce anxiety in humans (Charney et al., 1987). Benjamin et al. (1990) found that mCPP and TFMPP decreased entries and time spent in the open arms of the maze in a manner similar to that shown with PTZ.

Finally, although evidence exists for anxiolytic-like effects of 5-HT_3 antagonists in other procedures, these drugs do not produce anxiolytic-like actions in the elevated plus maze (File and Johnston, 1989). Thus, at present, the overall pattern of results with the elevated-plus maze are inconsistent and warrant systematic evaluation of the relevant parameters to achieve better reproducibility across laboratories and to ascertain what this procedure is measuring.

Defensive Burying: Under conditions where there is material available, rats will push or spray that material over a source of aversive stimulation. The burying response appears to be a species-typical behavior to a variety of aversive stimuli (Treit, 1985). In the procedure used to evaluate anxiolytic drug effects, the source of aversive stimulation is an electrified probe and the number of probe contacts and/or time spent burying the probe are measured. Meert and Colpaert (1986) have investigated a procedure in which no bedding material is available and only probe contacts are measured. Chlordiazepoxide and diazepam suppress the burying response or increase probe contacts, whereas other compounds, such as d-amphetamine, morphine and PTZ do not. These results are not always consistent, however, since Craft et al. (1988) found a reduction in burying with chlordiazepoxide only at doses that reduced general activity. In addition, burying was reduced by imipramine and morphine at doses that did not reduce general activity. There were, however, a number of differences between the Craft et al. (1988) study and those conducted by Treit and his colleagues (e.g., route of administration, period of habituation, source of rats and shock procedure) that require further analysis. However, Craft et al. concluded that the defensive burying procedure lacks therapeutic class specificity.

Effects of buspirone in the conditioned burying procedure have also been mixed, with Craft et al. (1988) and Meert and Colpaert (1986) reporting no reduction in burying or increases in probe contacts, respectively, except at doses that suppressed activity. Treit and Fundytus (1988), using the same procedure as that of Craft et al. reported an increased number of probe contacts and a decrease in burying behavior with buspirone; Treit (1990) has extended this work to other drugs. At present, it is difficult to draw any definitive conclusions about the sensitivity and selectivity of this procedure because of the inconsistent findings. It may be that, as Rohmer et al. (1990) suggest, additional behavioral measures are necessary to modify the sensitivity and selectivity of this procedure.

Social Interaction: The effects of drugs on the social interaction of rats under levels of high illumination in a novel environment has been used frequently as a measure of anxiolytic drug effects (reviews by File, 1985, 1988). BZs increase social interaction, particularly after a few days of chronic administration during which the sedative effects diminish. Although File (1985, 1988) did not find anxiolytic-like effects with buspirone, Guy and Gardner (1985) have reported increases in social interaction with buspirone and Dunn et al. (1989) have reported increased social interaction after buspirone, gepirone and 8-OH-DPAT. The main difference in the procedures is that social interaction in the latter studies was measured in a familiar environment and used the novelty of the rats as the main variable. In the Dunn et al. study, increases in social interaction with gepirone and buspirone occurred at doses that decreased motor activity. The $5\text{-HT}_{1B/C}$ agonists mCPP and TFMPP reduced social interaction without reducing locomotor activity (Kennett et al., 1989). Ritanserin did not increase social interaction in this test (Gardner, 1986).

Results from social interaction tests with the 5-HT_3 antagonists have been inconsistent. Although GR38032F increased social interaction in one study (Jones et al., 1988), increases did not occur with GR38032F, zacopride or BRL43964 in another (File and Johnston, 1989). When increases have occurred, they have been small. File and Johnston (1989) discuss the possible role of a rate-dependent effect in these experiments, suggesting that when social interaction scores were low, increases with these drugs were more likely to occur. Chronic administration of the 5-HT_3 antagonists ICS205-930 and BRL434694 produced an increase in selected components of social behavior of gerbils (Cutler, 1990).

As with the elevated plus maze, however, data obtained in the social interaction procedure appear to depend to a large extent on the specific methods employed. Further evaluation of a range of compounds under standard conditions would seem warranted.

Distress Vocalization: A number of conditions such as cold or separation from the mother will elicit ultrasonic vocalizations in rat pups; these are suppressed by BZs at non-sedating doses (Gardner, 1985; Insel et al., 1989). Mos and Olivier (1989) have investigated a variation of this procedure that measures vocalizations induced by separation from the mother and where the pups are tested under either warm or cold conditions. A number of 5-HT_{1A} agonists, such as 8-OH-DPAT, buspirone, flesinoxan and ipsapirone, reduced vocalizations under both conditions at doses that did not disrupt negative geotaxic behavior - a measure of an overall sedative effect. Similar effects with 8-OH-DPAT were reported by Hård and Engel (1988). Drugs that act at other 5-HT receptor subtypes (e.g., 5-HT_{1B}, 5-HT_2 and 5-HT_3), or that block the reuptake of 5-HT, reduced calls under the cold condition only or had no effect (Mos et al., 1989). Further use of this procedure will answer necessary questions that have arisen with regard to specificity, the role of hypothermia induced by the drugs, as well as problems or special considerations of drug action in the neonate. Although these remain to be examined more thoroughly, at present, this procedure appears quite promising.

Other behaviors: There are a number of other models that have been examined for which there is insufficient space for discussion. Some of these procedures appear to be of limited value for a variety of reasons (e.g., neophagia), whereas others are too new to have achieved sufficient verification and validation [e.g., stress-induced hyperthermia, Lecci et al., (1990)]. Considerable work has been devoted to examining the effects of drugs on various aggressive/defensive behaviors (see Blanchard et al., 1989 and Olivier et al., 1989 for a review of these procedures with special reference to 5-HT). These procedures provide a wide range of behavioral measures with considerable face validity and may be capable of subtle differentiation between drugs and drug effects on selected components of different aggressive and defensive responses. Because of the complexity of the measures and sophistication involved in monitoring and recording behavior, these procedures may not be used widely as screening methods.

SUMMARY AND CONCLUSIONS

A wide range of procedures are available for evaluating the effects of drugs in animal models of anxiolytic drug action. Until recently, effects in all of these procedures with known anxiolytics were reasonably consistent. Efforts were directed mainly towards the development of more rapid methods for evaluating potential anxiolytic compounds. However, the introduction of buspirone and the discovery of the potential role of 5-HT in mediating its anxiolytic effects has resulted in considerable disarray in results obtained with these models. As shown in this review, most of the procedures yield inconsistent effects with compounds acting at those 5-HT receptor subtypes believed to play some role in anxiety. Of the compounds with known or suggested anxiolytic effects in humans (BZs, 5-HT_{1A} agonists and 5-HT_2 antagonists), only data with the pigeon conflict procedure appear to be uniformly consistent in predicting anxiolytic actions of these drugs. Data from clinical trials with the 5-HT_3 antagonists will be of special interest, since these do not appear to increase punished responding in the pigeon conflict procedure but show anxiolytic-like effects under some conditions in other models (light-dark crossing and social interaction). It may be the case that each of these procedures effectively models different types of anxiety disorders and that they will be sensitive and selective screens for the development of future compounds with greater specificity.

Suitable and reliable models require extensive validation and a balance must be struck between the proliferation of newer models and the refinement of existing ones. Progress in the development of valid models for psychiatric disorders can come from close interaction between preclinical and clinical research but also depends critically on careful evaluation and control of the variables affecting behavior under each of the models discussed in this chapter.

Acknowledgements: This work was supported by PHS Grant DA-02873. I thank Dr.S.Gleeson for comments on the manuscript.

REFERENCES

Ator, N.A., Cook, J.M., Griffiths, R.R. (1989) Drug Dev. Res. 16, 257-267.

Barrett, J.E., Witkin, J.M., Mansbach, R.S., Skolnick, P., Weissman, B.A. (1986) J. Pharmacol. Exp. Ther. 238, 1009-1013.
Barrett, J.E., Gleeson, S., Nader, M.A., Hoffmann, S.M. (1989) J. Psychopharmacol. 3, 64-69.
Barrett, J.E., Witkin, J.M. (1990) In: Buspirone: Mechanisms and Clinical Aspects (G.Tunnicliff, A.S.Eison and D.P.Taylor, Eds), Academic Press/Tampa, FL.
Benjamin, D., Lal, H., Meyerson, L.R. (1990) Life Sci. 47, 195-203.
Blanchard, D.C., Rodgers, R.J., Hendrie, C.A., Hori, K. (1988) Pharmacol. Biochem. Behav. 31, 269-278.
Brady, L.S., Barrett, J.E. (1985) J. Pharmacol. Exp. Ther. 234, 106-112.
Brocco, M.J., Koek, W., Degryse, A.-D., Colpaert, F.C. (1990) Behav. Pharmacol. 1.
Broekkamp, C.L.E., Berendsen, H.H.G., Jenck, F., Van Delft, A.M.L. (1989) Psychopathology 22 (suppl. 1), 2-12.
Cappell, H.D., LeBlanc, A.E. (1973) J. Comp. Physiol. Psychol. 83, 97-104.
Carli, M., Prontera, C., Samanin, R. (1989) Br. J. Pharmacol. 96, 829-836.
Carli, M., Samanin, R. (1988) Psychopharmacology 94, 84-91.
Chopin, P., Briley, M. (1987) Trends Pharmacol. Sci. 8, 383-388.
Charney, D.S., Woods, S.W., Goodman, W.K., Heninger, G.R. (1987) Psychopharmacology 92, 14-24.
Colpaert, F.C., Meert, T.F., Niemegeers, C.J.E., Janssen, P.A.J. (1985) Psychopharmacology 86, 45-54.
Costall, B., Domeney, A.M., Gerrard, P.A., Kelly, M.E., Naylor, R.J. (1988a) J. Pharm. Pharmacol. 40, 302-305.
Costall, B., Jones, B.J., Kelly, M.E., Naylor, R.J., Tomkins, D.M. (1989) Pharmacol. Biochem. Behav. 32, 777-785.
Costall, B., Kelly, M.E., Naylor, R.J., Onaivi, E.S. (1988b) J. Pharm. Pharmacol. 40, 494-500.
Craft, R.M., Howard, J.L., Pollard, G.T. (1988) Pharmacol. Biochem. Behav. 30, 775-780.
Crawley, J.N. (1985) Neurosci. Biobehav. Rev. 9, 37-44.
Critchley, M.A.E., Handley, S.L. (1987) Psychopharmacology 93, 502-506.
Cutler, M.G. (1990) Neuropharmacol. 29, 515-520.
Davis, M. (1988) Anim. Models Psychiat. Disord. 1, 61-89.
Davis, M., Cassella, J.V., Kehne, J.H. (1988) Psychopharmacology 94, 14-20.
Dunn, R.W., Corkett, R., Fielding, S. (1989) Eur. J. Pharmacol. 169, 1-10.
Engel, J.A., Hjorth, S., Svensson, K., Carlsson, A., Liljequist, S. (1984). Eur. J. Pharmacol. 105, 365-368.
Ervin, G.N., Cooper, B.R. (1988) J. Pharmacol. Exp. Ther. 245, 137-146.
Ervin, G.N., Soroko, F.S., Cooper, B.R. (1987) Drug. Dev. Res. 11, 87-95.
Fernández-Guasti, A., López-Rubalcava, C. (1990) Psychopharmacology 101, 354-358.
File, S. (1985) Neuropsychobiology 13, 55-62.
File, S. (1988) Anim. Models Psychiat. Disord. 1, 151-166.
File, S., Johnston, A.L. (1989) Psychopharmacology 99, 248-251.
Gardner, C.R. (1985) J. Pharmacol. Meth. 14, 181-187.
Gardner, C.R. (1986) Pharmacol. Biochem. Behav. 24, 1479-1485.
Gardner, C.R. (1988) Gen. Pharmac. 19, 347-356.
Geller, I., Hartmann, R.J. (1982) J. Clin. Psychiatry 43, 25-32.
Geller, I., Seifter, J. (1960) Psychopharmacologia 1, 482-492.
Gleeson, S., Ahlers, S.T., Mansbach, R.S., Foust, J.M., Barrett, J.E. (1989) J. Pharmacol. Exp. Ther. 250, 809-817.
Gleeson, S., Barrett, J.E. (1990) Pharmacol. Biochem. Behav. 37, 335-337.
Glenn, B., Green, S. (1989) Behav. Pharmacol. 1, 91-94.
Goa, K.L., Ward, A. (1986) Drugs 32, 114-129.

Goldberg, M.E., Salama, A.I., Patel, J.B., Malick, J.B. (1983) Neuropharmacology 22, 1499-1504.
Gower, A.J., Tricklebank, M.D. (1988) Eur. J. Pharmacol. 155, 129-137.
Graeff, F. (1987) In: Advances in Behavioral Pharmacology (T.Thompson, P.B.Dews and J.E.Barrett, Eds), Lawrence Erlbaum/Hillsdale, NJ, pp 129-156.
Guy, A.P., Gardner, C.R. (1985) Neuropsychobiology 13, 194-200.
Hård, E., Engel, J. (1988) Neuropharmacol. 27, 981-986.
Hodges, H., Green, S., Glenn, B. (1987) Psychopharmacology 92, 491-504.
Howard, J.L., Nanry, K.P., Perkins, A.N., Thomas, J.D. (1990) Faseb J. 4, A616.
Howard, J.L., Pollard, G.T. (1990) Drug Dev. Res. 19, 37-49.
Insel, T.R., Gelhard, R.E., Miller, L.P. (1990) Dev. Psychobiol. 22, 509-525.
Jenck, F., Broekkamp, C.L.E., Van Delft, A.M.L. (1989a) Psychopharmacology 97, 489-495.
Jenck, F., Broekkamp, C.L.E., Van Delft, A.M.L. (1989b) Eur. J. Pharmacol. 161, 219-221.
Jenck, F., Broekkamp, C.L.E., Van Delft, A.M.L. (1990a) Psychopharmacology 100, 372-376.
Jenck, F., Broekkamp, C.L.E., Van Delft, A.M.L. (1990b) Eur. J. Pharmacol. 197, 201-204.
Jones, B.J., Costall, B., Domeney, A.M., Kelly, M.E., Naylor, R.J., Oakley, N.R., Tyers, M.B. (1988) Br. J. Pharmacol. 93, 985-993.
Kennett, G.A., Whitton, P., Shah, K., Curzon, G. (1989) Eur. J. Pharmacol. 164, 445-454.
Kostowski, W., Plaznik, A., Stefanski, R. (1989) Eur. J. Pharmacol. 168, 393-396.
Lal, H., Emmett-Oglesby, M.W. (1983) Neuropharmacol. 22, 1423-1441.
Lecci, A., Borsini, F., Volterra, G., Meli, A. (1990) Psychopharmacology 101, 255-261.
Lister, R.G. (1990) Pharmac. Ther. 46, 321-340.
Mansbach, R.S., Geyer, M.A. (1988) Eur. J. Pharmacol. 156, 375-383.
Mansbach, R.S., Harrod, C., Hoffmann, S.M., Nader, M.A., Lei, Z., Witkin, J.M., Barrett, J.E. (1988) J. Pharmacol. Exp. Ther. 246, 114-120.
Mason, S.T., Skinner, J., Luttinger, D. (1987) Psychopharmacology 92, 30-34.
McCloskey, T.C., Paul, B.K., Commissaris, R.L. (1987) Pharmacol. Biochem. Behav. 27, 171-175.
Meert, T.F., Colpaert, F.C. (1986) Psychopharmacology 88, 445-450.
Meert, T.F., Janssen, P.A.J. (1989) Drug Dev. Res. 18, 119-144.
Merlo Pich, E.M., Samanin, R. (1986) Psychopharmacology 89, 125-130.
Mos, J., v.d. Heyden, J., Olivier, B. (1989) In: Behavioural Pharmacology of 5-HT (P.Bevan, A.R.Cools and T.Archer, Eds), Lawrence Erlbaum/Hillsdale, N.J., pp 389-395.
Mos, J., Olivier, B. (1989) In: Behavioural Pharmacology of 5-HT (P.Bevan, A.R.Cools and T.Archer, Eds), Lawrence Erlbaum/Hillsdale, N.J., pp 361-366.
Moser, P.C. (1989) Psychopharmacology 99, 48-53.
Moser, P.C., Tricklebank, M.D., Middlemiss, D.N., Mir, A.K., Hibert, M.F., Fozard, J.R. (1990) Br. J. Pharmacol. 99, 343-349.
Oakley, N.R., Jones, B.J. (1983) Eur. J. Pharmacol. 87, 499-500.
Olivier, B., Mos, J., Tulp, M., Schipper, J., Bevan, P. (1989) In: Behavioural Pharmacology of 5-HT (P.Bevan, A.R.Cools & T.Archer, Eds), Lawrence Erlbaum/Hillsdale, N.J., pp 89-115.
Onaivi, E.S., Martin, B.R. (1990) Prog. Neuro-Psychopharmacol. & Biol. Psychiat. 13, 963-976.
Papp, M., Przegalinski, E. (1989) J. Psychopharmacol. 3, 14-20.
Pellow, S. (1986) Meth. Find. Exptl. Clin. Pharmacol. 8, 557-565.
Pellow, S., Chopin, P., File, S., Briley, M. (1985) J. Neurosci. Meth. 14, 149-167.
Pellow, S., File, S.E. (1986) Pharmacol. Biochem. Behav. 24, 525-529.

Pollard, G.T., Howard, J.L. (1990) Pharmacol. Ther. 45, 403-424.
Przegalinski, E., Tatarczynska and Chojnacka-Wójcik (1989) J. Psychopharmacol. 3, 180-185.
Rohmer, J.-G., Di Scala, G., Sander, G. (1990) Behav. Brain Res. 38, 45-54.
Sanger, D.J., Joly, D., Zivkovic, B. (1985) J. Pharmacol. Exp. Ther. 232, 831-837.
Schefke, D.M., Fontana, D.J., Commissaris, R.L. (1989) Psychopharmacology 99, 427-429.
Sepinwall, J. (1985) In: Behavioral Pharmacology: The Current Status (L.S.Seiden and R.L.Balster, Eds), Alan R. Liss/New York, pp. 181-203.
Shimizu, H., Hirose, A., Tatsuno, T., Nakamura, M., Katsube, J. (1987) Jap. J. Pharmacol. 45, 493-500.
Söderpalm, B., Hjorth, S., Engel, J.A. (1989) Pharmacol. Biochem. Behav. 32, 259-265.
Stephens, D.N., Kehr, W., Schneider, H,H, and Schmiechen, R. (1984) Neurosci. Lett. 47, 333-338.
Treit, D. (1985) Neurosci. Biobehav. Rev. 9, 203-222.
Treit, D. (1990) Pharmacol. Biochem. Behav. 36, 203-205.
Treit, D., Fundytus, M. (1988) Pharmacol. Biochem. Behav. 30, 1071-1075.
Vellucci, S.V., Martin, P.J., Everitt, B.J. (1988) J. Psychopharmacol. 2, 80-93.
Vogel, J.R., Beer, B., Clody, D.E. (1971) Psychopharmacologia 21, 1-7.
Weissman, B.A., Barrett, J.E., Brady, L.S., Witkin, J.M., Mendelson, W.B., Paul, S.M., Skolnick, P. (1984) Drug Dev. Res. 4, 83-93.
Wettstein, J.G. (1988) Eur. J. Pharmacol. 151, 341-344.
Winter, J.C. (1988) Pharmacol. Biochem. Behav. 29, 193-195.
Winter, J.C., Rabin, R.A. (1989) Drug. Dev. Res. 16, 327-333.
Witkin, J.M., Barrett, J.E. (1986) Pharmacol. Biochem. Behav. 24, 751-756.
Witkin, J.M., Perez, L.A. (1990) Behav. Pharm. 1, 247-254.
Young, R., Urbancic, A. Emrey, T.A., Hall, P.C., Metcalf, G. (1987) Eur. J. Pharmacol. 143, 361-371.

Animal Models in Psychopharmacology
Advances in Pharmacological Sciences

DRUG EFFECTS ON SPEED OF CONFLICT RESOLUTION IN THE SKINNERBOX

P. Koene and J.M.H. Vossen

Department of Comparative and Physiological Psychology, Psychological Laboratory, University of Nijmegen, 6525 HR Nijmegen, The Netherlands

SUMMARY: An approach-avoidance conflict test was designed to measure speed of conflict resolution in rats. In a Skinnerbox explicit approach (light) and avoidance (tone) stimuli were present to induce the conflict. Dependent on the intensities of both light and tone the animal was rewarded or punished after a criterion leverhold. Because the animal had complete control over the presentation of stimuli and reinforcers animals kept responding and the effects of several drugs on speed of conflict resolution could be measured. As expected, diazepam increased the number of punished leverholds and amphetamine showed an anorectic effect in a diminished number of responses on food rewarded leverholds. However, both FG7142, a putative anxiogenic, and haloperidol - sometimes claimed to be anxiolytic - had no specific influence. It is claimed that the paradigm differs essentially from other conflict paradigms in that it measures effects of chronic conflict, fast acquisition and extinction of conflict behaviour, and potentially could reveal putative anxiolytics, different from those revealed by paradigms measuring acute conflict.

INTRODUCTION

Anxiolytic agents diminish anxiety. Tests to find putative anxiolytic agents are mainly done using so-called conflict paradigms, for instance the Geller-Seifter test and the Vogel punished drinking test (Geller and Seifter, 1960; Vogel et al., 1971). Problems inherent to these tests are long training duration, direct (primary) punishment and shock titration to keep the subjects in conflict. A number of these problems has been resolved in other tests. The incremental shock paradigm overcomes the shock titration problem (Pollard and Howard, 1979) and the conditioned punishment paradigm enables the subject control over the presentation of the aversive CS (Koene and Vossen, 1990). We designed a conflict paradigm as a combination of primary and secondary punishment in which rats could be kept in conflict for a long time and in which subjects had complete control over presentation of stimuli and reinforcers, enabling to test the characteristics of drugs with respect to the speed of conflict resolution.

METHODS

Subjects: Ss were experimentally naive male rats, reciprocal crosses of RHA/kun and RLA/kun inbred rats, 90 days old at the start of the experiment. They were housed separately on a reversed day/night light cycle. Animals were food deprived at 85% of their predeprivation free-feeding bodyweight and had free access to water.

Apparatus: The apparatus consisted of eight operant conditioning boxes (23x24x20 cms) connected to a PDP 11/03 computer, which was used to control the experiments and to record the results. Each box contained a speaker, a stimulus light, a reinforcement compartment (magazine) and a grid floor. A retractable lever was mounted at the left side of the reinforcement compartment. Food pellets were presented in the magazine by a pellet feeder. Scrambled electric shock of 0.3 mA could be delivered through the grid floor. The conditioned stimuli were a light of maximal 400 Lux and a tone of 4 kHz of max 60 dB SPL, measured 3 cm in front of the lever.

Procedure: Animals received the following pretraining in the Skinnerbox: magazine training, CRF-training and shaping to 5 seconds leverholding. Subjects were then trained in a discrete-trial discrimination procedure in which a light or tone stimulus was presented contingent on leverholding. When holding down the lever the presence of light signalled a food reward after 5 seconds of uninterrupted leverholding and the presence of the tone stimulus signalled a foot shock. Subjects quickly learned to abort responding when the tone was presented and to continue leverholding when the light was presented. Drugs were tested in units of five days. The first two days a baseline level was established by the discrimination paradigm only. In the three following days subjects received trials in which both the light and the tone stimulus were presented contingent on leverholding; the intensity of the stimuli was systematically varied in ten steps. For each of both stimuli, the ten steps were given values from 1 to 10. The value of 10 was defined as the intensity used during original discrimination training. Food reward was given after 5 seconds uninterrupted leverholding when the value of the light intensity was larger than the value of the tone intensity ($L>T$), while the animal was punished after a 5 second leverhold when the value of tone intensity was larger than the value of light intensity ($L<T$). In this way subjects could stop or continue leverholding dependent on the information presented by the light and tone intensity. In addition to this procedure the stimulus combinations could be presented in gradual order and in random order. The test sequence was the following: on the third day control injection, on the fourth day the drug and on the fifth day again a control injection.

Drugs: The agents tested were an anxiolytic (diazepam (Diaz): 1.0 mg/kg i.p.), an anxiogenic (FG7142 (FG): 10.0 mg/kg), a stimulatory (d-amphetamine (Amph): 1.5 mg/kg) and a neuroleptic drug (haloperidol (Hal): 0.05 mg/kg). All drugs were intraperitoneally injected.

RESULTS

All animals received the same treatment on each day. Drug effects in this within-subjects design were analyzed with respect to weeks in which only injections were given (fig. 1), using the response measure;
Drug effect $[=\sqrt{}\#Responses_{drug} - \sqrt{}\#Responses_{injection}]$ for each separate day.

A significant linear trend in the repeated measures was an indication of after effects of the drug, and a significant quadratic trend indicated effects of the drug, comparing drug day with the preceding and following control injection day (tables 1 and 2).

Animals responded in almost all trials, although there was a trend to stop responding in the predictable (gradual) schedule when the tone stimulus was more intense than the light stimulus; this effect was not present in the unpredictable schedule. On average animals received between 25-30 food pellets and 1-3 shocks during the testing phase. Differences with these control data were computed to measure the effects of the drugs (tables 1 and 2).

Diazepam decreased the total number of responses in both schedules (table 1; multivariate $F=8.82$), but increased the number of shocks received in the predictable and the unpredictable schedule, while it did not affect the number of food pellets eaten (table 2). FG7142 diminished the number of responses (table 1; multivariate $F=4.54$), but had no influence on the relative number food pellets or shocks received.

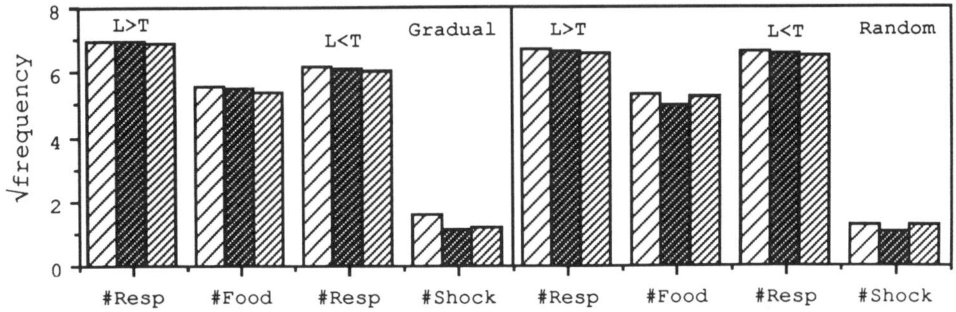

Fig. 1 Data of the three test days (control-drug-control). Number of responses in approach (Light > Tone) and avoidance (Light < Tone) trials and the number of rewarded and punished responses in a schedule with control.

Amphetamine decreased the number of responses in both schedules (table 1; multivariate F=8.29) and decreased on top of that the relative amount of food pellets eaten in both schedules (table 2). Haloperidol had no effect on the number of responses (table 1) and did not influence the number of food pellets and shocks earned in both schedules. Data suggest some aftereffects of haloperidol, concerning the total number of responses in the predictable schedule and a decrease of food pellets earned in the predictable schedule the day after drug administration.

Table 1 Drug effects and after effects on the difference in total number of responses in the 45 approach trials (Light > Tone) with food as reward and the 45 avoidance trials (Light < Tone) with shock as punishment. + $p<0.10$; * $p<0.05$; ** $p<0.01$

Total Responses	Drug	Diaz	after	FG	after	Amph	after	Hal	after
Schedule	Mul.F	8.82**	0.30	4.54**	1.16	8.29**	2.25+	0.90	2.00
predictable	Food	-1.35**	-.37	-1.17**	-.26	-1.01+	-.17	-.18	.08*
predictable	Shock	-0.94*	-.47	-1.15**	-.30	-0.88	-.29	-.28	.03**
unpredictable	Food	-1.10*	-.16	-0.06*	0.09	-1.08**	0.04	-.08	.04
unpredictable	Shock	-1.31**	-.07	-0.73**	0.03	-0.92**	0.07	-.09	.01

Table 2 Drug effects and after effects on the difference in reinforced number of responses in the 45 approach trials (Light$>$Tone) with food as reward and the 45 avoidance trials (Light$<$Tone) with shock as punishment. + $p<0.10$; * $p<0.05$; ** $p<0.01$

5-sec Responses	Drug	Diaz	after	FG	after	Amph	after	Hal	after
Schedule	Mul.F	7.43**	1.51	1.32	0.96	6.59**	0.91	1.22	1.77
predictable	Food	-.61	-.36*	-1.44	-.10+	-1.61**	-.38	0.54	0.92
predictable	Shock	0.94**	0.03	-.20	0.13	-.12	0.32	1.50	1.29
unpredictable	Food	-.34	-.87	-.35+	-.50	-1.66*	-.44	-.59	0.05*
unpredictable	Shock	0.89**	-.01	-..36	0.22	0.06	0.09	0.24	0.11

DISCUSSION

The anxiolytic profile of diazepam and the anorectic profile of amphetamine indicate that this approach-avoidance conflict paradigm can be used as a screening test for anxiolytic properties of agents. The anxiogenic effect of FG7142 (File and Baldwin, 1987) is not specifically found as in diazepam; only the nonspecific response suppression points to a possible anxiogenic effect. The effect of haloperidol is not in agreement with other reports (Pich and Samanin, 1986). However, the decrease in responding found after administration of the other three agents is not found after haloperidol treatment, indicating that the agent is at least not anxiogenic. Furthermore, the agreement between both schedules (gradual and random) in which the same drug is used is promising for the validity of the paradigm. Dose-response studies are required to further exploit this promising paradigm and compare it in more detail with the currently existing conflict paradigms.

REFERENCES

File, S.E., Baldwin, H.A. (1987) Brain Res. Bull. 19, 293-299.
Geller, I., Seifter, J. (1960) Psychopharmacologia 1, 482-492.
Koene, P., Vossen, J.M.H. (1990) Behaviour Genetics. (In press).
Merlo Pich, E., Samanin, R. (1986) Psychopharmacology 89, 125-130.
Pollard, G.T., Howard, J.L. (1979) Psychopharmacology 62, 117-121.
Vogel, J.R., Beer, R., Clody, D. (1971) Psychopharmacologia 21, 1-7.

Animal Models in Psychopharmacology
Advances in Pharmacological Sciences

A POTENTIAL ANIMAL MODEL FOR THE STUDY OF ANTI-PANIC TREATMENTS

R.L. Commissaris and D.J. Fontana

Department of Pharmaceutical Sciences, College of Pharmacy,Wayne State University, Detroit, Michigan 48202 USA.

SUMMARY: The effects of long-term chronic post-test treatment with numerous typical (e.g., imipramine, desipramine, phenelzine) or atypical (alprazolam, trazodone, mianserin, buproprion) antidepressants on conflict behaviour in rats were determined. Chronic treatment with most agents (exceptions: mianserin, buproprion) resulted in a gradually-developing anti-conflict effect, with a latency to onset of 3-5 weeks. These data suggest that animal conflict paradigms might be effective animal procedures for studying the anxiolytic and/or anti-panic effects of antidepressant agents.

Following the original observations by Klein (1964), many investigators have noted that tricyclic antidepressants (TCAs), monoamine oxidase inhibitors (MAOIS) and many (but not all) atypical antidepressants exert anxiolytic/anti-panic effects when administered chronically (Charney et al., 1986; Lydiard & Ballenger, 1987; Mavissakalian et al., 1987; Sheehan, 1983; Sheehan et al.,1983). There have been very few studies on the effects of these anti-panic treatments on behaviour in animal models for anxiety. In the present studies, the effects of long-term chronic treatment with typical (TCA and MAOI) and atypical antidepressants as well as several non-antidepressants on behaviour in the Conditioned Suppression of Drinking (CSD) conflict paradigm were investigated.

MATERIALS AND METHODS

Subjects: Female Sprague Dawley rats were housed in groups of four in a standardanimal colony room (lights on 0700 - 1900 hours). The subjects had unlimited access to food; water access was restricted during the period of behavioural testing.

Apparatus and Procedure: The apparatus was a standard rodent behavioural test chamber equipped with a tone generator and a drinking tube. An optical lickometer circuit was used tomonitor tube contacts (licks). Over the course of 4 weeks, subjects were trained as described previously (Fontana et al., 1989) to drink their daily water ration during 10-minute sessions; these sessions were characterized by alternations of silence (unpunished periods) and the presence of a tone (punished periods). Tube contact during the tone periods resulted in a 0.25 - 0.5 mA shock delivered to the mouth of the animal for the duration of the tube contact. Subjects were tested a minimum of 4 days/week, with unrestricted access to water provided on non-test days.

After 4 weeks of CSD training, subjects in each experiment were divided into two groups with comparable baselines for punished responding. One group of subjects received chronic post-test treatment with the antidepressant agent under investigation; controls received chronic post-test vehicle treatment. Drugs and drug vehicles were administered IP twice daily, seven days/week. Drug injections in the chronic post-test treatment regimens were made immediately post-test and again 12 hours later. Thus, CSD testing was conducted 12-14 hours after the most recent drug injection in the chronic treatment regimen. The purpose of this post-test chronic treatment was to minimize the effects of the acute actions of the drug at the time of behavioural testing.

Drugs and Drug Doses Examined: The following drugs were examined at the doses indicated: imipramine (2.5 mg/kg), amitriptyline (5 mg/kg), desipramine (5 mg/kg), phenelzine (4 mg/kg), pargyline (15 mg/kg), alprazolam (5 mg/kg), trazodone (40 mg/kg), buproprion (10 mg/kg), clonidine (40 μg/kg), phenobarbital (40 mg/kg) and chlordiazepoxide (10 mg/kg). Drug doses were based on literature reports of behavioural and/or neurochemical changes observed following chronic treatment.

RESULTS

In control (i.e., non-drug) conflict sessions, trained subjects reliably made approximately 2000-2500 unpunished licks (4-5 licks/second) and 30-40 punished licks (0.3-0.4 licks/second). Acute pre-test administration of traditional anxiolytics (benzodiazepines, barbiturates) dramatically increases the number of punished tube contacts made in a dose-dependent manner (McCloskey et al., 1987). In contrast, acute pre-test administration of several TCAS, MAOIs and atypical antidepressants (exception: alprazolam) do not exert such anxiolytic-like effects (Commissaris et al., 1990; Fontana & Commissaris, 1988; Fontana et al., 1989).

Table 1 Chronic Drug Treatments: Clinical Anti-Panic Efficacy -v- Anti-Conflict Efficacy in CSD Paradigm

Agent(s)	Clinical Anti-Panic Efficacy	Anti-Conflict Efficacy in CSD
Section A: Typical Antidepressants		
Imipramine	yes	yes
Desipramine	yes	yes
Amitriptyline	yes	yes
Phenelzine	yes	yes
Pargyline	-?-	yes
Section B: Atypical Antidepressants		
Alprazolam	yes	yes
Trazodone	weak	weak
Buproprion	none	none
Fluoxetine	yes	-?-
Mianserin	-?-	none
Section C: Non-Antidepressants		
Clonidine	yes	yes
Chlordiazepoxide	none	none
Phenobarbital	none	none
Clonazepam	yes	-?-

-?- Anti-panic or anti-conflict efficacy unknown.

Section A of Table 1 summarizes the effects of chronic post-test treatment with several TCAs and non-selective MAOIs on CSD conflict behaviour. As can be seen, in all cases chronic post-test treatment resulted in anxiolytic like-effects. The "onset" for this anxiolytic-like effect was approximately 3-4 weeks for all agents examined. Section B of Table 1 summarizes the effects of chronic post-test treatment with several atypical antidepressant agents on CSD conflict behaviour. As can be seen, there is an excellent correlation between the clinical anti-panic efficacy of an antidepressant agent and its effects on conflict behaviour following chronic post-test treatment. Finally, Section C of Table 1 summarizes the effects of chronic post-test treatment with several non-antidepressant agents on CSD behaviour. As can be seen, the clinically-effective anti-panic agent clonidine (Hoehn-Saric et al., 1981) also exerts anxiolytic-like effects when administered chronically, but not acutely (Commissaris et al., 1990). The agents chlordiazepoxide and phenobarbital did not result in anxiolytic-like effects when administered in a chronic post-test regimen.

DISCUSSION

In the present studies, many antidepressant agents were found to exert anxiolytic-like effects following chronic, but not acute treatment. The anti-conflict efficacy of these agents paralleled their clinical efficacy as anti-panic treatments. Chronic post-test treatment with the clinically-effective anti-panic (but not antidepressant) agent clonidine also resulted in an anxiolytic-like effect. Finally, chronic post-test treatment with phenobarbital or chlordiazepoxide failed to exert "anxiolytic-like" effects. These data suggest that rodent conflict paradigms such as the CSD may be effective animal models for the study of anti-panic treatment effects.

Bodnoff et al. (1988) have reported that chronic TCA and MAOI treatment results in anxiolytic-like effects in a novelty-suppressed feeding conflict task. In contrast to the present findings and those of Bodnoff et al., chronic antidepressant treatment has been reported not to exert anxiolytic-like effects in several other animal models for anxiety (Cassella and Davis, 1985; File et al., 1985; File and Johnston, 1987). It should be noted that the duration of chronic drug treatment (3 weeks maximum) may not have been sufficient for the production of an anxiolytic-like effect in those studies. On the other hand, chronic treatment with TCA or MAOI antidepressants for 7-12 weeks did not result in anxiolytic-like effects in the defensive burying paradigm (Beardslee et al., 1990). Thus, it appears that not all animal models for the study of benzodiazepine-like anti-anxiety effects are effective animal models for anti-panic treatment effects.

Acknowledgements: Supported in part by USPHS MH-42501 to RLC; DJF supported by the Department of Pharmaceutical Sciences, WSU.

REFERENCES

Beardslee, S.L., Papadakis, E., Fontana, D.J., Commissaris, R.L. (1990) Pharmacol. Biochem. Behav. 35, 451-455.
Bodnoff, S.R., Suranyi-Cadotte, B., Aitken, D.H., Quirion, R., Meaney, M.J. (1988) Psychopharmacology 95, 298-302.
Cassella, J.V., Davis, M. (1985) Psychopharmacology 87(3), 278-282.

Charney, D.S., Woods, S.W., Goodman, W.K., Rifkin, B., Kinch, M., Aiken, B. Quadrino, L.M., Henninger, G.R. (1986) J. Clin. Psychiatry 47, 580-586.
Commissaris, R.L., Ellis, D.M., Hill, T.J., Schefke, D.M., Becker, C.A., Fontana, D.J. (1991) Pharmacol. Biochem. Behav. (In press).
Ellis, D.M., Fontana, D.J., McCloskey, T.C., Commissaris, R.L. (1991) Pharmac. Biochem. Behav. (In press).
File, S.E., Johnston, A.L. (1987) Neuropsychobiology 17, 187-192.
File, S.E., Pellos, S., Chapin, P. (1985) Neurosci. Abstr. 11, 273.
Fontana, D.J., Carbary, T.J., Commissaris, R.L. (1989) Psychopharmacology 98, 157-162.
Fontana, D.J. Commmissaris, R.L. (1988) Psychopharmacology 95, 147-150.
Hoehn-Saric, R., Merchant, A.F., Keyser, M.L., Smith, V.K. (1981) Arch. Gen. Psychiatry 38, 1278-1282.
Klein, D.F. (1964) Psychopharmacology 5, 397-408.
Lydiard, R.B., Ballenger, J.C. (1987) J. Affective Disord. 13(2), 153-68.
Mavissakalian, M., Perel, J., Bowler, K., Dealy, R. (1987) Am. J. Psychiatry 144, 785-787.
McCloskey, T.C., Paul, B.K., Commissaris, R.L. (1987) Pharmacol. Biochem. Behav. 27, 171-175.
Sheehan, D.V. (1986) Psychosomatics 27(11), 10-16.
Sheehan, D.V., Davidson, J., Manschreck, T. (1983) J. Clin. Psychopharmacol. 3, 28-31.

A NEW ANIMAL MODEL FOR ANTICIPATORY ANXIETY?

T.J.J. Zethof*, J.A.M. van der Heyden and B. Olivier

CNS Pharmacology, Duphar B.V., P.O.Box 900, 1380 DA Weesp, The Netherlands

SUMMARY: Mice removed last from their home cage have a higher rectal temperature (hyperthermia) compared to those removed first from the same cage. This phenomenon was interpreted as a state of anxiety: anticipatory anxiety. This increase in rectal temperature was prevented in a dose dependent way by diazepam and serotonin-1A agonists. Several non-anxiolytic compounds did not affect the hyperthermia.

INTRODUCTION

In many animal models for anxiety the effects of some form of aversive stimulus on behaviour is measured. This aversive stimulus can be a footshock coupled to water drinking as in the Vogel test (Vogel et al., 1971) or a footshock coupled to exploration as in the 4-plate test (Aron et al., 1971; Bossier et al., 1968). In the light/dark anxiety model (Crawley and Goodwin, 1980) the natural aversion of rodents to avoid brightly lit places is used. Treatment with anxiolytic drugs should ideally facilitate punished (suppressed) behaviour without affecting unpunished behaviour.

In a recently described model (Borsini et al., 1989) it was observed that mice removed last from their home cage had a higher rectal temperature (hyperthermia) compared to those removed first from the same cage. This phenomenon was interpreted as a state of anxiety due to the expectation of a known or unknown event: anticipatory anxiety.

This increase in rectal temperature was prevented in a dose dependent way by prior treatment with diazepam, a well known anxiolytic compound. Several non-anxiolytic compounds (imipramine, haloperidol) did not affect the hyperthermia.

Because of its attractiveness as a new model for anticipatory anxiety, we tried to replicate and extend this model.

MATERIALS AND METHODS

Male NMRI mice (Charles River, Sulzfeld, Germany) weighing ± 25 gram were housed in groups of 10 animals per cage (34-22-15 cm) at normal 12 h light-12 h dark cycle conditions (light on 07.00-19.00 h).
The procedure used is adapted from that published by Borsini et al. (1989). The animals were handled and randomly marked (1-10) one day prior to the temperature measurement. All ten mice in each cage were treated orally with a test compound 60 minutes before the temperature measurement. The cages were allocated following a balanced design (e.g. treatment, technician, daytime). Each test group involved 8 cages.

Experiments were carried out between 9.00 a.m. and 3 p.m., at least 1 week after arrival of the animals. The temperature from all ten mice was measured sequentially. The basal temperature and increase in rectal temperature was calculated for each cage. Basal temp. = mean temp. mouse (1+2+3). Increase = mean temp mouse (8+9+10) - mean temp mouse (1+2+3). Each cage is considered as an experimental unit and between the four test groups homogeneity of variance for the variables basal temp. and increase was checked before data processing. Results were evaluated by a threeway analysis of variance (ANOVA) according to the experimental design.

RESULTS

The hyperthermic response was found to be determined only by the order in which the animals were removed from the cage (not shown). Reversing the order of measurement produced the same hyperthermic response. The response was also not dependent on whether the number of animals was reduced or maintained constant in the cage (not shown). We also found that handling did not affect the increase in temperature as revealed by the fact there was no difference between the groups handled for 1, 2, 3 or 5 consecutive days prior to testing (not shown). We essentially replicated Borsini's findings but found that only minimal handling is sufficient.

The increase in rectal temperature of mice removed last was prevented by prior treatment with diazepam in a dose dependent way (table 1). Several other non-anxiolytic drugs (imipramine, fluvoxamine, haloperidol) did not affect the hyperthermia. Interestingly flesinoxan, a 5-HT_{1A} receptor agonist, shows an anxiolytic profile in this model. 8-OH-DPAT shows a similar profile, but this is masked by the temperature lowering effects of the drug itself (basal).

Table 1 Effect of oral or subcutanious administration of various drugs on hyperthermia in mice removed at a later time

Treatment (route)	Dose mg/kg	Rectal temperature ± SEM	
		Basal (°C)	Increase (Δ°C)
Diazepam (p.o.)	0	38.0 ± 0.1	1.1 ± 0.2
	3	37.8 ± 0.1	0.3 ± 0.2 *
	6	36.6 ± 0.1 *	0.2 ± 0.2 *
	12	36.2 ± 0.1 *	- 0.2 ± 0.2 *
Fluvoxamine (p.o.)	0	37.9 ± 0.1	1.3 ± 0.1
	3	38.2 ± 0.1	0.9 ± 0.1 *
	10	38.2 ± 0.1	0.9 ± 0.1 *
	30	38.2 ± 0.1	0.8 ± 0.1 *
Imipramine (p.o.)	0	38.0 ± 0.1	0.9 ± 0.1
	3	37.8 ± 0.1	1.0 ± 0.1
	10	37.8 ± 0.1	1.0 ± 0.1
	30	37.4 ± 0.1 *	1.2 ± 0.1
Haloperidol (p.o.)	0	38.0 ± 0.1	1.1 ± 0.1
	0.3	38.1 ± 0.1	1.0 ± 0.1
	1	38.0 ± 0.1	0.9 ± 0.1
	3	37.8 ± 0.1	0.9 ± 0.1
Flesinoxan (p.o.)	0	37.9 ± 0.1	1.1 ± 0.1
	0.3	38.3 ± 0.1	0.7 ± 0.1
	1	38.2 ± 0.1	0.4 ± 0.1 *
	3	38.1 ± 0.1	0.3 ± 0.1 *
8-OH DPAT (s.c.)	0	38.4 ± 0.1	0.8 ± 0.1
	0.03	38.3 ± 0.1	0.8 ± 0.1
	0.1	38.4 ± 0.1	0.5 ± 0.1
	0.3	38.0 ± 0.1 *	0.3 ± 0.1 *

Values represent mean ± SEM from 8 cages. The time between drug administration and temperature recording was 30 min (diazepam and 8-OH-DPAT) or 60 min (others). Basal refer to the rectal temperature of the first three mice and increase refer to the difference between the last three mice and the first three mice. *$p<0.05$ ANOVA statistics compared with control treatment.

DISCUSSION

The increase in rectal temperature could be demonstrated even after reversal the order of removal, which means that the hyperthermia depends only on the order of removal and is not a characteristic of any particular animal. We extended Borsini's findings (Borsini et al., 1989) by showing that prior handling is not necessarily to measure the hyperthermia, so minimal handling is sufficient.

The fact that the increase in rectal temperature was prevented by diazepam (anxiolytic), but not by haloperidol (neuroleptic) or imipramine (antidepressant) would suggest that mechanisms related to anxiety are involved. For this reason it is very encouraging that 5-HT_{1A} agonists, like 8-OH-DPAT (Kennett et al., 1987) and flesinoxan (Olivier et al., 1990), are also active in this model. Lecci et al. (1990) have also shown that 5-HT_{1A} agonists (8-OH-DPAT and buspirone) antagonized the rise in body temperature (stress-induced hyperthermia (SIH). Although at the higher doses 5-HT_{1A} agonists decrease bodytemperature (8-OH-DPAT), this effect does not explain the decrease in the stress-induced enhanced temperature of the last removed mice. Therefore this anticipatory anxiety model seems to detect 5-HT_{1A} as potential anxiolytics (besides the benzodiazepines). Recent studies (Lecci et al., 1990) have indicated that the model does not pick up other potential anxiolytics like 5-HT_2 or 5-HT_3 antagonists. Although further studies are clearly needed, this model seems to represent a new tool for studying the psychobiological basis of anxiety, in particular the "alarm reaction" due to emotional stimuli. In this model we can demonstrate anxiolytic activity of compounds with different mechanisms.

REFERENCES

Aron, C., Simon, P., Larousse, C., Boissier J.R. (1971) Neuropharmacology 10, 459-569.

Borsini, F., Lecci, A., Volterra, G., Meli, A. (1989) Psychopharmacology 98, 207-211.

Boissier, J.R., Simon, P., Aron C. (1968) Eur. J. Pharmacol. 4, 145-151.

Crawley, J., Goodwin, F.K. (1980) Pharmacol. Biochem. Behav. 13, 167-170.

Kennett, G.A., Dourish, C.T., Curzon, G. (1987) Eur. J. Pharmacolo. 134, 265-274.

Lecci, A., Borsini, F., Mancinelli, A., D'Aranno, V., Stasi, M.A., Volterra, G., Meli, A. (1990) J. Neural Transm. 82, 219-230.

Olivier, B., Mos, J., Tulp, M.Th.M., Van der Heyden, J.A.M., Ybema, C., Slangen, J. (1990) In: Stress, biological rhythms and psychiatric disorders. Medidact, Houten, pp 79-87.

Vogel, R.A., Beer, B., Clody, D.E. (1971) Psychopharmacology 21, 1-7.

Animal Models in Psychopharmacology
Advances in Pharmacological Sciences

PHARMACOLOGY OF FEAR, INDUCED BY PERIAQUEDUCTAL GRAY STIMULATION IN THE RAT

C.L. Broekkamp, C. Dortmans, H.H.G. Berendsen and F. Jenck

CNS-Pharmacology Department, Organon, P.O.Box 20, Oss, The Netherlands

SUMMARY: The effect of drugs on escape induced by aversive stimulation of the periaqueductal gray (PAG) matter is reviewed. Strongest influences were observed with compounds influencing GABA-ergic and serotonergic neurotransmission. Activation of the 5-HT_{1C}-receptor is an effective way towards inhibition of the aversive effect of brain stimulation. An opposite effect was obtained by stimulation of 5-HT_{1A} and 5-HT_2-receptors. Dopaminergic D_2-receptor activation seems to have weak enhancing effects on the stimulation induced aversive state. In view of these pharmacological characteristics it is not possible to conclude that there is a straight forward parallelism between brain stimulation-induced fear and a particular anxiety disorder.

INTRODUCTION

Many animal models for anxiety disorders are proposed to date and most have in common that the environment of the animal is manipulated. As a result the anxiety-like behaviour in such procedures is directly related to fear-provoking stimuli in the environment. A notable exception is anxiety-like behaviour provoked by aversive brain stimulation. From the phenomenological point of view clear fear-like behaviour is provoked with electrical stimulation of the periaqueductal gray matter (PAG).

The 'fear' which seems to be elicited, arises for the animal suddenly and completely unrelated to environmental changes. The study of brain stimulation-induced fear with therapeutic agents was proposed in particular as it seems a realistic simulation of fear in generalized anxiety disorders and panic disorder (Bovier et al., 1982; Broekkamp et al., 1989). We have investigated the effect of many drugs on this behaviour. The results reveal that the pharmacological characteristics sometimes correspond with therapeutic effects of drugs in anxiety disorders and sometimes not. It is therefore not possible at this stage to link brain stimulation induced avoidance to a particular type of anxiety disorder. Nevertheless, insights are gained into the neurotransmitter involvement in fear-like behaviour provoked by stimulation of the periaqueductal gray matter.

METHODS

Methods are described in detail in previously published reports. (Jenck et al., 1989 a,b; 1990 a,b). We have trained rats to terminate electrical brain stimulation via permanently implanted electrodes by moving to the other half of the experimental cage, passing a 2 cm high barrier which divides the cage into two halves. The bipolar electrodes were stereotaxically implanted under anaesthesia in the dorsal part of the periaqueductal gray substance. After a recovery period of at least one week the rats were trained to terminate electrical brain stimulation in daily 30 min training sessions.

Stimulation consisted of trains of square wave pulses (0.1 msec) passing to the implanted electrode via a constant current unit. Initially, an effective current intensity was determined for each rat at a fixed frequency of 50 Hz. Each minute stimulation was started and there was no warning signal for brain stimulation onset. A positive response was noted when an escape occurred within 20 sec. Once an effective current level at 50 Hz was determined, this current intensity was held constant for each animal and throughout the sessions. Changes in the sensitivity for brain stimulation by drug treatments was determined by stepwise variations in the frequency. The frequency was increased or decreased depending on the response to the previous brain stimulation trial. In this way an average stimulation threshold frequency was determined. Drug effects were tested by first determining a pre-injection threshold for escape in a 30 min session. Immediately afterwards a drug was administered. Fourty min later, rats were placed again into the experimental cage and threshold frequency was determined again during a 25 min session. Between each trial a time of at least 40 sec passed. During this time the animal could freely move in the experimental cage and could pass the barrier. Such spontaneous passages were counted and provided a measure for spontaneous activity in the cage.

RESULTS

We have organized our results in tables with selectively active compounds, non-selectively active compounds and other compounds.

In table 1 the compounds are listed which increased or decreased the frequency of aversive brain stimulation and, at the indicated doses, (dose range) were devoid of effects on spontaneous activity in the experimental cage as would have been reflected in the number of spontaneous passages during stimulation intervals.

Table 1 Compounds with reducing or enhancing effects on brain stimulation-induced escape responses without affecting spontaneous locomotor activity.

Escape reducing compounds	(mg/kg)	Escape enhancing compounds	(mg/kg)
Diazepam	3.2	8-OH-DPAT	0.2 - 1.0
Alprazolam	1.0 - 5	Cyproheptadine	3.2 - 10
mCPP	0.2 - 4.6	Ritanserin	32
TFMPP	0.45 - 1.0	Mianserin	1.0 - 32
DOI	1.0 - 4.6	Metergoline	0.1 - 10
Ketanserin	1.0 - 10	Mirtazapine	0.1 - 32
Pirenperone	0.1 - 0.32	Nomifensine	3.2
Trazodone	1.0 - 22	Quinpirole	
Sertraline	1.0 - 10	(= LY 171555)	0.2 - 2.2
Fluvoxamine	1.0 - 22	Pergolide	0.32
Haloperidol	0.01 - 0.032	RU 24969	1.0

In table 2 the compounds are listed which did not have selective effects on escape behaviour. These compounds did change the frequency thresholds of escape but there were also effects on spontaneous passages.

Table 2 List of compounds with non-selective effects on brain stimulation induced escape responses.

Compounds, reducing escapes and having decreasing effects on spontaneous locomotor activity	mg/kg	Compounds enhancing escapes and having increasing effects on spontaneous locomotor activity	mg/kg
Prazosin	5.0	Bupropion	1.0 - 10
Buspirone	0.03 - 10	Dexamphetamine	0.1 - 0.45
Ipsapirone	0.1 - 32	Isoprenaline	0.1 - 0.5
SKF 38393	1.0 - 10		
SCH 23390	0.02 - 0.1		
Metoclopramide	0.1 - 10		

In table 3 compounds are listed which did not influence behaviour in these experiments.

Table 3 List of compounds tested that do not affect brain stimulation induced escape responses nor the spontaneous locomotor activity.

Compound	Dose range	Unit
ICS 205-930	0.01 - 10	mg/kg
Ondansetron	0.1 - 10	mg/kg
MDL 72222	0.1 - 22	mg/kg
Clomipramine	10 - 22	mg/kg
Sulpiride	15 - 68	mg/kg
(-) Pindolol	0.32 - 3.2	mg/kg
Phenelzine	10 - 22	mg/kg
Imipramine	10 - 22	mg/kg

DISCUSSION

As shown in previous studies benzodiazepines are effectively reducing escape behaviour induced by periaqueductal gray stimulation (Bovier et al., 1982; Graeff, 1984). The role of GABA-ergic mechanisms was discussed in these reports. Other compounds can best be interpreted and discussed in relation to the receptor which seems to be responsible for the effect. Several of the conclusions described below are based on the interpretation of data with compounds which influence more than one receptor. A more complete justification for some of these conclusions is given in the original reports on this work (Jenck et al., 1989a,b; 1990a,b).

The clearest involvement in modulation of aversive PAG stimulation was found for the 5-HT_{1C}-receptor (Jenck et al., 1990a). Compounds with 5-HT_{1C} agonistic effect, such as mCPP, DOI and TFMPP, inhibited escape responses to brain stimulation and compounds which block the 5-HT_{1C}-receptor, such as mianserin, cyproheptadine, metergoline and ritanserine, enhance the sensitivity for PAG-stimulation.

The 5-HT_{1A}-receptor is also mediating effects on aversive brain stimulation although it is less dominant than the 5-HT_{1C} effect. 8-OH-DPAT, the selective 5-HT_{1A}-receptor agonist, and RU 24969, a compound with 5-HT_{1A} agonistic effects, facilitated escape from PAG stimulation. Blockade of the 5-HT_2-receptor with for example ketanserin, pirenperone and trazodone had effects opposite to the mixed 5-HT_{1C}-receptor antagonists and reduced escape responses.

We did not find any anxiolytic-like effect with 5-HT_3-receptor antagonists such as ICS 205-930, MDL 72222, ondansetron and metoclopramide.

Thus, the influence of serotonin, for which an important role was indicated in early studies with brain stimulation (Kiser et al., 1978) is effectuated via different pathways and receptors. The fear-like behaviour elicited with PAG stimulation appears to be inhibited by 5-HT_{1C}-receptor activation and this is opposed by activation of 5-HT_{1A} and 5-HT_2-receptors. It is therefore attractive to suppose that the inhibitory effects which were found by indirect activators of serotonergic transmission such as the serotonin re-uptake inhibitors sertraline and fluvoxamine, is mediated by 5-HT_{1C}-receptors. The re-uptake inhibitor of noradrenaline, imipramine, and the mixed serotonin-noradrenaline re-uptake inhibitor clomipramine, did not influence escape behaviour, nor did we find selective effects with agents which influence alpha-1 or beta noradrenergic receptors such as prazosin, isoprenaline or (-)pindolol.

The results with compounds which influence dopaminergic transmission were less clear. An indirect activator of dopamine receptors by dopamine re-uptake inhibition, nomifensine, enhanced escape for brain stimulation, but a compound with a similar pharmacological effect, bupropion, was not selectively active. The strong effects of agents influencing motor performance, which is often the case with compounds interacting with dopaminergic receptors, complicate proper evaluation of such compounds. Nevertheless, the selective escape facilitating effect of the D_2-receptor agonist quinpirole and the lack of selective effects of D_1-selective agents SKF 38393 and SCH 23390 suggests some fear-enhancing effect of D_2-receptor activation. In agreement with this, some reduction of escape behaviour which was selective with respect to other motor effects was observed with low doses of haloperidol. Dexamphetamine acted more like a stimulant and a facilitative effect on brain stimulation-induced escape responses was not found at doses which were free of effects on spontaneous passages in the experimental cage. Possibly, amphetamine's releasing effect of serotonin opposes a selective escape facilitating effect due to dopamine release.

The inefficacy of buspirone might also be related to a mixed effect whereby a possible escape reducing effect via 5-HT_{1A}-receptor blockade could have been opposed by its dopamine receptor blocking effect.

The activity of these compounds and observations with efficacy in anxiety disorders should enable a discussion the relevance of this experimental procedure as a model for some kind of anxiety disorder. However, there are still many differences in the kind of observations which make it premature to draw conclusions in this direction. Drug testing in medical practice is time consuming. Usually, the drugs only show activity after chronic administration and many drugs are not yet tested in specifically diagnosed anxiety disorders. Closest resemblance of this animal model should be expected with panic disorder. One can point at positive effects for serotonin uptake inhibitors both in panic disorder and in brain stimulation-induced panic (Den Boer and Westenberg. 1989) and on negative effects with buspirone also in panic disorder patients (Sheehan et al., 1990). However, mCPP and the serotonin releaser fenfluramine are reported to induce panic disorder in man after acute intravenous injections (Charney et al., 1987. Targum and Marshall, 1989), but mCPP is inhibiting fear induced by brain stimulation. Therefore, we think that, rather than studying this behaviour from the perspective of an animal model for a particular anxiety disorder, it is more useful to study it first from the perspective of gaining more insight into mechanisms and transmitter involvement in fear-like behaviour elicited from the periaqueductal gray substance, hoping that this will later fit into a broader scheme on anxiety and fear in mammals which should incorporate both animal and clinical data.

REFERENCES

Bovier, P., Broekkamp, C.L.E., Lloyd, K.G. (1982) Brain Res. 248, 313-320.
Broekkamp, C.L.E., Berendsen, H.H.G., Jenck, F., van Delft, A.M.L. (1989) Psychopathology 22 (suppl. 1), 2-12.
Den Boer, J.A., Westenberg G.M. (1989) Int. Clin. Psychopharmacol. 3, 59-74.
Charney, D.S., Woods, S.W., Goodman, W.K., Heninger, G.R. (1987) Psychopharmacology 92, 14-24.
Graeff, F.G. (1984) TIPS 5, 230-233.
Jenck, F., Broekkamp, C.L.E., van Delft, A.M.L. (1989a) Psychopharmacology 97, 489-495.
Jenck, F., Broekkamp, C.L.E., van Delft, A.M.L. (1989b) Eur. J. Pharmacol. 161, 219-221.
Jenck, F., Broekkamp, C.L.E., van Delft, A.M.L. (1990a) Psychopharmacology 100, 372-376.
Jenck, F., Broekkamp, C.L.E., van Delft, A.M.L. (1990b) Eur. J. Pharmacol. 177, 201-204.
Kiser, R.S., German, D.C., Lebovitz, R.M. (1978) Pharmacol. Biochem. Behav. 9, 27-31.
Sheehan, D.V., Raj, A.B., Sheehan, K.H. (1990) J. Clin. Psychopharmacol. 10, 3-11.
Targum, S.D., Marshall, L.E. (1989) Psychiatry Res. 28, 295-306.

Animal Models in Psychopharmacology
Advances in Pharmacological Sciences

ELECTROPHYSIOLOGICAL EVIDENCE FOR THE INVOLVEMENT OF 5-HT2 RECEPTORS IN THE ANTIAVERSIVE ACTION OF 5-HT IN THE DORSAL PERIAQUEDUCTAL GRAY

M.L. Brandao, J.A. Lopez-Garcia and M.H.T. Roberts

Department of Physiology, University of Wales College of Cardiff, Cardiff, CF1 1SS, U.K.

SUMMARY: The technique of microiontophoresis was employed to assess the identity of the receptors mediating the effects of 5-hydroxytryptamine (5-HT) upon neurones of the dorsal periaqueductal gray (DPAG). Most cells which responded were excited by 5-HT. These effects were clearly attenuated by ketanserin, a selective 5-HT_2 antagonist. Zimelidine, a specific 5-HT reuptake blocker significantly enhanced the excitatory effects of 5-HT. These results give electrophysiological support to the proposal of Graeff (1990) that 5-HT_2 mechanisms in the DPAG may mediate, at least part of the therapeutic effects of 5-HT uptake inhibitors in panic disorders.

INTRODUCTION

It has been suggested that activation of the DPAG is a valid animal model of panic attack (Panksepp, 1982, Graeff, 1987) and that 5-HT_2 mechanisms phasically inhibit the substrates of aversion in the DPAG (Graeff, 1990). In order to test this hypothesis 5-HT was applied by microiontophoresis to cells in the DPAG and challenged with ketanserin, a highly selective 5-HT_2 receptor blocker (Fozard, 1987). The 5-HT_{1A} receptor seems also to be involved in aversive responses in the DPAG (Jenck et al.,1989) but no specific antagonist for 5-HT_{1A} receptor is available. It has been shown, however, that 8-OH-DPAT is a powerful and selective agonist at this 5-HT receptor subtype (Middlemiss and Fozard, 1983). The effects on DPAG cells of 8-OH-DPAT and two anxiolytic drugs, buspirone and gepirone, which also have high affinity for 5-HT_{1A} receptors (Fozard, 1987) were studied. Zimelidine, a selective 5-HT reuptake blocker (Ross and Renyi, 1977) was also applied to cells responding to 5-HT.

MATERIALS AND METHODS

Experiments were conducted on male Wistar rats anaesthetized with halothane. Five-barrelled microelectrodes with a tip of 5 to 7 μm were aimed at the lateral parts of the DPAG. One barrel was filled with NaCl (4M) for the extracellular recording of action potentials. A second barrel was used for current balancing and the remaining three were filled with drug solutions.

The effects of ketanserin were studied on cells excited both by 5-HT and by DL-homocysteic acid (DLH). When consistent and repeatable responses to 5-HT and DLH were obtained, the antagonist was applied for 15 min from another barrel by the passage of a small current (lOnA) without interrupting the agonist cycle. In order to study the effects of zimelidine, 5-HT was repeatedly applied to a DPAC cell until the excitatory responses were stable. Zimelidine was then applied with a single iontophoretic pulse (lOOnA, 60s) without interrupting the sequence of 5-HT applications. Responses to drug administration were quantified by measuring the area of the response from the polygraph recordings and converting to total spike number (TSN) as described previously (Bradshaw et al., 1974).

The following drug solutions were used: 5-HT bimaleate (0.2M, pH 4.5, Sigma), DLH (0.2M, pH 8.0, Sigma), ketanserin tartrate (0.05M, pH 4.0; Janssen), buspirone HCl (O.O5M, pH 4.5, Sigma), gepirone HCl (0.2M, pH 4.0, Bristol Myers) zimelidine (0.2M, pH 4.0, Astra) and 8-hydroxy-2-(di-n-propylamino) tetralin HBr (8-OH-DPAT, 0.02M, pH 4.0, Res. Biochem. Inc.).

RESULTS

5-HT was applied to a total of 176 spontaneously active DPAG cells in this work. One hundred and thirteen of these cells were unresponsive while 45 were excited and 18 were depressed by microiontophoretically applied 5-HT. In general, cells giving excitatory responses to 5-HT had very low baseline firing rates whereas cells giving depressant responses to 5-HT showed high spontaneous firing activity at baseline.

Zimelidine depressed the activity of cells depressed by 5-HT but neither depressed nor excited cells excited by 5-HT. However, zimelidine strongly potentiated the excitatory effects of 5-HT (Fig. 1A). Ketanserin clearly attenuated the excitatory effects of 5-HT without altering the effects of the control agonist DLH (Fig. 1B). The 5-HT_{1A} agonists failed to modify the firing rate of cells excited by 5-HT but caused a significant reduction in the firing rate of cells depressed by 5-HT (Fig. 2).

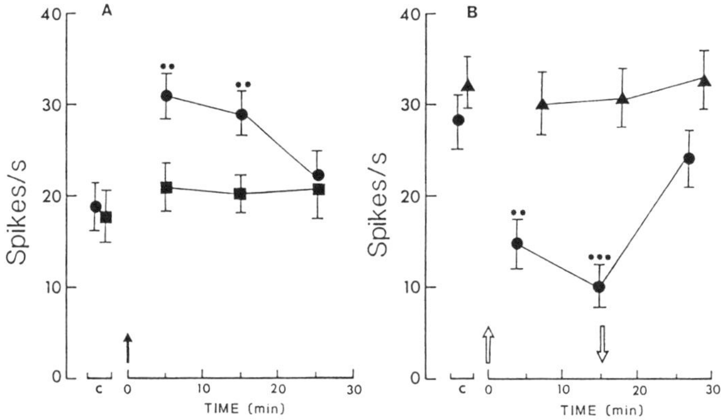

Fig. 1 A. Effects of zimelidine (ZIM) and gepirone(GEP) on the excitation induced by 5-HT. The 5-HT effects on DPAG cells tested with ZIM are shown in circles and on cells tested with GEP in squares. Single administration (lOOnA, 60s) of ZIM orGEP was made at the time indicated by the arrow. n=8 in both cases. B. Effects of ketanserin (lOnA) on DPAG cells excited by 5-HT (circles) and DLH (triangles). n=6 in both cases. Ketanserin was applied during the time indicated by the arrows. c=control. Spikes/s was calculated by dividing the total spike number (TSN) during the response by the time of drug application.*P<O.05, **P<0.01, ***P<0.001, paired Student's t test.

DISCUSSION

These results show that the majority of cells reacting to 5-HT in the DPAG are excited. The 5-HT_{1A} agonists failed to excite any cells but mimicked the depressant effects of 5-HT. Only the excitatory effects of 5-HT were blocked by ketanserin suggesting that excitation is mediated by 5-HT_2 receptors.

It has been suggested that generalized anxiety disorder may be the result of an overactivation of 5-HT_2 and/or an impairment of 5-HT_{1A} mechanisms in forebrain structures such as hippocampus, amygdala and frontal cortex (Deakin, 1988). An action on these receptors could explain the beneficial effects of 5-HT_2 antagonists and 5-HT_{1A} agonists in anxiety. However, this hypothesis does not seem to apply to the neurohumoral mechanisms underlying the aversive states related to panic attacks (Graeff,1990). In fact it has been found that 5-H_{1A} agonists may precipitate panic in

Animal Models in Psychopharmacology
Advances in Pharmacological Sciences

BEHAVIOURAL AND HORMONAL INDICES OF ANXIOLYTIC AND ANXIOGENIC DRUG ACTION IN THE SHOCK PROD DEFENSIVE BURYING/AVOIDANCE PARADIGM

Sietse F. de Boer, Jan van der Gugten and Jef L. Slangen

Department of Psychopharmacology, Faculty of Pharmacy, University of Utrecht, Sorbonnelaan 16, 3584 CA Utrecht, The Netherlands

INTRODUCTION

The bodily response of mammals to stressful situations is characterized by a complex pattern of behavioural, endocrine and autonomic changes (Fig. 1). A variety of these so-called "stress" indices are nowadays routinely employed as the dependent variable in animal models designed for studying the neuropharmacological mechanisms of stress/anxiety and, in particular, for evaluating the potential anti-stress or anxiolytic actions of new compounds. These include: suppressed operant behaviour (Geller & Seifter, 1960; Vogel et al.,1971), startle potentiation (Davis, 1979), decreased exploratory activity (Crawley, 1981; Meert & Colpaert, 1984; Yamamoto & Ueki, 1987), decreased social interaction (File, 1980), increased burying behaviour (Treit et al., 1981), and elevated plasma corticosteroid concentrations (Lahti & Barsuhn, 1974; Le Fur et al., 1979; Gram & Christensen; 1986). It is generally assumed that compounds with anti-stress or anxiolytic activity are characterized by their potency to inhibit or block these changes, while the ability to induce or enhance them is a reflection of anxiogenic or pro-stress properties. The usefulness and shortcomings of different animal models of stress/anxiety have been extensively reviewed by Treit (1985), and will not be discussed further in this chapter.

However, in contrast to the multimodality (Figure 1), complexity and heterogenity of the stress response (Bohus et al., 1987, De Boer, 1990), the above listed animal models make use of only one single (mostly unidirectional) behavioural or hormonal index of stress, fear and/or anxiety. As a result, the interpretation of anxiolytic drug action is complicated and may lead to "false negatives" and/or "false positives". Therefore it seems attractive to study anxiolytic drug effects in a test-situation that allows several different behavioural and autonomic/endocrine manifestations of stress/anxiety to be measured simultaneously.

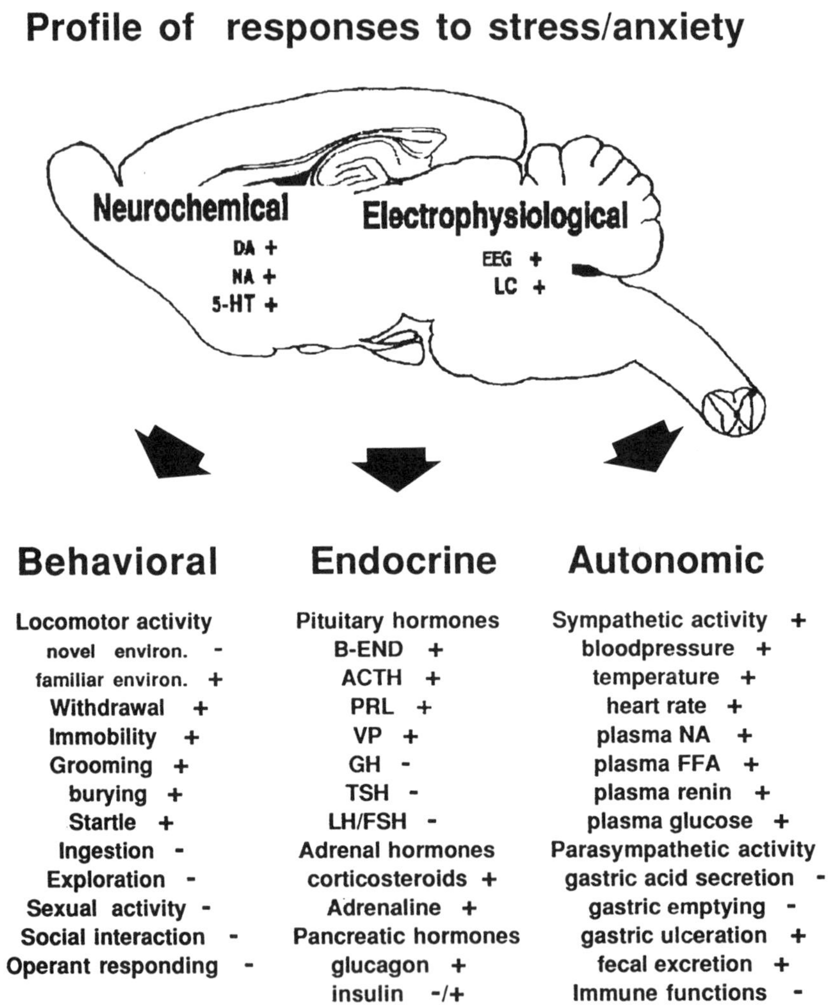

Fig. 1 A schematic overview of the various neurochemical, electrophysiological, behavioural, endocrine and autonomic parameters which are affected when an organism is exposed to stressful or anxiety-provoking stimuli. "+" indicates an increase in the measure. "-" a decrease.

In order to combine different behavioural indices of stress/anxiety with several endocrine/autonomic parameters within a single experimental setting, we recently adapted a fairly-well proven validated and relatively economic behavioural test for anxiolytic and anxiogenic action, i.e. the shock-prod defensive burying/avoidance paradigm. In this chapter we shortly review the use of this animal model in psychopharmacological research and describe results obtained in experiments which combine behavioural and hormonal measurements.

THE SHOCK-PROD DEFENSIVE BURYING/AVOIDANCE PARADIGM: GENERAL CHARACTERISTICS AND BEHAVIOURAL MEASURES

In 1978, Pinel, Treit and co-workers (Pinel & Treit, 1978) reported that rats, when tested in an environment with suitable bedding material on the floor, have a very strong innate tendency to bury well-localized sources of noxious stimulation, such as a wire-wrapped prod attached to the test-chamber wall from which they have received an electric shock upon contacting it. This prod-burying behaviour consists of very characteristic movements of the front paws and snout in a pushing manner so as to shove material (either forward or sideways) toward and over the shock-prod. Typically, the frequency and duration of burying behaviour as well as an intimately related measure i.e. the height of burying material accumulated around the shock-prod, are used as the dependent variable measures of fear/anxiety in this so-called defensive burying paradigm. Subsequent studies performed in a variety of laboratories over the past decade have expanded and delineated the parameters which affect this robust and species-typical active avoidance or coping response (see Pinel & Treit, 1983 for an extensive review of this literature). A summary of the research on the main factors influencing defensive burying behaviour is presented in table 1, and will not be discussed in detail. It shows that the degree of burying is both determined and modified by a variety of environmental, stimulus and organismic variables. This plasticity makes it feasible to manipulate "baseline" levels of burying behaviour by, for example, simply adjusting the size of the test chamber or the intensity of the prodshock.

Furthermore, it has been shown that in an environment without bedding material on the floor, where the burying option is eliminated, rats display a passive form of avoidance behaviour e.g., immobility in locations away from the continuously electrified prod (De Boer et al., 1990; Treit et al., 1986). In addition, exploration of the prod, which can be quantified as the number and duration of physical contacts with

TABLE 1

FACTORS AFFECTING DEFENSIVE BURYING.

Factors		References
ENVIRONMENTAL/STIMULUS VARIABLES		
- Size and shape of test chamber		Pinel *et al.*, 1980
		Moser *et al.*, 1983
	small > large	Treit *et al.*, 1986
	one-compartment > two-compartment	Davis *et al.*, 1982a
- Type and availability of burying material		Pinel *et al.*, 1979
	sand wooden blocks	Treit *et al.*, 1986
		Peacock *et al.*, 1982
	clean bedding > stress odor contaminated bedding	Williams, 1987a,b
- Type and intensity of the localized aversive stimulus		Treit *et al.*, 1980
	electrified prod electrified lever	Terlecki *et al.*, 1979
	flash bulb dead conspecifics	Goldberg *et al.*, 1983
	mouse trap noxious food/water	Wilkie *et al.*, 1979
		Davis *et al.*, 1981, 1982b
	high intensity > low intensity	Poling *et al.*, 1981
ORGANISMIC VARIABLES		
- Species		Whillans *et al.*, 1981
	only rats, mice and gerbils; hamsters not	Treit *et al.*, 1980
		Davis *et al.*, 1980
- Strain	PVG > Wistar	Tsuda *et al.*, 1988
	Fischer > Wistar	McKim *et al.*, 1979
	Long-Evans > Wistar	Tarte *et al.*, 1982
	MR = MNRA	Beardslee *et al.*, 1989
	CF1 > CB1 > BALB/c	Treit *et al.*, 1980
		Harder *et al.*, 1983
- Interindividual variability		Peacock *et al.*, 1982
- Age	old > young	Treit *et al.*, 1980
- Sex	no sex difference	Treit *et al.*, 1980
- Metabolic/physiological condition		
	satiated > food/water deprived	Davis *et al.*, 1981
	control > insulin-induced hypoglycemia	Davis *et al.*, 1980
- Previous experiences		Treit *et al.*, 1988
		Davis *et al.*, 1981
	number of habituation trials	Tarte *et al.*, 1982
	without prod > with prod	Oberdieck *et al.*, 1981
	controllable stress > uncontrollable stress	Williams, 1987a,b
NEUROBIOLOGICAL MANIPULATIONS		
- Brain lesions		
	anterior septum ↔	Gray *et al.*, 1981
	posterior septum ↓	Treit *et al.*, 1990a
	frontal cortex ↓	Kolb *et al.*, 1981
	neocortex ↓	
- Pharmacological drug treatments		see table 2

↓ decrease in burying.

it, is strongly reduced when the prod remains electrified (see figures 2 and 3). Both the occurrence of immobility (freezing and/or scanning) and the suppression of prod-exploration have been used as the dependent variable measure of fear and/or anxiety in this so-called "shock-probe conflict test" (De Boer et al., 1990; Meert & Colpaert, 1986; Treit et al., 1990), which uses a continuously electrified prod and no bedding material.

Thus, it may be clear that when a continuously electrified shock-prod is used in a chamber with bedding material, three concurrent behavioural measures of fear/anxiety can be registrated: 1) suppressed prod-exploration; 2) increased prod-burying, and 3) increased immobility. In addition, various other behavioural elements which are sensitive to stress can be scored as well, e.g., grooming, rearing and ambulation, eating and drinking, and approach-withdrawal responses (Moser & Tait, 1983; Peacock & Wong, 1982; Tsuda et al., 1988). Behavioural studies have also revealed that there is a reliable sequence or hierarchy in the rat's defensive response repertoire upon contacting the electrified prod, i.e. reflexive startle or leaping to the far end of the chamber and initial freezing/scanning in that location away from the prod followed by a series of approach-withdrawal movements and, finally, the partial or complete burying of the prod. The occurrence and amount of time spent on each of these defensive behavioural elements appears to be a function of organismic variables (individual coping style) as well as of environmental constraints (availability of bedding, escape possibilities) or experienced fear in a particular test situation (De Boer et al., 1990; Treit et al., 1986, Tsuda et al., 1988). Linked to this notion are the findings that anxiolytic drug action might vary as a function of these parameters (Treit et al., 1986) and/or that these drugs modulate the hierarchy in a rat's defensive coping repertoire (De Boer et al., 1990; Tsuda et al., 1988).

The shock-prod paradigm also allows to investigate the conditioned emotional consequence of former shock-punishment. One day after the electrified shock-prod test (conditioning test), the suppressed prod-exploration, increased burying and enhanced immobility responses can be reliably evoked when the rats are re-exposed to the same, but now unelectrified shock-prod (see Fig. 2). Viewing this experimental set-up within the Pavlovian scheme, prod-burying, prod-exploration and immobility behaviour may be seen as a conditioned response (CR) that results from the association of the stationary prod, or conditioned stimulus (CS) and the shock, or unconditioned stimulus (US).

Since it has been shown that burying is employed, and most easily elicited, in a familiar, "home-bound" environmental situation (Tarte & Oberdieck, 1982; Treit & Fundytus, 1988), we tested our rats in their own home cages (for details of the

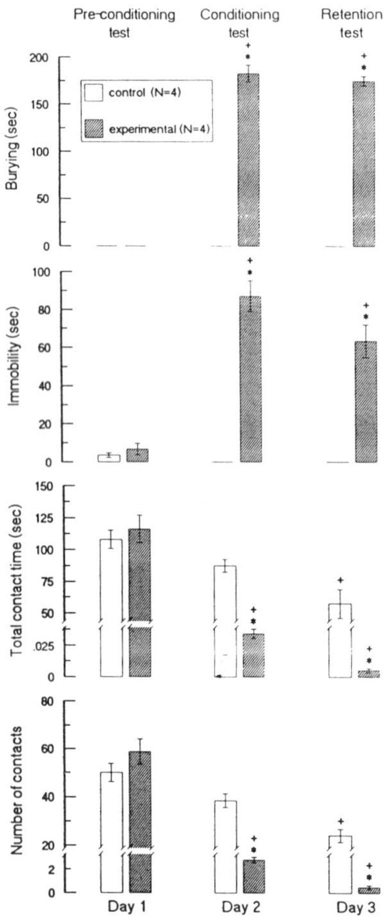

Fig. 2 Mean (± SEM) number and duration of prod contacts as well as mean (± SEM) duration of immobility and burying behaviour observed for the control (never shocked) and experimental group of rats at pre-conditioning (day 1), conditioning (day 2) and retention (day 3) sessions.
* Significantly different from control group of rats.
\+ Significantly different from the respective pre-conditioning value on day 1 ($p<0.05$).
During 3 consecutive days, control rats were confronted for 15 min/day with an unelectrified shock-prod. Note the gradual habituation in shock-prod exploratory behaviour over the 3 days. Experimental animals received the same procedure, except on day 2 where they were confronted with an electrified (2 mA) shock-prod. Note the pronounced conditioned burying (increase), immobility (increase) and prod-exploration (suppression) responses in the experimental animals on day 3 (retention test).

methodology see De Boer et al., 1990). Interestingly, it was frequently observed that the shocked rats covered/plugged the hole in the front wall of the cage through which the prod had been inserted during the conditioning test. These observations strenghten the assumption that burying behaviour represents a fundamental part of the rat's naturalistic coping repertoire (Pinel & Treit, 1983), serving to protect the individual from sources of aversive stimulation.

TABLE 2

SUMMARY OF DRUG EFFECTS IN THE DEFENSIVE SHOCK PROD BURYING TEST.

Drug	Route	Dose range (mg/kg)	ITI (min)	MED (mg/kg)	Reference
DRUG TREATMENTS THAT DECREASE BURYING					
Buspirone	SC	0.05–1.0	30	0.05	Treit *et al.*, 1988
Chlorpromazine	IP	1.0–4.0	30	1.0	Treit *et al.*, 1981
	IP	2.0	60		Davis *et al.*, 1981b
Chlordiazepoxide	IP	1.0–6.0	30	3.0	Treit *et al.*, 1981, 1990b
(CDP)		2.5–10.0	30	2.5	Treit *et al.*, 1988
		4.0–32.0	30	4.0	Craft *et al.*, 1988
		3.0–7.1	30	5.0	Beardslee *et al.*, 1990
	IG	9.0	60		De Boer *et al.*, 1990
Diazepam	IP	0.1–6.0	30	0.5	Treit *et al.*, 1981
		0.5–1.0	20	0.5	Blampied *et al.*, 1983
		0.5–2.0	30	0.5	Tsuda *et al.*, 1988
Diazepam + Nalaxone	IP	1/10	30/10		Treit, 1985
Ethanol	IP	0–2000	30	2000	Treit, 1990b
Imipramine	IP	4.0–16.0	30	8.0	Craft *et al.*, 1988
Ipsapirone	IP	0.065–10	30	2.5	Korte *et al.*, 1990
Midazolam	IP	1.0–3.0	30	1.0	Treit, 1990b
Morphine	IP	2.0–8.0	30	2.0	Craft *et al.*, 1988
Meprobamate	IP	75–125	60	75	Craft *et al.*, 1988
Oxprenolol	IP	10–20	20	10	Blampied *et al.*, 1983
Pentobarbital	IP	1.0–6.0	30	3.0	Treit *et al.*, 1981
		10–20	30	15	Treit, 1990b
Pentylenetetrazol	IP	20	10		Treit, 1987
Picrotoxine	IP	2	10		Treit, 1987
Pimozide	IP	1	240		Beninger *et al.*, 1980
DRUG TREATMENTS THAT DO NOT CHANGE BURYING					
Buspirone	IP	8.0–64	30		Craft *et al.*, 1988
CDP + Flumazenil	IP	5/5 and 5/10	30/10		Treit, 1987
CDP + Pentylenetetrazol	IP	5/5	30/10		Treit, 1987
CDP + Picrotoxine	IP	5/1	30/10		Treit, 1987
CDP + CGS 8216	IP	5/10	30/10		Treit, 1987
CGS 8216	IP	5.0–10.0	10		Treit, 1987
Chlorpromazine	IP	1.0–16.0	30		Craft *et al.*, 1988
	IP	0.5–5.0	30		Treit, 1990b
Desipramine (chronic)	IP	2.5 b.i.d.; 8 weeks	14–18 h		Beardslee *et al.*, 1990
Diazepam + Picrotoxine	IP	1/1	30/15		Treit *et al.*, 1982
Flumazenil	IP	5.0–10.0	30		Treit, 1987
Imipramine (chronic)	IP	2.5 b.i.d.; 7 weeks	14–18 h		Beardslee *et al.*, 1990
Morphine	IP	1.5	30		Treit *et al.*, 1981
Naloxone	IP	3.0	30		Whiteside *et al.*, 1985
		10.0	10		Treit, 1985
Pargyline (chronic)	IP	30 b.i.d.; 12 weeks	14–18 h		Beardslee *et al.*, 1990
Pentylenetetrazol	IP	20	30		Treit *et al.*, 1981
	IP	5.0–20	10		Treit, 1990b
Physostigmine	IP	2.0	30		
Picrotoxine	IP	0.5	30		Treit *et al.*, 1981
Scopolamine	IP	0.05–1.25	30		Treit, 1990b
Yohimbine	IP	0.5–2.0	30		Treit, 1990b
DRUG TREATMENTS THAT INCREASE BURYING					
Amitriptyline	IP	2.0	60		Davis *et al.*, 1981b
β-CCE	IP	0.1–5.0	0	0.5	Tsuda *et al.*, 1989
d-Amphetamine	IP	1.0	30		Treit *et al.*, 1981
Yohimbine	IP	0.5–2.0	30	0.5	Tsuda *et al.*, 1988

ITI = injection test interval; MED = minimal effective dose.

PHARMACOLOGICAL VALIDATION OF THE BEHAVIOURAL INDICES IN THE SHOCK-PROD DEFENSIVE BURYING/AVOIDANCE PARADIGM

Animal models for anxiolytic effects have to fulfil a number of pharmacological criteria as specified recently by Shephard (1984): 1) Specificity: anxiolytic effects in the model should be induced by more than one compound from one class of drugs with similar pharmacological actions. 2) Sensitivity: the model is expected to be sensitive for a low dose of a benzodiazepine. 3) Correlation with clinical potency. 4) Validity: the action of drugs in the model should not be based on a system unrelated to anxiety like feeding or anti-nociception. 5) Absence of tolerance: anxiolytic action of a drug should still be present after repeated daily administration of the drug.

Table 2 summarizes results from studies of drug effects on burying behaviour in rats. A variety of clinically proven "classical" antianxiety agents such as diazepam, chlordiazepoxide, and pentobarbital as well as the "novel" anxiolytics buspirone and ipsapirone have been shown to suppress prod-burying behaviour (the most frequently used index of fear/anxiety in this paradigm). This suppression of burying behaviour occurs in a dose-related manner and with a relative potency that is similar to that found in clinical settings. Tolerance to the suppressive effect of diazepam does not occur when rats are tested with low-intensity prod-shock although anxiolytic tolerance has been detected when tested with higher intensity shocks (Treit, 1985b). Anxiolytic-induced suppression of prod-burying can be distinguished from those of several nonanxiolytic agents (Beardslee et al., 1990; Treit, 1990; Treit et al., 1981), and do not appear to be secondary to behavioural sedation (Blampied & Kirk, 1983; Treit, 1990; Treit & Fundytus, 1988; Treit et al., 1981), analgesia (Treit, 1985), or associative deficits (Blampied & Kirk, 1983; Korte & Bohus, 1990). In addition, the benzodiazepine anxiolytic-induced suppression of burying behaviour can be reversed by benzodiazepine/GABA-receptor antagonists (Treit, 1987; Treit et al., 1982). Furthermore, it has been shown that some anxiogenic agents like d-amphetamine, amitryptiline, yohimbine and the benzodiazepine-receptor inverse agonist β-CCE enhance the amount of shock-prod burying in rats (Davis et al., 1981; Treit et al., 1981; Tsuda et al., 1988, 1989), indicating that the burying paradigm is also capable of detecting pro-stress or anxiogenic drug effects.

By using the suppression of prod-exploratory behaviour as the index of fear/anxiety in the closely related "shock-probe conflict test", Meert & Colpaert (1986) have demonstrated that anxiolytic agents reliably and dose-dependently block this inhibition, while most nonanxiolytics do not produce this effect. In addition, it was recently shown that the anxiolytic-induced decreases in prod-burying are concurrent with increases in prod-contacts using the (combined) shock-prod defensive burying/avoidance test (Craft et al., 1988; De Boer et al., 1990; Treit, 1990; Treit & Fundytus, 1988). Taken together, these studies have provided a reasonable pharmacological validation of reduced prod-exploration and defensive prod-burying as behavioural models for the study of anxiolytic agents.

In certain animal tests, immobility responses (i.e., passive avoidance, freezing) are also considered to be true indicators of fear and anxiety (Bolles, 1970; Bolles & Fanselow, 1980), and do seem to be selectively sensitive to anxiolytic drugs (see Treit, 1985). In the shock-prod defensive burying/avoidance paradigm, however, no decrease in immobility is seen after benzodiazepine anxiolytic treatment but, paradoxically, an increase is observed concurrent with suppression of burying (De Boer et al., 1990; Treit et al., 1986; Tsuda et al., 1988). This non-uniform attenuation of different behavioural indices of fear/anxiety by a prototypical anxiolytic makes the interpretation of drug effects in this paradigm problematic. Moreover, it has been found that increasing a rat's fear, as a result of prior exposure to uncontrollable electric footshock or the presence of conspecific stress odors, potentiates freezing and disrupts burying behaviour (Williams, 1987), indicating a complex relationship between levels of fear and the occurrence of burying and freezing behaviour in the rat. Assessment of additional autonomic/endocrine measures within such a setting may resolve this problem as discussed in the next section.

AUTONOMIC AND ENDOCRINE MEASURES IN THE SHOCK-PROD DEFENSIVE BURYING/AVOIDANCE PARADIGM

By employing a chronic vascular cannulation technique for repetitive blood collection in unaenesthetized and unrestrained animals (Steffens, 1969), combined with highly sensitive and specific assay methods for determination of plasma catecholamines and glucocorticoids (Van der Gugten et al., 1977), the dynamics of plasma noradrenaline, adrenaline and corticosterone concentrations were measured in connection with behavioural responses to shock-prod exposure under different experimental test conditions (De Boer et al., 1990). Since the individually housed rats were tested in their own home cages, no habituation trials and handling of the animals prior to testing is needed, thus preventing disturbance of the neuroendocrine measures under study. Two different experimental housing conditions were used, i.e., either with or without bedding material on the cage floor, and exposure (for 15 min) to either a non-electrified or constantly electrified (2 mA) prod. The results of this study are shown in figures 3 and 4. Both in the presence and absence of bedding material, rats explored the non-electrified prod and showed a small (novelty-induced) increase in plasma noradrenaline and corticosterone contents. Exploration of the prod was strongly reduced when the prod was electrified. In the presence of bedding material, shocked rats typically displayed burying responses (active form of avoidance or coping behaviour), whereas in the absence of bedding (i.e., burying option eliminated) shocked rats engaged in immobility (freezing and scanning) behaviour (passive form of avoidance) (see figure 3).

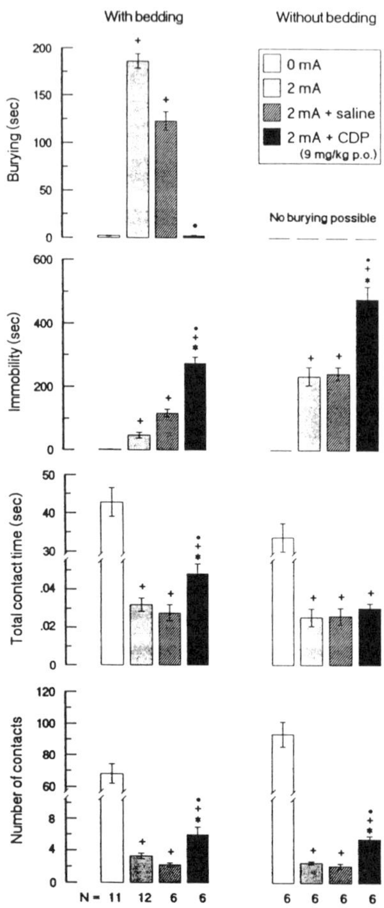

Fig. 3 Mean (+SEM) number and duration of prod contacts and mean (+SEM) duration of immobility and burying behaviour performed by non-shocked (0 mA) and shocked (2 mA) untreated control rats as well as by shocked vehicle- and CDP-treated subjects tested in the presence (left panels) or absence (right panels) of bedding material (adapted from De Boer et al., 1990).
* Significantly different from saline-treated group.
\+ Significantly different from untreated shocked (2 mA) group.
. Significantly different from untreated non-shocked (0 mA) group.

Although actively and passively coping rats received similar amounts of electric shock through the stationary prod, their profiles of changes in plasma noradrenaline, adrenaline and corticosterone concentrations differed substantially from each other: The passive avoidance situation was accompanied by larger adrenaline and corticosterone increases but a lower noradrenaline rise as compared to the hormonal responses associated with the active avoidance situation (see figure 4).

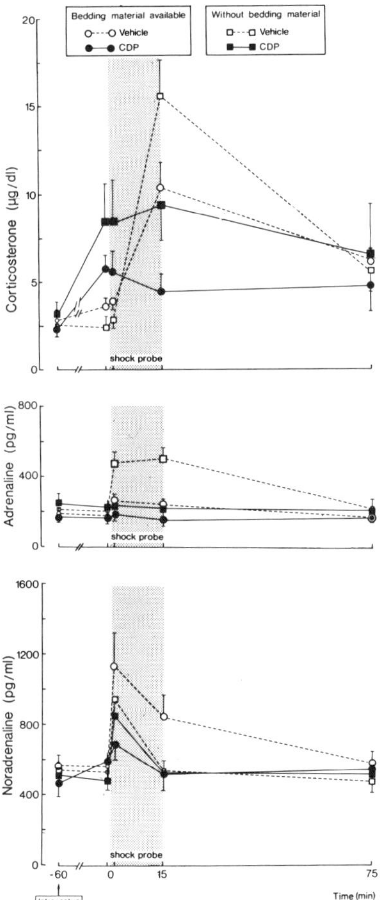

Fig. 4 Plasma corticosterone, adrenaline and noradrenaline concentrations before, during and after 15 min. exposure to an electrified shock-prod in vehicle (open symbols) and chlordiazepoxide (closed symbols) treated rats tested either with (circles) or without (squares) bedding material on the floor of their home cages. Data are expressed as mean ± SEM for 6 rats per condition. (From De Boer et al., 1990, with permission).

Concentrations of noradrenaline and adrenaline in circulating blood reflect neurosympathetic and adrenomedullary outflow, respectively, whereas plasma corticosterone concentrations serve as a reliable indicator of pituitary-adrenocortical activity (Axelrod & Reisine, 1984; Goldstein et al., 1983). Therefore, burying the shock-prod is accompanied by a relatively high neurosympathetic activation. Accordingly, by using biotelemetry equipment, it was recently shown that sympathetically regulated autonomic indices like blood-pressure, heart-rate and core-temperature also increase substantially during shock prod-burying (Korte & Bohus, 1990; Croiset & De Zwart, 1990). On the other hand, passive avoidance of the shock-prod, when no bedding is available, is accompanied by a relatively high adrenomedullary and adrenocortical activation. These observations reinforce the stress concept of Henry (1986) that different behavioural coping styles to a challenge are associated with distinct patterns of neurosympathetic, adrenomedullary and adrenocortical activation. Furthermore, the data provide empirical support for the view that the activity of sympathetic nerves, as reflected in plasma noradrenaline content, is principally related to conditions involving actual skeletal muscle exertion (i.e., either organized or disorganized postures and movements, physical exercise) regardless of the emotional connotations of the challenge, whereas adrenomedullary and adrenocortical stimulation occurs primarily during emotional stress, fear or anxiety-provoking situations in which coping capabilities are limited or abolished.

Administration of the prototypical anxiolytic chlordiazepoxide (CDP) attenuated the shock-induced suppression of prod exploration, decreased prod-burying behaviour but, as already mentioned in the preceding section, increased immobility behaviour. Thus it seems that benzodiazepine anxiolytics do not uniformly antagonize all behavioural manifestations of the rat's fear/anxiety state. However, during both active coping (i.e., burying behaviour in the presence of bedding) and passive coping (i.e., immobility behaviour in the absence of bedding), the prod shock-induced elevations in plasma corticosterone and adrenaline (the "emotional stress" hormones) were completely abolished after CDP-treatment, indicating that the fear/anxiety state is reduced. The burying-related rise in plasma noradrenaline (the "physical stress" hormone) was attenuated only in CDP-treated rats tested on a bedding-floor (Fig. 4). Recently, it was demonstrated that the "new" 5-HT_{1A} anxiolytic compound ipsapirone is effective in reducing the conditioned defensive burying and accompanying tachycardiac response (Korte & Bohus, 1990). To which extent the anxiolytic-induced decrease in sympathetic activation (as indicated by alevased plasma noradrenaline and tachycardia) results from the concurrent diminished vigour of burying (i.e., somatic-autonomic coupling) is not known yet. Based on these preliminary findings, it is proposed that the combined assessment of behavioural and autonomic/endocrine indices of the animal's fear/anxiety state within the same experimental setting may facilitate the interpretation of anxiolytic drug actions.

Table 3 Effects of different experimental testing conditions on behavioural, autonomic and endocrine measures in the shock-prod defensive burying/avoidance paradigm.

Measure	With bedding material		Without bedding material	
	UCS	CS	UCS	CS
BEHAVIOURAL				
burying	++	++		
height of pile	++	++		
prod-exploration	--	--	--	--
approach-withdrawal	+	++	++	++
immobility	+	+	++	++
AUTONOMIC				
heart rate	++	+	?	?
blood pressure	+	+	?	?
core temperature	+	+	?	?
ENDOCRINE				
noradrenaline	++	?	0/+	?
adrenaline	+	?	++	?
corticosterone	0/+	?	++	?

"+" indicates an increase, "0" indicates no change and, "-" indicates a decrease in the measure as compared to the non-shocked controls. "?" indicates not known/tested yet. UCS = unconditioned shock-prod test (conditioning session). CS = conditioned shock-prod test (extinction or retention session).

SUMMARY AND CONCLUDING REMARKS

The shock-prod defensive burying/avoidance paradigm serves well as an animal test which allows several behavioural and autonomic/endocrine manifestations of stress/anxiety to be measured simultaneously. A summary of the different behavioural, autonomic and endocrine measures of stress, fear and/or anxiety in this paradigm are listed in table 3. Also indicated are the effects of different experimental testing conditions on these indices. Simply by manipulating the availability of bedding material, one can titrate the animal to adopt an active (i.e., burying in the presence of bedding) or passive (i.e., immobility in the absence of bedding) behavioural strategy in order to cope with the shock-prod. The neuroendocrinological consequences of these behavioural response styles differ substantially: the passive mode of avoidance response being accompanied by a predominant activation of the pituitary-adrenocortical and adrenomedullary systems resulting in pronounced elevations of corticosterone and adrenaline, respectively, whereas the active mode of response is accompanied by a preferential neurosympathetic activation, resulting in increased plasma noradrenaline levels, tachycardia, and hyperthermia. Furthermore, by incorporating an additional retention/extinction test the conditioned emotional consequence of former prod-shock punishment can also be assessed. Clearly, by using this relatively simple animal model it is possible to study the effects of drugs with potential anti-stress and/or anxiolytic activity on several behavioural and hormonal measures under four different experimental conditions, e.g., unconditioned and conditioned active avoidance of a shock-prod, and unconditioned and conditioned passive avoidance of a shock-prod. Particularly the conditioned test situation may prove to be sensitive for detecting drugs with a presumed anxioselective profile of action (e.g. the novel 5-HT putative anxiolytics). Thus by employing a multidisciplinary approach in the shock-prod defensive burying/avoidance test not only the interpretation of anxiolytic effects of drugs in this particular test may be facilitated, the understanding of the neuronal mechanisms of stress and anxiety in general may be furthered as well.

Acknowledgements: This research was supported in part by the Netherlands Organization For Scientific Research (NWO-MEDIGON, Grant No. 900-548-087). Our sincere thanks are extended to mrss. W.E. van der Wal and A.B. van Woerkom for their technical and analytical assistance.

REFERENCES

Axelrod, J., Reisine, T. D. (1984) Science 224, 452-459.

Beardslee, S.L., Papadakis, E., Altman, H.J., Harrington, G.M., Commissaris, R.L. (1989) Physiol. Behav. 45, 449-451.

Beardslee, S.L., Papadakis, E., Fontana, D.J., Commissaris, R.L. (1990) Pharmacol. Biochem. Behav. 35, 451-455.

Beninger, R.J., McLennan, A.J., Pinel, J.P.J. (1980) Pharmacol. Biochem. Behav. 12, 445-448.

Blampied, N.M., Kirk, R.C. (1983) Life Sci. 33, 695-699.

Bohus, B., Benus, R.F., Fokkema, D.S., Koolhaas, J.M., Nyakas, C., van Oortmerssen, G.A., Prins, A.J.A., de Ruiter, A.J.H., Scheurink, A.J.W., Steffens, A.B. (1987) In: Neuropeptides and Brain Function (E.R.De Kloet, V.M.Wiegant and D.De Wied, Eds), Elsevier, Amsterdam, pp 57-70.

Bolles, R.C. (1970) Psychol. Rev. 77, 32-48.

Bolles, R.C., Fanselow, M.S. (1980) Behav. Brain Sci. 3, 291-323.

Craft, R.M., Howard, J.L., Pollard, G.T. (1988) Pharmacol. Biochem. Behav. 30, 775-780.

Crawley, J.N. (1981) Pharmacol. Biochem. Behav. 15, 696-699.

Davis, M. (1979) Psychopharmacology 62, 1-7.

Davis, S.F., Rossheim, S.A. (1980) Bull. Psychon. Soc. 16, 229-231.

Davis, S.F., Whiteside, D.A., Heck, D.G., Dickson, V.A., Tramill, J.L. (1981a) Bull. Psychon. Soc. 17, 45-47.
Davis, S.F., Hazelrigg, M., Moore, S.A., Petty-Zirnstein, M.K. (1981b) Bull. Psychon. Soc. 18, 325-327.
Davis, S.F., Moore, S.A., Cowen, C.L., Thurston, D.K., Maggio, J.C. (1982a) Anim. Learn. Behav. 10, 516-520.
Davis, S.F., Whiteside, D.A., Dickson, V.A., Heck, D.G., McKnab, P.A. (1982b) J. Gen. Psychol. 106, 101-108.
De Boer, S.F., Slangen, J.L., Van der Gugten, J. (1990) Physiol. Behav. 47, 1089-1098.
De Boer, S.F. (1990) Ph.D. Thesis, University of Utrecht, pp 11-18.
De Zwart, J.F. (1990) Unpublished doctoral report Rudolph Magnus Institute for Pharmacology, University of Utrecht.
Geller, I., Seifter, J. (1960) Psychopharmacologia 1, 482-492.
Goldberg, S.G., Ghezzi, P.M., Cheney, C.D. (1983) Psychol. Rec. 33, 191-199.
Goldstein, D.S., McCarty, R., Polinsky, R.J., Kopin, I.J. (1983) Hypertension 5: 552-559.
Gram, L.F., Christensen, P. (1986) Pharmacopsychiatry 19, 19-22.
Gray, D.S., Terlecki, L.J., Treit, D., Pinel, J.P.J. (1981) Physiol. Behav. 27, 1051-1056.
Harder, D.B., Maggio, J.C. (1982) Animal Learn. Behav. 11, 465-473.
Henry, J.P. (1986) In: Biological Foundations of Emotion (R.Plutchik, H.Kellerman, Eds), Academic Press, Orlando, pp 37-60.
Kolb, B., Wishaw, I.Q. (1981) J. Comp. Physiol. Psychol. 95, 468-483.
Korte, S.M., Bohus, B. (1990) Eur. J. Pharmacol.
Lahti, R.A., Barsuhn, C. (1974) Psychopharmacologia 35, 215-220.
Le Fur, G., Guilloux, F., Mitrani, J., Mizoule, J., Uzan, A. (1979) J. Pharmacol. Exp. Ther. 211, 305-308.
McKim, W.A., Lett, B.T. (1979) Behav. Neural Biol. 26, 76-80.
Meert, T.F., Colpaert, F.C. (1986) Psychopharmacology 88, 445-450.
Moser, C.G., Tait, R.W. (1983) J. Comp. Psychol. 97, 338-352.
Oberdieck, F.G., Cheney, C.D. (1982) Bull. Psychon. Soc. 19, 295-297.
Peacock, E.J., Wong, P.T.P. (1982) Anim. Learn. Behav. 10, 103-107.
Pinel, J.P.J., Treit, D. (1978) J. Comp. Physiol. Psychol. 92, 708-712.
Pinel, J.P.J., Treit, D., Ladek, F., McLennan, A.J. (1980) Animal Learn. Behav. 8, 447-451.
Pinel, J.P.J., Treit, D. (1983) In: Behavioural Approaches to Brain Research. (T.Robinson, Ed), Oxford University Press, New York, pp 212-234.
Poling, A., Cleary, J., Monaghan, M. (1981) J. Exp. Anal. Behav. 35, 31-44.
Shephard, R.A. (1984) Neurosci. Biobehav. Rev. 10, 449-461.
Steffens, A.B. (1969) Physiol. Behav. 4, 833-836.
Tarte, R.D., Oberdieck, F. (1982) Psychol. Rec. 32, 101-107.
Terlecki, L.J., Pinel, J.P.J., Treit, D. (1979) Learn. Motivat. 10, 337-350.
Treit, D., Terlecki, L.J., Pinel, J.P.J. (1980) Bull. Psychon. Soc. 16, 451-454.
Treit, D., Pinel, J.P.J., Fibiger, H.C. (1981) Pharmacol. Biochem. Behav. 15, 619-626.
Treit, D., Pinel, J.P.J., Fibiger, H.C. (1982) Pharmacol. Biochem. Behav. 17, 359-361.
Treit, D. (1985) Neurosci. Biobehav. Rev. 9, 203-222.
Treit, D. (1985a) Pharmacol. Biochem. Behav. 22, 47-52.
Treit, D. (1985b) Pharmacol. Biochem. Behav. 22, 383-387.
Treit, D., LoLordo, V.M., Armstrong, D.E. (1986) Pharmacol. Biochem. Behav. 25, 561-565.
Treit, D. (1987) Brain Res. Bull. 19, 401-405.
Treit, D., Fundytus, M. (1988) Pharmacol. Biochem. Behav. 30, 1071-1075.

Treit, D., Pesold, C. (1990a) Physiol. Behav. 47, 365-371.
Treit, D. (1990b) Pharmacol. Biochem. Behav. 36, 203-205.
Tsuda, A., Ida, Y., Tanaka, M. (1988a) Anim. Learn. Behav. 16, 354-358.
Tsuda, A., Ida, Y., Tanaka, M. (1988b) Psychobiology 16, 213-217.
Tsuda, A., Ida, Y., Nishimura, H., Tanaka, M. (1989) Psychobiology 17, 202-206.
Van der Gugten, J., Slangen, J.L. (1977) Pharmacol. Biochem. Behav. 7, 211-215.
Vogel, J.R., Beer, B., Clody, D.E. (1971) Psychopharmacologia 21, 1-7.
Whiteside, D.A., Devenport, L.D. (1985) Behav. Neurosci. 99, 436-440.
Whillans, K.V., Shettleworth, S.J. (1981) Anim. Learn. Behav. 9, 357-362.
Wilkie, D.M., McLennan, A.J., Pinel, J.P.J. (1979) J. Exp. Anal. Behav. 31, 299-306.
Williams, J.L. (1987a) Psychol. Rec. 37, 369-380.
Williams, J.L. (1987b) Anim. Learn. Behav. 15, 333-341.
Yamamoto, T., Ueki, S. (1987) Pharmacol. Biochem. Behav. 26, 111-117.

Animal Models in Psychopharmacology
Advances in Pharmacological Sciences

THE LACK OF TOLERANCE TO THE ANXIOLYTIC EFFECTS OF BENZODIAZEPINES IN THE GELLER/SEIFTER CONFLICT TEST

L.M. Broersen, F. Woudenberg and J.L. Slangen

Department of Psychopharmacology, Faculty of Pharmacy, University of Utrecht, Sorbonnelaan 16, 3584 CA Utrecht, The Netherlands

SUMMARY: The lack of tolerance to the anti-conflict effects of benzodiazepines in the Geller/Seifter conflict test was reinvestigated, and was found not to be related to any of four factors examined.

Tolerance to the anxiolytic action of benzodiazepines (BDZs) has not been documented reliably within the clinical situation (Lader, 1988). In contrast, tolerance to this drug effect has been found in a number of animal models of anxiety (Cooper et al., 1981; Davis & Gallager, 1988; Emmett-Oglesby et al.,1987; File et al., 1987; Treit, 1985; Vellucci & File, 1979), except in the Geller/Seifter conflict (GSC) test (Babbini et al., 1982; Geller & Seifter, 1960; Margules & Stein, 1968).

The present experiment investigated four factors, that may be responsible for the reported lack of tolerance development in the GSC-test: 1. dose and duration of drug administration; 2. habituation to shock; 3. contingency of drug administration with testing; 4. development of a conditioned anxiolytic response.

In addition, it was checked whether tolerance to a non-anxiolytic action. i.e. muscle-relaxation developed after BDZ administration.

MATERIALS AND METHODS

Animals: 32 male Wistar rats (CPB-TNO, Zeist, The Netherlands) weighing 250 g at the beginning of the experiment, were individually housed under a non-reversed 12 h light-dark cycle and were maintained at 85% of their free-feeding weight. Tap water was freely available.

Apparatus: Two ventilated rat chambers, equipped with two retractable levers and a grid floor, were used. A pellet dispenser delivered 45 mg pellets (Noyes) in a tray placed between the levers. Shocks were delivered through a constant current shock generator (Campden 521C) and scrambler (Campden 521S). A Digital Equipment Corporation PDP-ll micro computer and software (SKED) supplied by State Systems Incorporated (Kalamazoo, MI) were used to control session events and to record data.

Procedure: 32 rats were trained to press the left lever for food according to a multiple schedule derived from the one described by Geller and Seifter (1960). Sessions consisted of three 6 min periods under a VI-30sec, interspersed by two 3 min periods under a CRF schedule. VI-30sec periods, signalled by a white light, and CRF periods, signalled by a red light, were separated by a 5 sec dark period. Lever presses during CRF periods were punished by 0.5 sec foot-shocks. Shock intensities were adjusted until animals responded 2 to 5 times per conflict period. Animals were divided in 4 matched groups of 8 animals. These groups were administered either vehicle or one of three doses (1, 2.25 or 5 mg/kg) of diazepam (DZP) and tested for 25 consecutive days. Thereafter, dose-effect curves were constructed for DZP, chlordiazepoxide (CDP), baclofen (BAC) and methocarbamol (MCM). During the 25 day period, rats were also tested for muscle-relaxation with the pull-up test (Deacon & Gardner, 1984).

The vehicle and DZP 2.25 mg/kg groups were then used in a second experiment. They received 2.25 mg/kg of DZP and vehicle respectively for 25 days, but were tested with the GSC-procedure only before and after this chronic administration phase.

Drugs: Diazepam (OPG, Utrecht, The Netherlands) was dissolved in a vehicle containing 1.5% benzylalcohol, 8.5% ethylalcohol, 41.5% propylene glycol and 48.5% distilled water. Chlordiazepoxide-HCl (Bufa-chemie, Castricum, The Netherlands) was dissolved in 0.9% saline. Baclofen (Ciba-Geigy, Basle, Switzerland) and methocarbamol (A.H. Robins, Crawley, England) were suspended in a vehicle of distilled water and Tween 80 (2 drops/10 ml). Doses refer to the forms indicated. Drugs were administered i.p. 25 min prior to testing, in an injection volume of 2 ml/kg, except for the doses of DZP. The 1 and 2.25 mg/kg doses were given in a volume of 0.6 ml/kg; the 5 mg/kg dose in a volume of 1.25 ml/kg.

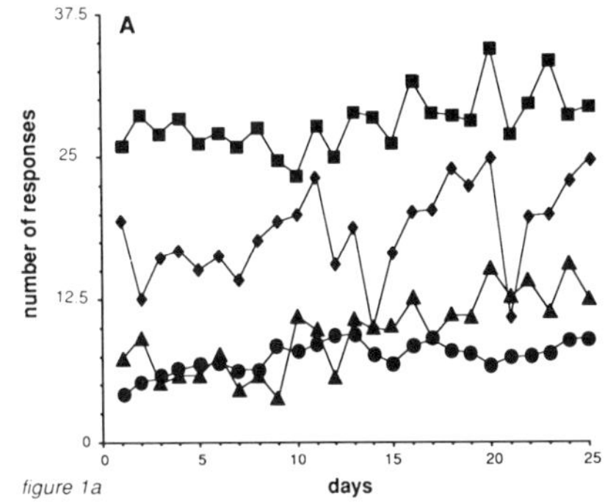

figure 1a

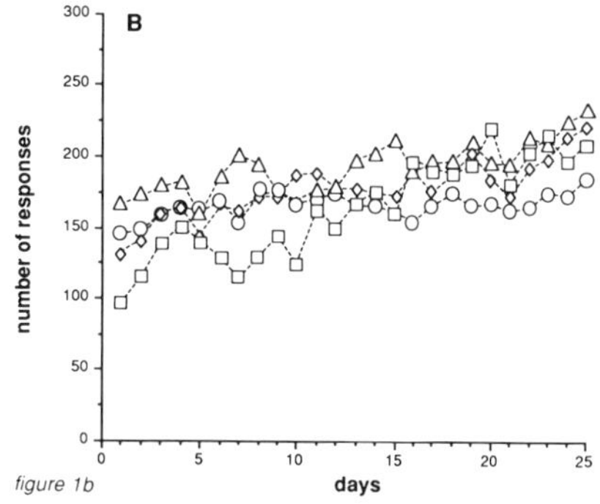

figure 1b

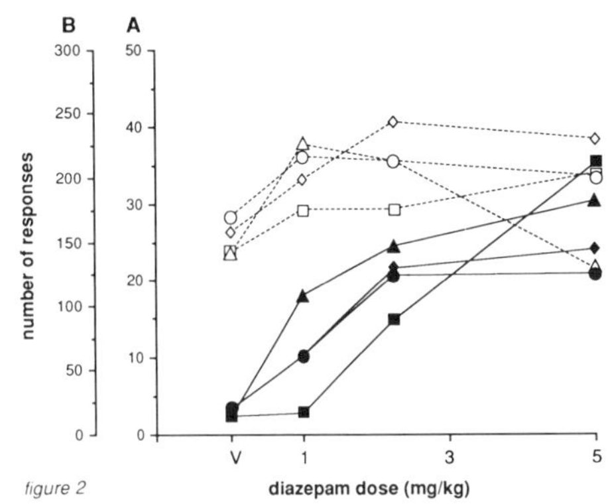

figure 2

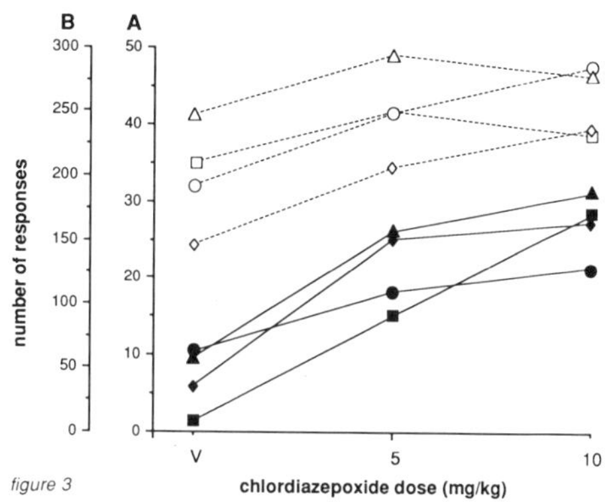

figure 3

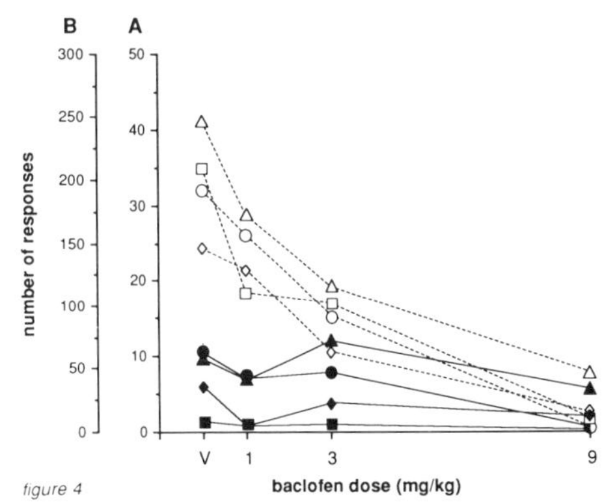

figure 4

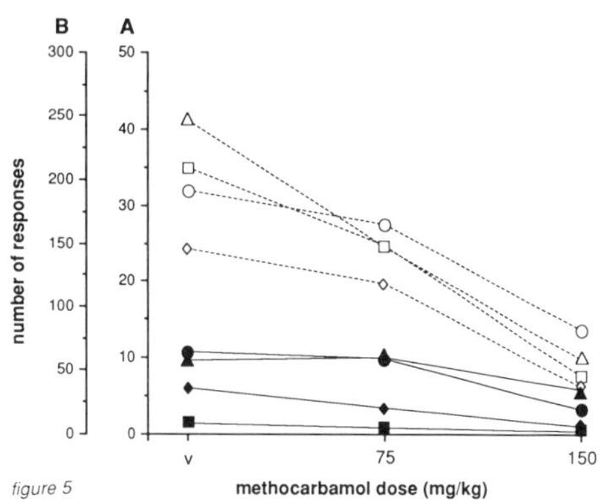

figure 5

Fig. 1-5 Mean number of punished (y-axis A, closed symbols) and unpunished (y-axis B, open symbols) responses obtained in rats treated chronically with vehicle (circles) or 1 (triangles), 2.25 (diamonds) or 5 (squares) mg/kg of diazepam (Figs. 1a, 1b) and subsequently tested with different doses of diazepam (Fig. 2) chlordiazepoxide (Fig. 3), baclofen (Fig. 4) and methocarbamol (Fig. 5).

Data analysis: Punished and unpunished responses were analyzed separately with a MANOVA. In addition, paired comparisons were performed. Pull-up latencies were analyzed with Student's t-test and a linear regression analysis.

RESULTS

During the 25 days of chronic DZP administration, the 2.25 and 5 mg/kg doses always had a clear anti-conflict effect (Fig.la). The effect of the 1 mg/kg dose was not different from the vehicle effect. The number of both punished (Fig. la) and unpunished (Fig. lb) responses increased over this period in all experimental groups ($F(1,28)=12.9$, $p<0.01$ and $F(1,28)=10.8$, $p<0.01$ respectively.

When tested after the chronic administration phase, DZP ($F(3,26)=87.1$, $p<0.01$) and CDP ($F(2,27)=40.0$, $p<0.01$) dose-dependently increased the number of punished responses (Figs. 2 and 3). DZP and CDP also increased unpunished responding ($F(3,26)=4.5$, $p<0.05$ and $F(2,27)=9.8$, $p<0.01$ respectively) as compared to vehicle. MCM (Fig. 5) decreased, but not dose-dependently, the number of punished responses ($F(2,27)=4.9$, $p<0.05$); BAC was without effect on punished responses. BAC ($F(3.26)=24.4$, $p<0.01$) and MCM ($F(2,27)=54.4$, $p<0.01$) dose-dependently decreased the number of unpunished responses (Figs. 4 and 5).

In experiment 2, concerning the administration of DZP without test sessions, the effect of DZP (2.25 mg/kg) on punished responding did not significantly differ from its effect in experiment 1 (data not shown).

On the first day of DZP administration, pull-up latencies were increased by the 2.25 ($T(14)=-2.8$, $p<0.05$) and 5 ($T(14)=-4.4$, $p<0.01$) mg/kg doses. Within 9 days, latencies returned to baseline values in all experimental groups.

DISCUSSION

The present results indicate that in the Geller-Seifter conflict-test tolerance does not develop to the anxiolytic effects of three different chronic doses of diazepam. The small increase in punished responding over the course of chronic administration is caused by habituation to shock and not by sensitization of the drug effect, for a similar increase was found in the vehicle treated group. In addition, both diazepam and chlordiazepoxide dose-dependently increased punished responding, irrespective of chronic pretreatment. In contrast, tolerance to the muscle-relaxant action of diazepam developed rather rapidly.

The experience of a drug state in the test apparatus can influence the development of tolerance to a drug action (Barrett & Leith, 1981; Chen, 1968). As opposed to the GSC-procedure, chronic benzodiazepine administration in other anxiolytic test procedures has often been dissociated from actual testing. However, still no tolerance to the anxiolytic action of diazepam was observed in the present experiment, when diazepam was chronically administered without subsequent testing.

It has been demonstrated, that a Pavlovian conditioned response that is a replica of the unconditioned drug response may develop during chronic drug administration (see Siegel,1989). Such a response might mask the development of tolerance. However, a vehicle injection after chronic benzodiazepine administration does not increase the number of punished responses in the GSC-test (Margules & Stein, 1968).

In the present study, baclofen and methocarbamol, which share internal stimulus effects with diazepam but exert no anti-conflict effects (Bennett & Amrick, 1986), were tested to see whether the anxiolytic response was maintained through an internal stimulus effect acting as conditioned stimulus and eliciting a conditioned anxiolytic response. However, neither baclofen nor methocarbamol increased punished responding in any group.

The lack of tolerance in the GSC-test can not be explained by any of the factors examined. With regard to the benzodiazepines it seems, that the GSC-test reflects the clinical situation more adequately than other tests of anxiolytic drug activity.

REFERENCES

Babbini, M., Gaiardi, M., Bartoletti, M. (1982) Pharmacol. Biochem. Behav. 17, 43-48.
Barrett, R.J., Leith, N.J. (1981) Psychopharmacology 20, 251-255.
Bennett, D.A., Amrick, C.L. (1986) Life Sci. 39, 2455-2461.
Chen, C.S. (1968) Psychopharmacologia 12, 433-440.
Cooper, S.J., Burnett, G., Brown, K. (1981) Psychopharmacology 73, 70-74.
Davis, M., Gallager, D.W. (1988) Eur. J. Pharmacol. 150, 23-33.
Deacon, R.M.J., Gardner, C.R. (1984) J. Pharmacol. Exp. Ther. 11, 119-124.
Emmett-Oglesby, M.W., Mathis, D.A., Lal, H. (1987) Drug Dev. Res. 11, 145-156.
File, S.E., Baldwin, H.A., Aranko, K. (1987) Brain Res. Bull. 19, 607-610.
Geller, I., Seifter, J. (1960) Psychopharmacologia 1, 482-492.
Lader, M. (1988) In: The psychopharmacology of addiction (M.Lader, Ed). Oxford University Press, New York, pp 1-14.
Margules, D.L., Stein, L. (1968) Psychopharmacologia 13, 74-80.
Siegel, S. (1989) In: Psychoactive drugs: tolerance and sensitization (A.J.Goudie and M.W.Emmett-Oglesby, Eds), Humana Press, Clifton, pp 115-180.
Treit, D. (1985) Pharmacol. Biochem. Behav. 22, 383-387.
Vellucci, S.V., File, S.E. (1979) Psychopharmacology 62, 61-65.

EXTINCTION OF CONFLICT BEHAVIOUR

C.E.J. Ketelaars* and J. Bruinvels

Department of Pharmacology, Workgroup Biological Psychiatry, Erasmus University Rotterdam, P.O.Box 1738, 3000 DR Rotterdam. *: Present address: Polikliniek Kinder- en Jeugdpsychiatrie A.Z.G., P.O.Box 30001, 9700 RB Groningen, The Netherlands

SUMMARY: The mechanism behind the anxiolytic effect of the $GABA_B$ receptor agonist baclofen was studied using a new animal model, namely extinction of conflict behaviour. It is shown that baclofen, benzodiazepines and several drugs which decrease the function of the serotonergic (5-HT) system enhance extinction of conflict behaviour. Baclofen does not affect extinction via activation of the $GABA_A$/benzodiazepine receptor complex, but probably via a decrease in function of the 5-HT system. Extinction of conflict behaviour might have predictive value for drugs active in anxiety disorders in which the memory for aversive events plays an important pathogenic role.

The $GABA_A$ receptor agonist baclofen has a beneficial effect on three components of panic disorder, namely panic attacks, anticipatory anxiety and phobic avoidance (Breslow et al., 1989; Pepplinkhuizen and Bruinvels, 1978). Since baclofen is not active in the Geller-Seifter conflict test, this paradigm was modified in such a way that the new test might be a more sensitive animal model for detecting anxiolytic drugs. The modification, (removal of foot-shock punishment during the test situation), causes a slow extinction of conflict behaviour in untreated animals (Cook and Davidson, 1973).

It is postulated that enhancement of this process might reflect a decrease of anticipatory anxiety and phobic avoidance in man.

MATERIALS AND METHODS

Extinction of conflict behaviour: In order to study extinction of conflict behaviour, the Geller-Seifter test (Geller and Seifter, 1960) was modified as described previously (Ketelaars et al., 1988). In brief: After being deprived of water for 16 hours, male

Wistar rats (starting weight: 70 - 90 g; after training was completed: 250 - 350 g) were placed in an experimental chamber measuring 20 x 20 x 20 cm, containing a response lever, milk delivery apparatus, an electrifiable grid floor and a stimulus light. The rats were trained to press a lever in order to obtain milk and were subsequently trained once a day for a duration of 42 minutes. Within these training (and test) periods, two schedules were used in alternation. During 5 minutes a 5% random interval (RI) schedule was offered. This schedule was alternated with 2 minutes of a fixed ratio 1 (FR1) schedule. A steady light signal during the FR1 schedule served as the discriminative stimulus. This combination of RI and FR1 schedules was presented 6 times to each rat per session. After stabilization of performance, response-contingent foot shock was given during the FR1 periods. Shock levels (100-160 V, 2 - 3.3mA) were individually adjusted until cumulative lever pressings during the FR1 periods were below 5 per 12 minutes, which represents about 5% of the number of lever pressings before foot shock was applied.

Drug effects on extinction of conflict behaviour were measured on the third day, always after 2 days of training in order to achieve a stable "base line" level of conflict behaviour. On the test days, foot shock was turned off. During the 2 days before the test days training continued in the presence of foot shock, and shock levels were adjusted when the animals took more than 5 shocks per 12 minutes. After at least 20 training sessions, spontaneous extinction was non-existent in control rats during the trial period; i.e. these animals responded with an average of 0 to 5 pressings per 12 minutes. Vehicle and different doses of each drug were administered to each rat of a group using a latin square design. The effects of drugs were tested once a week to allow for a sufficiently long wash out period.

RESULTS AND DISCUSSION

Table 1 shows a list of drugs which are active or inactive on extinction of conflict behaviour. Baclofen enhances extinction in a $GABA_A$-independent way, since the effect of baclofen on extinction is not antagonized by the $GABA_A$ antagonist bicuculline, while, in contrast, bicuculline antagonizes the oxazepam-induced extinction of conflict behaviour (data not shown, (Ketelaars, 1990)). Muscarinic receptor blockade (cyproheptadine, scopolamine) or muscarinic autoreceptor activation (oxotremorine) enhances extinction (Ketelaars and Bruinvels, 1989). Impairment of 5-HT function via biosynthesis inhibition (PCPA), autoreceptor activation (8-OH-DPAT) or postsynaptic

Table 1 Drugs active or inactive in extinction of conflict behaviour. Each drug was tested in 3 to 5 doses of which the maximal dose is shown. (+) means enhancement of extinction, (-) means not active. Data from Ketelaars et al. (1988), Ketelaars and Bruinvels (1989) and Ketelaars (1990).

Drug	Dose (mg/kg)	Activity
Baclofen	2.5	+
Oxazepam	10	+
Chlordiazepoxide	10	+
Haloperidol	1	+
Chlorpromazine	3	-
Thioridazine	5	-
Imipramine	10	-
Amytryptiline	1	-
Protryptiline	1	-
Mianserine	1	-
Fluvoxamine	20	-
Fluvoxamine (long-term treatment)	20/day	+
Cyproheptadine	10	+
Scopolamine	0.2	+
Physostigmine	0.25	-
Oxotremorine	0.08	+
Ritanserine	20	-
ICS 205-930	0.0625	-
PCPA	300	+
8-OH-DPAT	0.1	+
Propranolol	40	+
Butoxamine	10	-
Metoprolol	8	-

5-HT_1 receptor blockade (propranolol) also enhances extinction of conflict behaviour. Blockade of 5-HT_3 receptors, 5-HT_2 receptors and β-adrenoceptors does not affect extinction of conflict behaviour. In addition, it is shown that in PCPA pretreated rats, baclofen and 8-OH-DPAT do not enhance extinction above the level of extinction induced by PCPA alone (data not shown, Ketelaars, 1990). These data suggest that impairment of the 5-HT system, or more precisely: reduced postsynaptic 5-HT_1 receptor stimulation, plays a role in the baclofen-induced extinction. Long-term administration of fluvoxamine enhances extinction of conflict behaviour, while acute administration of the selective 5-HT reuptake inihibitor fluvoxamine counteracts both the oxazepam and baclofen-induced extinction (data not shown, Ketelaars, 1990). This supports the suggestion that the latter effect of baclofen is caused by an impairment of the 5-HT system. However, the 8-OH-DPAT-induced extinction was not counteracted with the same dose of fluvoxamine.

We consider extinction of conflict behaviour a sensitive animal model for anxiety. However, the nature of this procedure implies that it is not inconceivable that a mixture of anxiolytic effects and/or effects on retrieval of memory for aversive events is responsible for a drug-enhanced extinction. In fact, extinction of conflict behaviour resembles models such as passive or active avoidance which are used to test memory functions, but are also sensitive to anxiolytic drugs (Gray, 1982; Traber et al., 1984). It is tempting to draw an analogy between the design of the extinction procedure on one hand and the history of patients suffering from posttraumatic stress disorder or panic disorder on the other hand, since the memory for an aversive event plays an important role both in the animal model and in the pathogenesis of the latter anxiety disorders.

REFERENCES

Breslow, M.F., Fankhauser, M.P., Potter, R.L., Meredith, K.E., Misiaszek, J., Hope, D.G. (1989) Am. J. Psychiatry 146, 353-356.
Cook, L., Davidson, A.B. (1973) In: The Benzodiazepines. Raven Press, New York, pp 327-345.
Geller, I., Seifter, J. (1960) Psychopharmacologia (Berlin) 1, 482-492.
Gray, J.A. (1982) The neuropsychology of anxiety: An enquiry into the functions of the septo-hippocampal system. Clarendon Press, Oxford.
Ketelaars, C.E.J. (1990) PhD Thesis, Rotterdam.
Ketelaars, C.E.J., Bollen, E.L., Rigter, H., Bruinvels, J. (1988) Life Sci. 42, 933-942.
Ketelaars, C.E.J., Bruinvels, J. (1989) Life Sci. 44, 1743-1749.
Pepplinkhuizen, L., Bruinvels, J. (1978) Abstr. IInd World Congr. Biol. Psychiatry, p 156.
Traber, J., Davies, M.A., Dompert, W.U. (1984) Brain Res. Bull. 12, 741-744.

IS CLONIDINE ANXIOGENIC IN RAT PUPS?

A.M. van der Poel, E. Molewijk, J. Mos and B. Olivier

CNS-Pharmacology, DUPHAR b.v., P.O.Box 900, 1380 DA Weesp, The Netherlands

SUMMARY: Compared to increasing the level of distress by lowering the ambient temperature, clonidine affects the vocal output of rat pups and its temporal patterning to a much larger extent. Moreover, the normal quieting effect of contact with littermates does not operate in clonidine-treated animals. Clonidine therefore either is much more distressful (anxiogenic) than the most distressing manipulation applied to untreated rat pups (cold plate) or acts via a completely different mechanism (e.g. directly upon vocal motor output neural circuits). The latter hypothesis was rejected, since it was found that 2 anxiolytic drugs with different mechanism of action (the 5-HT_{1A} agonist flesinoxan and the benzodiazepine oxazepam) suppressed the clonidine-stimulated ultrasonic vocalizations at their normal anxiolytic dose levels. The ultrasonic distress vocalization of rat pups as an anxiolytic model does not seem to be apt for testing noradrenergic compounds as putative anxiolytics.

INTRODUCTION

Rat pups below the age of 16-20 days emit ultrasonic cries in a frequency range of 35-55 kHz (Sales & Smith, 1978), when separated from nest and littermates. Stressful stimulation like exposure to cold (Allin & Banks, 1971; Okon, 1972), tactile stimuli (Noirot, 1972; Oswalt & Meier, 1975), and unfamiliar olfactory stimuli (Oswalt & Meier, 1975) all enhance calling, whereas stimuli emanating from the mother and/or littermates have a quieting effect (Hofer & Shair, 1978, 1980). Rat pup ultrasounds can therefore be taken to reflect a state of distress. They serve as key stimuli for evoking retrieval behaviour in the mother (Smotherman et al., 1974).

Gardner (1985a) was the first to use the calling of isolated rat pups as an animal model for anxiety. Benzodiazepines (Engel & Hård, 1987; Gardner, 1985[b]; Gardner & Budhram, 1987; Insel et al., 1986) and non-benzodiazepine anxiolytics like buspirone and ipsapirone (Mos & Olivier, 1989) suppress distress vocalizations at dose levels that do not induce sedation or respiratory depression, whereas a number of other psychotropic substances devoid of anxiolytic properties was not active in this model (Gardner, 1985a).

However, there is at least one notable exception to the rule that agents with anxiolytic actions relieve distress and suppress ultrasonic calling in isolated rat pups. Clonidine, claimed to diminish anxiety in humans, both in patients suffering from withdrawal symptoms during abstinence from morphine or alcohol (Björkqvist, 1975; Ahgajanian, 1978) and in patients with anxiety disorders (Liebowitz et al., 1981; Kolb et al., 1984; Hollander et al., 1988), potently stimulates rat pup distress vocalizations (Hård et al., 1988; Kehoe & Harris, 1989).

This paper addresses the question whether this seemingly paradoxic effect of clonidine in rat pups may be considered anxiogenic. The question was tackled in two different ways. Firstly, the effects of increasing dosages of clonidine were compared with increased distress by manipulating the ambient temperature in untreated animals. Secondly, the effects of increasing dosages of two types of anxiolytic drugs on clonidine-stimulated ultrasonic vocalizations were studied.

It has been demonstrated in a number of rodent species (infant mice: Elwood & Keeling, 1982; infant Woodmice: Pontet et al., 1989; adult rats: Van der Poel et al., 1989) that ultrasonic vocalizations are not emitted in a random fashion, but come in bouts. It was therefore deemed necessary not to limit the analysis solely to calling rate, but to include other temporal characteristics of the sounds like bout rate and duration by applying log survivor analysis to intervals between calls (Machlis, 1977; Slater, 1974).

MATERIALS AND METHODS

Subjects were groups of 10 infant Wistar rats (Harlan/CPB, Zeist, the Netherlands) derived from 2 batches of 10 litters bred in our own laboratory. Within 24 hours after birth (Day 0) litters were culled to eight regardless of sex. The animals were tested for ultrasonic vocalization at 9-11 days of age.

On the experimental day the pups were separated from their mother and transported in their home cage to the experimental room. The cage was placed on a warm plate (37°C) and left undisturbed for at least 15 min. Pups were then weighed, marked and injected intraperitoneally with vehicle or clonidine and put back in the home cage. Untreated pups (exp. 1) were weighed and marked but did not receive an injection.

Pups pretreated with one of the anxiolytic drugs, were injected intraperitoneally with either flesinoxan or oxazepam 5 min before treatment with 0.3 mg/kg clonidine. Pups were randomly allocated to treatments according to a balanced block design.

Thirty minutes after the last treatment the animals were placed singly on one of two aluminum plates, which were kept at a temperature of either 18°C (cold plate) or 37°C (warm plate) by circulating water through a reservoir below the plate, similar to the set up described by Gardner (1985a,b). A plexiglass cylinder of 25 cm height was placed on the plate and covered by a plexiglass top on which the microphone was mounted. Animals remained on the plate for 5 min and were then returned to their home cage.

Ultrasounds were recorded with a Bruel & Kjaer 2619 preamplifier, connected to a Bruel & Kjaer 2608 amplifier using a Bruel & Kjaer 4135 condensor microphone. The signal was filtered using 2 Interelectronic AF 173 filters (settings: 30 kHz high pass, 60 kHz low pass) before being fed into a customized electronic device with variable trigger level and further processed by computer. The trigger level was set just above the ambient noise level, allowing the registration of the beginning of each call when signal intensity surpassed the trigger level. When signal intensity dropped below the trigger level, the beginning of a new call was only registered if it was separated from the previous one by a gap of at least 20 ms. Onset and end of each call could thus be timed to the nearest ms. Bouts of calls were defined by a method to discriminate between short intervals separating calls within bouts and longer intervals separating bouts in a formal way. The method was originally proposed by Machlis (1977) and applied to the temporal analysis of rat ultrasonic vocalizations by Van der Poel et al. (1989), and makes use of log survivor analysis. Log survivor plots of intervals between calls characteristically display a biphasic pattern: a steep initial part followed by a less steep second part. In most cases the second part is reasonably linear, indicating that the longer intervals form a homogeneous, exponential distribution and therefore have a constant termination rate. A goodness of fit method (the Cramer-von mises test) as proposed by Haccou (1987) was applied to determine the minimal interval (criterion interval; see Fig. 5^E) between bouts. The average minimal interval ranged from about 200 ms in untreated animals to over 600 ms in animals receiving the combined treatment of oxazepam and clonidine.

In the first experiment 5 groups of 10 animals received an intraperitoneal injection of saline or clonidine (0.01, 0.03, 0.1 or 0.3 mg/kg). They were all tested on the warm plate. Two other groups were not treated. One of the latter groups was tested on the warm plate, the other on the cold plate. In the second experiment groups of 10 animals were injected intraperitoneally with saline, flesinoxan (0.03, 0.1 or 0.3 mg/kg) or oxazepam (0.3, 1 or 3 mg/kg) 35 min before the test, 5 min later followed by a second intraperitoneal injection of 0.3 mg/kg clonidine. Results were analyzed by ANOVA, followed by Scheffé's multiple comparisons test.

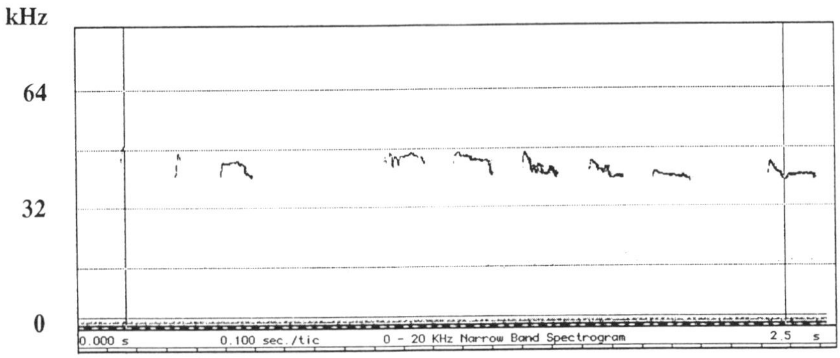

Fig. 1 0-80 kHz narrow-band sonagram of distress calls of a 9 day-old rat pup.

RESULTS

Experiment 1:

Narrow band sonagrams show that distress calls of rat pups consist of bouts of less than 20 to 200 ms, 35-50 kHz highly modulated whistles (see Fig. 1).

Fig. 2 depicts 30 s-samples of recordings from each experimental condition. Special care was taken to select examples of "average" animals. The recordings highlight the most prominent features of the ultrasonic vocalizations under the various experimental conditions. Untreated (Fig. 2A) or saline-treated animals (Fig. 2C) call at low rates. Bouts are short and highly irregular. Single calls are frequent. Pups placed on the cold plate (Fig. 2B) call at a higher rate without much change in irregularity and length of the individual bouts. With increasing dosages of clonidine (Fig. 2D-G) this pattern changes. Bouts become increasingly more regular and longer, while call duration clearly increases.

Fig. 3 illustrates these changes even more clearly in a quantitative way. Compared to exposition to a warm environment pups exposed to cold stress increase their vocal output (Fig. 3A) primarily by increasing the bout rate (Fig. 3D) without changes in call duration (Fig. 3E) and bout duration (Fig. 3F). The increase in calling rate (Fig. 3C) completely parallels the increase in bout rate. There is a modest, insignificant decrease in call latency.

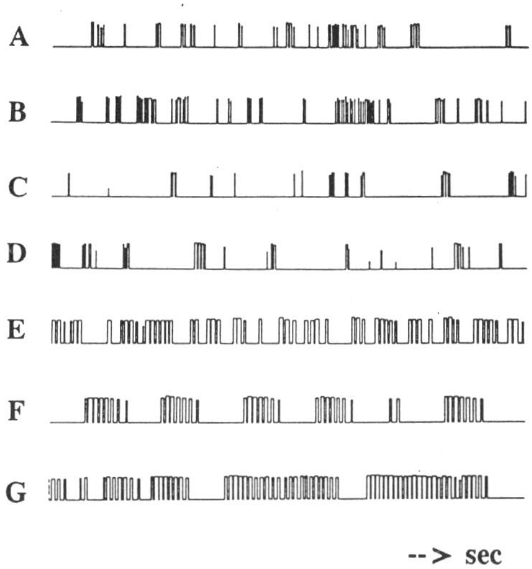

Fig. 2 Representative 30 s-samples of ultrasound recordings:
A. untreated pup on warm floor (37°C).
B. untreated pup on cold floor (18°C).
C. saline-treated pup on warm floor.
D. 0.01 mg/kg clonidine on warm floor.
E. 0.03 mg/kg clonidine on warm floor.
F. 0.1 mg/kg clonidine on warm floor.
G. 0.3 mg/kg clonidine on warm floor.
Each deflection represents one call. The deflection amplitude reflects the mean pitch of each call (kHz).

Clonidine increased the time spent vocalizing too (Fig. 3A). The increase was dose-dependent and much larger than the increase obtained following cold exposure in untreated pups. It was due not only to an increased number of bouts (Fig. 3D), but also to significantly longer bouts (Fig. 3F) of more numerous and longer calls (Fig. 3E). At the 2 highest doses bouts became so long that the bout rate decreased again (Figs. 3D and 2G). In contrast to the effect of manipulating the level of distress in untreated animals clonidine decreased the first call latency dose-dependently (Fig. 3B). It was noted that the clonidine-treated animals already were calling before the test while still with their littermates in the nest and continued to do so after the test when returned to the home cage.

Experiment 2:

Fig. 4 depicts representative 30 sec-samples of ultrasound recordings from clonidine-treated pups pretreated with either saline, flesinoxan or oxazepam.

Fig. 5 illustrates the dose-response relationships. Saline-pretreated clonidine-treated animals again showed high vocal output, producing large number of long bouts consisting of long calls. The mean criterion interval of 400 ms is slightly above the values usually obtained in saline-treated or untreated animals (200-300 ms; unpublished observations).

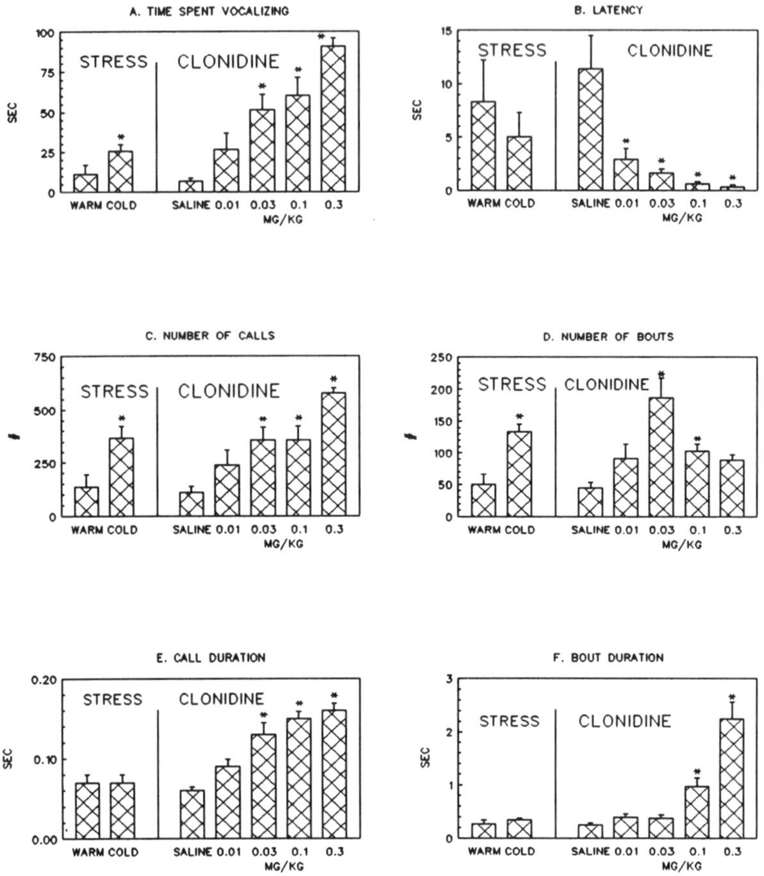

Fig. 3 Temporal characteristics of rat pup ultrasonic distress calls following manipulation of level of distress (left part of the graphs) or increasing doses of clonidine (right part).

Both flesinoxan and oxazepam dose-dependently suppressed clonidine-stimulated calling (see Fig. 5A), though each in a slightly different way. Following pretreatment with flesinoxan vocal output already is significantly decreased at the lowest dose. Pups almost completely ceased vocalizing at the 2 highest dosages due to concurrent reduction of call duration (Fig. 5B), bout rate (Fig. 5D) and bout duration (Figs. 5C and 4B). The bout structure of the ultrasonic calls is not affected however, since the criterion interval is not significantly changed by flesinoxan.

Following pretreatment with oxazepam the clonidine-stimulated vocalization was diminished significantly due to a dose-dependent reduction of call duration (Fig. 5B) and bout rate (Fig. 5D) in addition to spacing out of calls (Fig. 4C). The spacing out of calls, which is a prominent feature of the vocalization of most oxazepam-pretreated animals, is responsible for the observed increase of the criterion interval, which is significant at the highest dose (Fig. 5E). Interestingly, it is coupled with an amazing increase in mean bout duration (Figs. 5C and 4C).

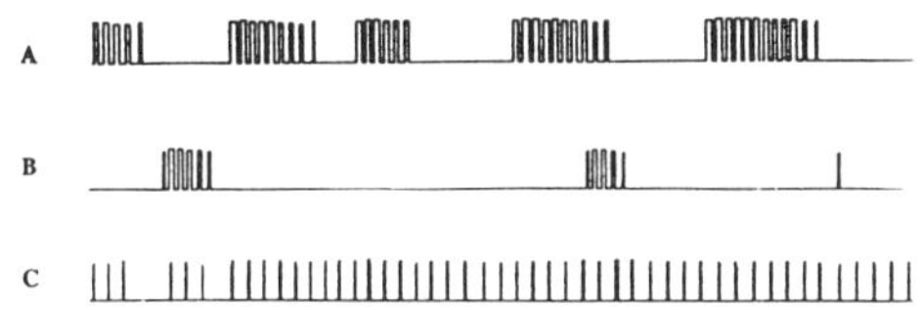

Fig. 4 Representative 30 s-samples of ultrasound recordings of pups treated with 0.3 mg/kg clonidine:
A. saline pretreatment.
B. flesinoxan pretreatment.
C. oxazepam pretreatment
Each deflection represents one call. The deflection amplitude reflects the mean pitch of each call (kHz).

DISCUSSION

The effects of clonidine on the temporal patterning of rat pup ultrasonic distress vocalizations are different from the effects of manipulating the level of distress by cold exposure. Both clonidine and cold exposure admittedly enhance vocal output, but clonidine does so to a much larger extent. Animals exposed to cold spend more time vocalizing primarily by increasing the bout rate, whereas clonidine-treated animals in addition shorten their latency and increase call duration concurrently. At higher doses they also increase bout duration. At the highest dose tested (0.3 mg/kg IP) the vocal output of clonidine-treated pups is 3-4 times higher than that of cold-exposed animals. Moreover, the normal quieting effect of contact with littermates does not operate in clonidine-treated animals. Clonidine therefore either is much more distressful than the most distressing manipulation applied to untreated pups (cold floor) or acts via a completely different mechanism (e.g. directly upon vocal motor output neural circuits).

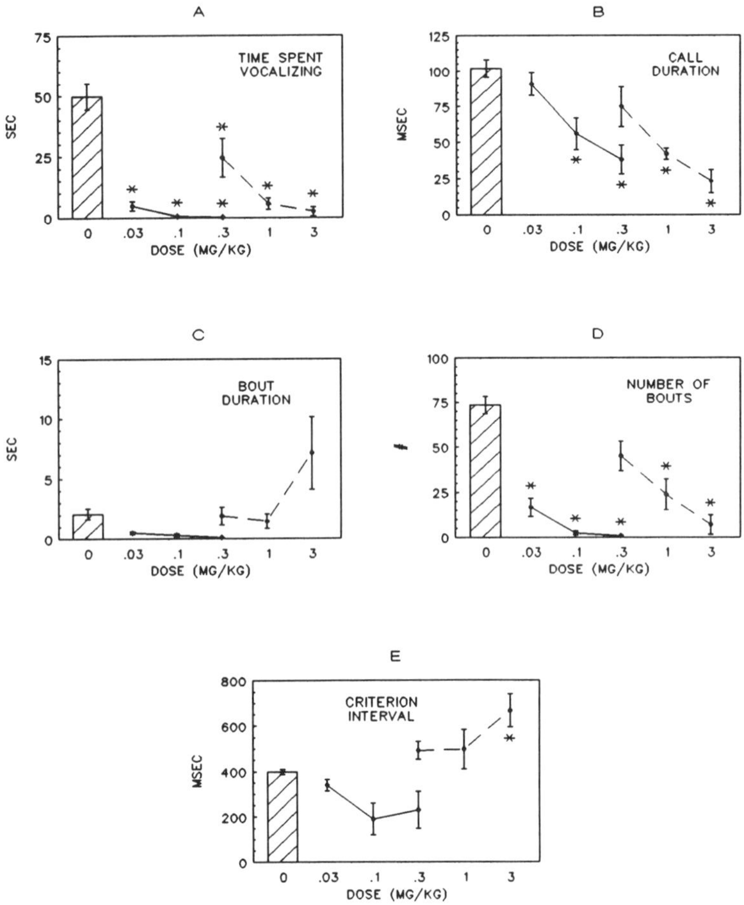

Fig. 5 Temporal characteristics of rat pup distress calls following treatment with 0.3 mg/kg clonidine IP. Bar: saline pretreatment: continuous lines: flesinoxan pretreatment; broken lines: oxazepam pretreatment.

Since benzodiazepines and 5-HT_{1A} agonist drugs are known to suppress rat pup distress vocalizations presumably by their anxiolytic action, it was of interest to examine the effects of both types of drugs on clonidine-stimulated vocalization. Oxazepam as a representative of a benzodiazepine-agonist and flesinoxan as an example of full agonists at the 5-HT_{1A} receptor were selected since both drugs have been studied before in the pup vocalization test in our laboratory. The lowest dose of oxazepam which significantly lowered distress calling on both warm and cold plate was shown to amount to 1 mg/kg (Mos & Olivier, 1989; 1990). The present study shows that oxazepam is similarly effective in suppressing clonidine-stimulated vocalizations. Flesinoxan also is at least as effective in suppressing clonidine-stimulated vocalizations (present study: less than or equal to 0.03 mg/kg) as normal distress vocalizations (unpublished observations: 0.1 mg/kg). It is concluded that anxiolytic drugs acting via different mechanisms of action, are able to counteract the stimulatory effect of clonidine on rat pup distress vocalization. This effect of clonidine therefore may be labelled anxiogenic.

In animal models of anxiety which use adult rats, biphasic dose-response curves have been obtained (Handley & Mithani, 1984; Söderpalm & Engel, 1988), low doses of clonidine exerting anxiolytic, high doses anxiogenic effects. Since the effects of high doses may be confounded by the well-known sedative action of clonidine (Drew et al., 1979), these results may be interpreted as to confirm its anxiolytic action in man.

Apparently, there is a radical change in the effects of clonidine on anxiety during ontogeny. This has been confirmed by an ontogenetic study of rat pup distress vocalization by Kehoe & Harris (1989). They found that clonidine enhanced vocal output at ages up to day 16, while reducing it at day 18. At the critical age of day 17 low doses enhanced and high doses reduced vocal output. Most likely, maturational changes in the noradrenergic system are responsible for this change. Hård et al. (1988) found that the α_2-antagonist idazoxan antagonized the effects of clonidine on rat pup distress vocalizations at all ages studied, whereas the suppressant effect of the α_1-antagonist prazosin disappeared somewhere between the ages of day 12 and day 16. It has been shown that postnatally the numbers of both α_1- and α_2-receptors increase continuously and concurrently up to day 20 and then show a slight regression (Morris et al., 1980). The development of the noradrenergic system appears to progress at unequal speeds in different parts of the brain. Kellogg & Wennerström (1974) found functional noradrenaline-containing neurons in cortex and brain stem already by 4 days of age, whereas such neurons and their receptors were not apparent in the diencephalon until 28 days. Reinstein & Isaacson (1977) suggested that differential development of various parts of the noradrenergic system is responsible for the observed ontogenetic change in the clonidine response on locomotor activity (hyperactivity at first, followed by hypoactivity and catalepsy). Since this change seems to occur at about the same age (between day 14-21) as the change in the clonidine response on distress vocalization, it is tempting to attribute both changes to the same maturational process.

The results of this study therefore suggest that the ultrasonic distress vocalization of rat pups as an anxiety model is not apt for testing the putative anxiolytic effects of noradrenergic compounds, since in contrast to the serotonergic and benzodiazepine/GABA systems, the noradrenergic system does not yet seem to function at its normal, adult level at this young age in pups.

REFERENCES

Aghajanian, G.K. (1978) Nature 276, 186-188.
Allin, J.T., Banks, E.M. (1971) Devl. Psychobiol. 4, 149-156.
Björkqvist, S-E. (1975) Acta Psychiatr. Scand. 52, 256-263.
Drew, G.M., Gower, A.J., Marriott, A.S. (1979) Br. J. Pharmacol. 67, 133-141.
Elwood, R.W., Keeling, F. (1982) Devl. Psychobiol. 15, 221-227.
Engel, J., Hård, E. (1987) Alcohol & Alcoholism Suppl. 1, 709-712.
Gardner, C.R. (1985[a]) J. Pharmacol. Meth. 14, 181-187.
Gardner, C.R. (1985[b]) Drug Devl. Res. 5, 185-193.
Gardner, C.R., Budhram, P. (1987) Eur. J. Pharmacol. 134, 275-283.
Haccou, P. (1987) Thesis, Leiden, pp 19-44.
Handley, S.L., Mithani, S. (1984) Naun. Schmiede. Arch. Pharmacol. 327, 1-5.
Hård, E., Engel, J., Lindh, A.S. (1988) J. Neural Transm. 73, 217-237.
Hofer, M.A., Shair, H. (1978) Devl. Psychobiol. 11, 495-504.
Hofer, M.A., Shair, H. (1980) J. Comp. Physiol. Psychol. 94, 271-279.
Hollander, E., Fay, M., Liebowitz, M.R. (1988) Am. J. Psychiatry 145, 388-389.
Insel, T.R., Hill, J.L., Mayor, R.B. (1986) Pharmacol. Biochem. Behav. 24, 1263-1267.
Kehoe, P., Harris, J.C. (1989) Behav. Neurosci. 103, 1099-1107.
Kellogg, C., Wennerström, G. (1974) Brain Res. 79, 451-464.
Kolb, L.C., Burris, B.C., Griffiths, S. (1984) In: Post-Traumatic Stress Disorder: Psychological and Biological Sequelae (B.A. Van der Kolk, Ed), Washington, DC, American Psychiatric Press.
Liebowitz, M.R., Fyer, A.J., McGrath, P., Klein, D.F. (1981) Psychopharmacol. Bull. 17, 122-123.
Machlis, L. (1977) Behaviour 63, 1-70.
Morris, M.J., Dausse, J-P., Devynck, M-A., Meyer, P. (1980) Brain Res. 190, 268-271.
Mos, J., Olivier, B. (1989) In: Behavioural Pharmacology of 5-HT (P.Bevan, A.R.Cools and T Archer, Eds), pp 361-366.
Mos, J., Olivier, B. (1990) In: Stress, biological rhythms and psychiatric disorders (H.G.M.Westenberg, Ed), Medidact, Houten, pp 89-96.
Noirot, E. (1972) Devl. Psychobiol. 5, 371-387.
Okon, E.E. (1972) J. Zool. Lond. 168, 139-148.
Oswalt, G.L., Meier, G.W. (1975) Devl. Psychobiol. 8, 129-135.
Pontet, A., Gyger, M., Schenk, F. (1989) Behaviour 108, 241-261.
Reinstein, D.K., Isaacson, R.L. (1977) Brain Res. 135, 378-382.
Sales, G.D., Smith, J.C. (1978) Devl. Psychobiol. 11, 595-619.
Slater, P.J.B. (1974) An. Behav. 22, 506-515.
Smotherman, W.P., Bell, R.W., Starzec, J., Elias, J., Zachman, T.A. (1974) Behav. Biol. 12, 55-66.
Söderpalm, B., Engel, J.A. (1988) Pharmacol. Biochem. Behav. 30, 471-477.
Van der Poel, A.M., Noach, E.J.K., Miczek, K.A. (1989) Psychopharmacology 97, 147-148.

Animal Models in Psychopharmacology
Advances in Pharmacological Sciences

RISK ASSESSMENT AND ANIMAL MODELS OF ANXIETY

D.C. Blanchard[1], R.J. Blanchard[2] and R.J. Rodgers[3]

1 Békesy Laboratory of Neurobiology and John A. Burns School of Medicine, University of Hawaii, Honolulu, HI 96822
2 Department of Psychology, University of Hawaii, Honolulu, HI 96822
3 Department of Psychology, University of Leeds, Leeds LS2 9JT, England

SUMMARY: Analysis of the pattern of defensive behaviours in semi-natural situations suggests a long-duration process in which high level freezing/movement arrest gradually gives way to active risk assessment activities, terminating with a gradual return to nondefensive behaviours. Risk assessment is conceptualized as the key feature in this process, with the information obtained by these oriented scanning and exploratory activities providing feedback to reduce the subject's initial high level of defensiveness. Risk assessment measures derived from this analysis are selectively responsive to classic anxiolytics. Examination of a range of commonly used measures of anxiolytic action suggest that many of these contain clear elements of the risk assessment process. Furthermore, variation in the results obtained in these tests might be profitably analyzed in terms of some of the features of risk assessment and other aspects of the defense pattern seen to partial or potential threat stimuli.

THE FUNCTION OF ANIMAL MODELS IN RESEARCH

A good animal model of anxiety must meet formidable criteria. The human phenomena encompassed by this rubric are both common, and complex: Anxiety is the most frequent psychiatric diagnosis made in America, with approximately 8.3% of the adult US population estimated as having suffered from anxiety disorders during the past 6 months (Rickels & Schweizer, 1987). About 15% of the population are estimated to suffer some anxiety disorder on a lifetime basis (Barlow, 1988). Anxiety disorders are associated with a number of different behavioural symptomatologies, typically rather consistent within individuals (APA, 1987) and the variety of these, in addition to their differential responsivity to drug therapies, increases the difficulty of determining an adequate animal model or models for anxiety (Lister, 1990).

A survey of current animal models of anxiety (for reviews: File, 1986; Gray et al., 1981; Lal & Emmett-Oglesby, 1983; Lister, 1990) reveals a bewildering diversity of procedures. Broadly speaking, however, all models involve exposure of animals to either exteroceptive (e.g., cues previously paired with footshock) or interoceptive (e.g., internal drug states) stimuli which appear capable of causing anxiety in humans. Behavioural responses to such stimuli that are selectively antagonized by anxiolytic drugs are generally accepted as analogues of anxiety. Although the criterion of pharmacological validation is not unimportant, a major drawback of exclusive reliance upon this criterion is the potential danger of excluding compounds with novel actions.

Tests employing exteroceptive stimuli comprise a broad spectrum, at one end of which are the classic conflict procedures and their derivatives (e.g., Geller-Seifter, Vogel and Bossier) while at the other are the more recently developed models such as the rat social interaction (File & Hyde, 1978), and plus-maze (Handley & Mithani, 1984) tests and the mouse novelty (Crawley & Goodwin, 1980) and stretched attention posture tests (Kaesermann, 1986). The classic models involve training of subjects and the use of painful training stimuli, whereas the more recent models are based upon the study of spontaneously emitted behaviour. While animal models of anxiety have proliferated and expanded in recent years, there is considerable agreement that no single existing model is capable of providing an adequate view of the range of anxiety-related disorders (Lister, 1990). In fact, understanding of the mechanisms and treatment of anxiety states is currently in a state of flux (File & Johnston, 1987).

AN ALTERNATIVE APPROACH TO THE INVESTIGATION OF ANXIETY THROUGH USE OF ANIMAL MODELS

A classic distinction in the analysis of relationships between behaviour patterns in different species is between homologous and analogous behaviours (Lorenz, 1974), with the former involving derivation from a common ancestral source with little subsequent divergence. It is obviously too much to ask of an animal model of anxiety that it be homologous in terms of this definition, as the common ancestor of any nonhuman animal (including primates), and man, is simply not available for examination, particularly for the study of neurobehavioural mechanisms. While it may be possible to exclude such homologies, in the case of very distant divergence from an ancestor with primitive characteristics, a reasoned acceptance of the view that a particular neurobehavioural system is homologous for humans and some other nonhuman species, in terms of this definition, remains difficult if not impossible.

However, an implicit feature in the concept of an homology, that the same mechanisms be involved, remains an important, and assessable, criterion. Such a view, we maintain, should give weight to each of the key factors involved in understanding behaviour patterns: the stimuli and situations that produce them; how they are organized and modulated in the brain; the behaviours they encompass; the typical function of these behaviours in the normal environment of the animal.

From this perspective, animal modelling of anxiety should ideally involve systems which bear a broad-based correspondence to the phenomena encompassed by the rubric "anxiety".

NEUROBEHAVIOURAL SYSTEMS OF DEFENSE, FEAR, AND ANXIETY

The pioneering work of Darwin (1872) laid a foundation for regarding the defensive response of lower animals to predators, attacking conspecifics and environmental threats and hazards, as essential evolutionary precursors to human fear and anxiety reactions. The behavioural and experiential complexities of human fear and anxiety reactions have been increasingly acknowledged and detailed in various classifications of fear or anxiety disorders in recent years (APA, 1987; Marks, 1987). However, there has been no systematic attempt to provide a correspondingly detailed description of defensive reactions in laboratory and wild animals in order to analyze the possible relationships between these reactions and the behavioural characteristics of human anxiety. It is this specific gap that we have been attempting to fill.

Defensive behaviours are significant, first, because they are likely to be the most pervasive and common behaviour patterns for any vertebrate species, encompassing the reactions of animals to the host of life-threatening dangers encountered in every natural environment (Endler, 1986). An additional significance is that they limit and control all of the personal- and genetic-fitness enhancing activities of these animals. As the consequences of incorrect, inadequate, or misplaced defense are so immediate and (often) final for the individual, defensive behaviours have been submitted to a particularly stringent selection process. Consequently, they are particularly well represented in terms of complex brain systems not only providing for the form of the response, but also linking specific defensive behaviours to features of both the threat source, and the situation in which it is encountered (Blanchard, R.J. et al., 1989; Blanchard, R.J. & Blanchard, D.C. in press).

These views are well supported by previous research indicating that a range of discrete defensive behaviours occurs in laboratory and other rodent species (Adams, 1980), even when the subjects are maintained under conditions in which previous experience of threat stimuli is minimal. Several specific behaviours, particularly freezing (Blanchard, R.J. & Blanchard, D.C., 1969b; Fanselow & Sigmundi, 1986; 1987), startle (Blanchard, R.J. et al., 1986; Brown et al., 1951), flight/avoidance (Blanchard, R.J. & Blanchard, D.C., 1969a; Mackintosh, 1974) and defensive threat/attack (Blanchard, R.J. et al., 1978; Flynn, 1972) have long been recognized as components of a behavioural defense system.

Investigators concerned with the pharmacological and neural control of fear have typically concentrated on the neurobiological correlates of a subset of these behaviours. Thus Fanselow and his coworkers (Fanselow & Sigmundi, 1986, 1987) have studied opiate effects on freezing following foot shock or exposure to a predator; LeDoux and his coworkers (Iwata et al., 1986; LeDoux, 1986; LeDoux et al., 1986) have primarily worked with neural correlates of conditioned freezing and startle; Siegel and his coworkers (Brutus et al., 1988, 1989; Fuchs et al., 1985; Fuchs & Siegel, 1984; Shaikh et al., 1985; Siegel & Pott, 1988), and Bandler and his colleagues (Bandler & Carrive, 1988; Bandler & Depaulis, 1988) have investigated neural systems involved in defensive threat, or flight; and Davis and his colleagues (Davis et al., 1987; Hitchcock et al., 1989) the neural systems involved in conditioned startle reactions. The resulting literature on the neuroanatomic and neurotransmitter correlates of defense (above: also Adams, 1979; Blanchard, D.C. & Blanchard, R.J. 1988) indicates that each of these (freezing, flight, defensive threat/attack, startle) involves a complex of forebrain, midbrain and hindbrain systems which may be differentiated from the others at a number of levels. However, a more systematic analysis of the relationship between defensive behaviour and neural systems requires a thorough understanding of the extended range of defensive behaviours, the circumstances in which they occur, and their organization with reference to each other.

DEFENSIVE BEHAVIOURS TO A PRESENT, DISCRETE, THREAT STIMULUS

The immediacy of threat and its embodiment in a discrete physical stimulus appears to be an extremely important organizational factor in defense (Blanchard, R.J. et al., 1989). When an immediate (i.e. present) , discrete and complete threat stimulus (i.e.

predator) is available, it elicits flight, freezing, or defensive threat/attack on a very predictable basis as a function of the defensive distance (distance between the subject and the threat), and as a function of the escapability of the situation (Blanchard, R.J. et al., 1986). While freezing in an inescapable situation, measured directly, is high and stable over a range of defensive distances, it is notable that startle (elicited by sound or movement) increases as the predator approaches, suggesting an increase in the muscle tension and readiness to react associated with freezing. In contrast, when the situation permits escape flight speed increases as the predator approaches (Ydenberg & Dill, 1986). When contact becomes imminent, regardless of the availability of escape, defensive threat, followed shortly by defensive attack, is seen.

The simplicity of this analysis is facilitated by the lack of specialized structural defenses (toxic sprays, armor, spines, etc.) in laboratory rodents, and animals possessing such features typically show additional, species-specific, behaviours in their use. However, the focus of this analysis is on species-typical behaviours (i.e. typical of a range of mammalian species: see Adams (1980), for description of between-species generality of intraspecies defense in muroid rodents) and the lack of specialized structural defenses in rodents, as in humans, thus makes them more appropriate subjects for such analyses.

The foregoing description of flight, freezing and defensive threat/attack is based on wild rat reactions to an approaching experimenter. However, the same pattern of reactions occurs in naive wild rats bred in a laboratory, as in wild-trapped wild rats (Blanchard, R.J. et al., 1986), indicating that prior experience with these behaviours in a predatory context is not necessary for their appearance in a highly organized manner with reference to relevant features of the threat stimulus and situation. Even the laboratory rat, an animal long recognized as less reactive than its wild congeners (Stone, 1932; Yerkes, 1913) will show consistent defensive behaviour to the presence of a cat (Curti, 1935). While laboratory rats tend to show the same pattern of relationships of defense to threat elements as do wild rats, it is notable that flight, and especially defensive threat/attack, are much reduced in laboratory rats (Blanchard, R.J. et al., 1986) with freezing a more prominent component of the defensive repertory for these animals (Takahashi & Blanchard, 1982).

DEFENSIVE BEHAVIOURS TO SITUATIONS WITH A THREAT POTENTIAL

Anxiety has long been conceptualized in many psychological views as a reaction to impending danger (Freud, 1936) or as a response to a threat which, while not clearly present, is signalled by either a cue stimulus or situation (Estes & Skinner, 1941). While this stimulus-based distinction is widely accepted, and there have been suggestions that specific behaviours may be associated with anxiety as opposed to fear (Blanchard, D.C. & Blanchard, R.J., 1988), systematic analysis of behaviour patterns to situations involving ambiguous or potential threat as opposed to discrete and present threat sources has only just begun.

The Visible Burrow System (VBS) is a seminatural habitat with transparent Plexiglas tunnels and chambers (under infrared light illumination) designed to resemble actual burrows constructed in this laboratory by rats in a dirt substrate. It has also an open "surface" area which is supplied with a 12:12 hr light/dark cycle. Behaviours can be videotaped in both the burrow system and in the adjacent "surface" area. Laboratory rats living in this system show consistent diurnal variations in activity, with females and dominant males emerging on the surface shortly after lights out. Normal ongoing activities such as eating, drinking, and social, sexual and aggressive behaviours provide a background against which the behaviour changes associated with defense can be measured. Use of the VBS thus provides a relatively unstructured situation in which a full range of natural defensive reactions can be measured to partial threat stimuli, or after brief presentation of a predator.

The following data and analyses are more fully described in Blanchard, R.J. & Blanchard, D.C. (1989). Presentation of a noncontacting cat for 15 minutes in the "surface" area shortly after lights out systematically changes the behaviours of the animals of the group for 24 hours or more. The initial reaction to presentation of the cat in the open area consists of immediate flight to the burrows, a finding consistent with the effects of a proximal predator in other situations allowing escape (Blanchard, R.J. et al., 1986).

Movement Inhibition: Inside the burrow systems, immobility (presumably freezing) occurred next: Dominant males (typically less defensive than subordinates or females) remained totally motionless in the burrows for an average of about 12 minutes, and the first episode of grooming for these animals did not occur for over 30 minutes. During the next several hours, subjects gradually resumed movement in the burrow system.

Ultrasonic Vocalization: During this period there is intense ultrasonic vocalization, largely 20-24 kHz in frequency, by the rats in the burrow system (Blanchard, R.J. et al., in press b).

Risk Assessment: It is notable that during the periods of total (and then partial) movement inhibition subjects tended to maintain orientation toward the close tunnels through which a predator from the surface would have to pass in order to approach. As movement resumed, it consisted largely of approaches to the close tunnels and then to the openings to the surface. Typically, subjects would stand in a close tunnel, head poked through an opening to the surface area, scan, followed then by withdrawal into the tunnels. After several such episodes, subjects animals eventually emerged on the surface, where they showed thigmotaxis and constrained movement. The latter involved a flattening of the back and stretching of the animal's body in a "stretch attend" posture. These stretched-out postures or movements have also been described (Grant & Mackintosh, 1963; Pinel & Mana, 1989) for rats oriented to/approaching attacking conspecifics or conditioned threat stimuli. Thus "risk assessment" activities involve orientation (toward the potential threat source) , proxemic (approach to the threat source) and specific movement characteristics such as the lowered back and cautious movement seen in "stretch attend" and "stretch approach."

Inhibition of nondefensive behaviour: Nondefensive behaviours such as eating, drinking, mounting of females and offensive attack were relatively common during the pre-cat day, but declined abruptly with cat presentation and remained low during the 24-hr. period after cat exposure.

These results suggest a very clear pattern of reactions to potential threat in a semi-natural situation affording a wide range of activities. However, not all of the reactions seen here are to potential as opposed to actual and present, threat. Flight in this situation almost always occurs when the cat is presented, not after the cat has been removed. Flight also occurs when subjects are approached by a predator in the F/DTB. Thus, it seems likely that this component of the defense pattern is more a reaction to present, discrete threat stimuli than to potential threat. The same argument could be made concerning movement arrest/freezing, but with less force, in that these reactions continue to occur long past the removal of the cat, often in animals which have already scanned the open area no longer containing the cat. The inhibition of nondefensive behaviours, although of long duration in this situation, is also not likely to be specific to potential, as opposed to actual, threat, as it is even more marked while the cat is present (unpublished observations from this laboratory).

The reactions which most clearly are not an immediate response to the presence of the cat, but occur at high levels to situations or objects with threat potential, are the active, information-gathering risk assessment behaviours. Analysis of these suggests several important points. First, risk assessment is both a long-latency, and a long-duration process when high levels of potential threat are involved. To some degree the latency to risk assesment may be misleading, since even the most profoundly freezing rat maintains an orientation toward the possible threat, suggesting a risk assessment component even in the early and more intense stages of the defense process. However, movement arrest is clearly incompatible with the active risk assessment behaviours, and the latter appear with increasing frequency as the former decline. Risk assessment also continues much past the time when movement inhibition is no longer evident, and may appear as the initial and primary reaction to a very mildly threatening stimulus, attesting to its status as an independent behaviour category rather than some artifact of the breakup of movement arrest.

Thus, there is a complex relationship between movement arrest/freezing, and the various risk assessment activities which follow it, as the defense pattern progresses over time. We have suggested (Blanchard, R.J. et al., 1989) that risk assessment is, in fact, the central feature in promoting the sequence through the defense process. The information obtained from orientation, scanning, and approach activities act (in the continued absence of the cat) to decrease the subject's level of defensiveness, reducing movement inhibition and permitting more active and effective risk assessment to occur. Eventually, given a normal subject, this information reduces defensiveness to the point where risk assessment itself declines, and a return to normal, relatively nondefensive, behaviour occurs.

A TIME/INTENSITY PROGRESSION FOR DEFENSIVE BEHAVIOURS TO POTENTIAL THREAT: A CHANGING BEHAVIOURAL BASELINE FOR DEFENSE

A simplified representation of this process is shown in Figure 1, which presents active defenses, freezing/movement arrest, and risk assessment as a function of time after exposure to the predator.

The time line for this defense sequence presumably also represents an initial maximal level of defensive motivation while the terminal time is associated with a much lower value of defensiveness. A rat, 1 hour after exposure to the cat (time "A" on graph) might thus be presumed to be more defensive than the same animal at time "B" or "C" and any factor reducing defensiveness by changing the activity of brain systems involved in maintaining this long-term process might be expected to produce a behavioural shift toward portions of the defense process which typically occur later. Over most of the defense sequence such a shift is best described in terms of the magnitudes of both freezing/movement arrest, and, risk assessment. If an animal is at time "A" in the process, with a predominantly freezing baseline, reductions in defensiveness would be associated with decreased movement arrest and an increase in risk assessment, represented here by time "B". If intervention is made when the animal is at point "B", when risk assessment is the predominant defensive response, then the result would be a shift toward point"C" with a decrease in risk assessment. This analysis suggests that the effects of an anxiolytic compound on risk assessment must be evaluated against the initial behavioural baseline, with increases or decreases predictable for exactly the same compound/dose, depending on initial levels of defensiveness.

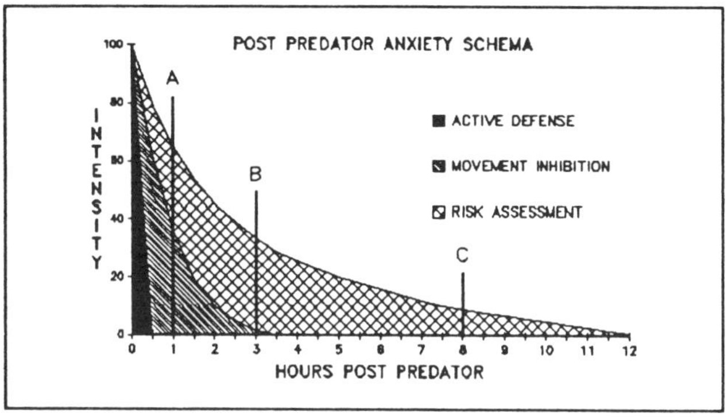

Fig. 1 Active defense, movement inhibition and risk assessment are shown as a function of time after exposure to a predator.

A slightly more complex view is that, although the components of this process do reflect the intensity of defensive motivation, they are also independently responsive to relevant stimulus features, and, to underlying neural processes. This view reflects the considerations (outlined above) suggesting that risk assessment, in particular, occurs primarily in the context of uncertainty concerning the threat characteristics of the stimulus. In contrast, the other components, flight, freezing/movement arrest, and interference with nondefensive behaviours may occur at equally high or higher rates in the presence of a clearly threatening object, the predator. This view also receives some support from studies of drug effects on various components of this process.

EFFECTS OF ANXIOLYTIC DRUGS ON COMPONENTS OF THE DEPENSIVE PROCESS

The time course for defense changes in the VBS is such that this situation is not suitable for use with drug manipulations. We have therefore created a set of more specific tests taking various measures of freezing/movement arrest, risk assessment, and interference with nondefensive behaviours in situations associated with the actual presence of a cat, or to a less intensely threatening cat odor stimulus. The results of these tests indicate a profound and specific alteration of risk assessment activities with diazepam (Blanchard, R.J. et al., 1990; Blanchard, D.C. et al., 1990), while ethanol (Blanchard, R.J. et al., 1990a,b) reduces freezing/movement arrest and also alters risk assessment in a manner similar to (though more variable than) that of diazepam.

What is particularly interesting about these results is that the direction of the effect of diazepam on risk assessment (and of ethanol, when ethanol effects are obtained) depends on the initial baseline activity (i.e. for vehicle control subjects) in the particular situation. When freezing is the dominant initial reaction in the situation (in all tests using presentation of the cat), risk assessment increases with diazepam, but when risk assessment is initially present in control animals (cat odor test), it decreases with diazepam. This is not a ceiling effect since, in the latter tests control subjects averaged about 25% risk assessment time in the situation.

The baseline activities in these tests were congruent with the behaviour sequences obtained in the VBS to an actual cat, as opposed to cat odor stimuli, and suggest the possibility that baseline behaviours can be precisely predicted for animals of a given sex (see sex differences, below) in a particular situation. However, prediction from the VBS or some other situation is less important than is the recording of baseline behavioural measures in a given test. In addition, the multiple measures taken in the four tests in this series can provide profiles of drug effects for rather detailed comparison and analysis of the actions of different classes of compounds.

SEX DIFFERENCES

These same tests have shown an unexpected, but consistent, pattern of sex difference in laboratory rat subjects. In two parallel studies involving situations in which a cat had been presented and then removed (Blanchard, R.J. et al., 1990b; Blanchard, D.C. et al., 1990), each resulting in a baseline of high freezing/movement arrest, female rats showed higher levels of crouching, and lower locomotion, drinking, eating and lying, with reliable sex effects or sex interactions in each case. Males showed reliably higher levels of risk assessment. It is notable that these differences were invariably in the opposite direction to the effect of cat exposure (i.e. males showed less of the behaviours increasing with cat exposure) and in the same direction as the effects of anxiolytic compounds (diazepam, and ethanol) where these produced reliable differences in these tests. In the cat odor test where diazepam and ethanol tended to reduce risk assessment, against a high (circa 25%) risk assessment back ground, females tended to show higher risk assessment than males (Blanchard, R.J. et al., 1990a). Thus the pattern of sex effects on these reactions to partial threat stimuli or situations associated with threat, indicate that females show higher baseline defensiveness than do males, with these higher baselines interacting with the effects of drug treatment on the tests.

In contrast, no sex differences were found in reactions to an approaching, contacting, discrete threat stimulus (Blanchard, R.J. et al., 1986). In the latter situation, there was also a remarkable lack of benzodiazepine effects on most measures. This conjunction of findings suggests that female rats may show higher levels of anxiety-related defensiveness, but no difference from males in fear-related defenses to a present and complete threat stimulus. This is congruent with indications of higher rates of anxiety-related disorders in females (Barlow, 1988; APA, 1987) and strongly suggests the necessity for consideration and investigation of sex differences in animal models of anxiety.

This program, still in early stages of development, demonstrates ready responsivity of measures derived from a purely behavioural analysis of defense, to some commonly used anxiolytic drugs. It therefore provides a possible framework for a more comprehensive analysis of other behavioural measures showing good responsivity to anxiolytic and anxiogenic drugs.

ADDITIONAL COMPONENTS OF THE DEFENSE SYSTEM

Examination of other defense-related behaviours, some already used in tests of anxiolytic drug action, suggests a possible correspondence with the sequence outlined above.

Defensive Burying: Pinel, Treit, and their collaborators have extensively investigated a "defensive burying" behaviour, typically in the context of a well-defined but inanimate object associated with shock or other noxious stimulation (Pinel & Treit, 1978). A number of features of defensive burying suggest that it may be a component of the risk assessment pattern:

a. Flight is typically the initial reaction to shock from a shock prod stimulus, followed by varying degrees of immobility, and by a stretch-attend/stretch approach return to the shock prod, with the subject often pushing bedding material ahead of itself during this approach (Pinel & Mana, 1989).

b. Post-shock conditioning effects suggest an information-gathering role for these stretch-attend/approach activities (Pinel et al., 1986).

c. Pinel & Wilkie (1983) have interpreted burying of a hole through which the shock prod was presented as an aid to recognition of its dangerousness.

d. Defensive burying increases with increasing shock intensity, and then declines (Pinel, personal communication), and, may be suppressed by manipulations which further increase subjects' defensiveness (Williams, 1987; Williams & Scott, 1989)

e. Burying does occur in the VBS, when a novel stimulus or a cloth saturated with cat odor is placed in the open area (Blanchard, R.J. et al., 1989): Subjects show flight, brief (to a novel stimulus) or longer (to a cat odor cloth) freezing in the burrows, followed by typical risk assessment approaches and eventual burying of the object. The least threatening, novel, stimuli are, in fact, often dug up and reburied several times in the course of a typical 3-hour test.

"Exploratory behaviour" tests of anxiolytic drug action: A number of the most commonly used tests for anxiolytic and anxiogenic action of drugs involve some form of exploratory behaviour (see Lister, 1990, for a review). Each of these tests measures either specific actions (Holeboard Test; head dipping into holes: Staircase Test: rearing on a staircase) or orientation/approach to specific potential danger stimuli (Plusmaze; open arm exploration: Light-Dark Transitions Test: transition between light and dark areas) which can be differentiated from more general locomotory aspects of exploration. These particular measures, with the possible exception of rearing in the Staircase Test (which also does not correlate with head dipping in the holeboard test, File, 1985) appear to be very similar to many of the risk assessment behaviours seen in the unstructured VBS tests, and a number of features of behaviour change in these tests further suggest a relationship to the risk assessment pattern. Many such features are described in Lister (1990), including the following findings relating to repeated testing effects:

1. When the subject is unfamiliar with the test apparatus, head dipping in the holeboard test increases with low doses of benzodiazepines, but decreases with high doses (Nolan & Parkes, 1973), an effect very congruent with the analysis of risk assessment as a behaviour which initially increases, then declines, with anxiety reduction from an initial freezing/movement inhibition baseline.

2. When the subject has previously been exposed to the test apparatus, benzodiazepine doses that produce an effect in unfamiliar animals fail to alter head-dipping (Nolan & Parkes, 1973). Since the only potential threat element in this situation is the novelty of the apparatus (and specifically of the holes, which are not so amenable to rapid visual inspection and re-inspection as is the small open chamber) this threatening element is presumably reduced by prior familiarization, reducing the anxiety baseline of the test.

3. Rapid reductions in defensiveness based on information obtained through exploration/head dipping activities should limit the time period in which head-dipping responds to anxiolytic effects: Benzodiazepine-induced increases in exploration are most visible during the initial moments of the test period (Nolan & Parkes, 1973).

4. In apparent contrast, repeated exposure to the plus-maze test either does not alter the behaviours of undrugged animals (Lister, 1987), or may even give an anxiogenic profile of decreased open arm activity (Lee & Rodgers, in press). Repeated testing in the Light-Dark test after diazepam administration also fails to alter the drug effect in increasing transitions. Notably, the threat sources in these situations (including the possibility of falling off the open arms of the elevated plus maze, and the aversive effect of the brightly lit Light chamber), are not eliminated and may even be increased by information-gathering activities, in contrast to the novelty-based threat potential of the holeboard situation. Some defensiveness due to the novelty of the test situation should also be a factor in the plus-maze and light-dark tests, but each of these situations combines two potential threat elements while those of the holeboard and staircase tests have no threat elements other than the novelty of the situation and its features.

One additional feature of these tests relevant to their relationship to the defense process outlined earlier, is that each attempts to provide some locomotory control measure in the same or a different test situation, with an interpretation of specific anxiolytic action based on changes (generally increases, but note the high dose benzodiazepine effects on head dipping, above) in a criterion measure with no accompanying locomotory alteration. These analyses, implicitly focusing on risk assessment activities as a specific measure of anxiety, have produced a wide variety of results, with common findings of concomitant increases in both the criterion measure and in locomotory activity, even for drugs at doses which in other situations do dissociate the two measures (see Lister, 1990 for review).

The variety in these findings is consonant with the view that freezing/movement arrest precedes risk assessment in the neurobehavioural defense process, given a sufficiently high level of defensiveness. In sofar as freezing is an important component of a given test situation (generally, in the more threatening situations), and the subject's behaviours are more or less limited by the test situation to freezing, locomotion, and the criterion measure, some degree of locomotory increase might be expected to accompany decreased defensiveness, in addition to increased risk assessment. When freezing/movement arrest is not a feature (i.e.in less threatening situations), or when the subjects used represent a strain or species in which freezing is less potent (e.g. wild as compared to laboratory rats; particular mouse strains), specific anxiolytic effects on such criterion measures might be easier to obtain.

Finally, as noted previously, both increases and decreases in the criterion measures for these tests can appear with drugs of undoubted anxiolytic potential. As these tests are presently conceptualized, there is no consistent interpretation of these effects and no basis on which specific predictions might be made about the circumstances in which they are likely to appear. However, insofar as these criterion measures do tap components of the risk assessment pattern, this pattern may be useful in the development of a basis for predicting the precise form that anxiolytic action should take, in a specific test one particular feature of the defense process analysis is that it utilizes the baseline behaviours in a particular test, in order to predict/evaluate drug effects. Another possibly useful feature is that it differentiates between the results of risk assessment in situations where novelty (easily habituated) is the threat source, as compared to threat sources (i.e. high places, light, a cat odor) for which the feedback from risk assessment is much less likely to produce rapid reductions in defensiveness.

Vigilance: Vigilance, searching/scanning activity, often seen in alternation with foraging or feeding, is common in virtually all higher animals (Caine, 1984; Wawra, 1988). While vigilance may involve additional or alternative activity patterns such as searching for consummatory objects, for the most part it appears to be a function of defense-related factors such as group size (Jarman, 1987), size of individuals (Berger & Cunningham, 1988), group social organization (Caine & Marra, 1988), position near the center or the periphery of a group (Underwood, 1982) and proximity to places of concealment (Carey & Moore, 1986; Holmes, 1984).

Most field studies of vigilance appear to reflect the alternation of eating (a nondefensive behaviour) with searching/scanning activity, (risk assessment), suggesting the utility of a finer-grained analysis of these activities in laboratory tests of defensiveness. We (Blanchard et al., in press) thus examined several characteristics of eating patterns for rats in four 1-male + 4-female groups maintained in VBS, prior to any presentation of a cat, and, following cat exposure. Notably, the first two cat exposures resulted in no eating whatever during the hours of observation, so pre-post cat comparisons reflected the eating which occurred after the third cat presentation.

Eating instances were measured and classified as "bouts" separated by a minimum of 1 minute of no eating, and "episodes" which were shorter intervals of continuous eating within a bout. The frequencies and durations of the bouts, the episodes and the intervals between the episodes within a bout were measured individually for the various subjects. Mean bout duration and number of episodes within a bout declined dramatically, and significantly, from the pre-cat day to the post-cat period. However, the intervals between episodes of eating within bouts showed no apparent change with cat exposure, suggesting that these search/scan intervals have a relatively fixed duration in a particular situation, perhaps reflecting the time required to make a comprehensive visual search/scan of the potentially threatening area in order to determine that a potential danger is, in fact, still absent. This analysis indicates fixed, as well as variable, parameters within the "normal" vigilance pattern, which may be valuable in suggesting criteria for "abnormal" vigilance. This is important because animal models of anxiety are typically little concerned with what criteria might be used to determine whether the behaviour of subjects is typical for the species, or if it might represent dysfunction of the neurobehavioural systems involved in anxiety.

CLINICAL RELEVANCE OF "RISK ASSESMENT" MEASURES

The initial premise of the present approach to the modelling of anxiety through analysis of natural defensive behaviours of wild and laboratory rodents was that it might be possible to find some components of defense that are similar to anxiety reactions in terms of each of the key factors involved in a behaviour system: Eliciting stimuli and situations, organismic (physiological, neural, neurochemical, experimental) correlates, response characteristics, and function.

In terms of eliciting stimuli/situations, the risk assessment pattern appears to be rather specifically associated with potential, as opposed to actual and present threat. The essential features of generalized anxiety disorder (GAD) do not refer to any specific stressful associated or precipitating event (APA, 1987), nor is there a particular type of object or event (as with simple phobia) which serves as a focus for the subject's concern. Instead, the subject " .... continually feels anxious, worries, ruminates and anticipates that something bad will happen.... " (APA,1987, p. 232). This fits well with the view, influential in much psychological research, that anxiety is the reaction to anticipated or signalled threat, as opposed to fear, the reaction to present threat (Arnold, 1960; Estes & Skinner, 1939; Freud, 1936).

With reference to organismic variables, only the 3 studies with diazepam and ethanol described earlier have a direct bearing on similarity of drug effects on risk assessment, in comparison to clinical anxiety. However, if the present analysis is correct, that the "exploratory" tests of anxiety and defensive burying both tap aspects of the risk assessment pattern, then there is a considerable literature suggesting that this pattern responds differentially to clinically effective drugs.

With reference to response characteristics, many of the diagnostic criteria for GAD focus on autonomic reactivity and motortension. Only two strictly behavioural criteria are involved: Apprehensive expectation, and, vigilance and scanning, involving hyper attentiveness (APA, 1987, p. 233). The potential isomorphism between risk assessment, and these behaviours, is obvious.

The functional question is one of the most interesting. It is often, and reasonably, pointed out (Lister, 1990) that present animal models of anxiety do not generally deal with etiology; nor, directly, does a view that risk assessment is the central feature of anxiety. However, risk assessment is a behaviour that is normally self-limiting in that it provides feedback which, in the actual presence of danger, enables the subject to identify and localize the danger source in order to deal with it more effectively through specific defenses such as flight or defensive attack; or, in the absence of real danger, promotes a return to nondefensive behaviours through a reduction in the perceived threat potential of the situation. Even for normal rat subjects, however,this is a highly conservative system involving a lengthy process of active risk assessment for many hours after a single cat exposure in a small environment (VBS). The continuous worry and rumination that "something bad" will happen, seen in GAD, provides a clear cognitive/verbal counterpart to this process, different largely in an apparent failure of normal feedback systems to reduce the anxious defensiveness. What these systems might be, and specific causes of their failure to provide negative feedback in the absence of real danger, is an interesting focus of investigation.

Acknowledgements: The research reported in this paper was supported by NIMH Grant MH42803, by NIH Grant AA06220, and by RCMI Grants RRO3061 and RRO1825.

REFERENCES

Adams, D.B. (1979) Behav. Brain Sci. 2, 201-241.
Adams, D.B. (1980) Aggress. Behav. 6, 295-346.

American Psychiatric Association (1987) Diagnostic and statistical manual of mental disorders.' 3rd ed., Rev., Washington, D.C.
Arnold, M.B. (1960) Emotion and Personality (2 vols) New York: Columbia University Press.
Bandler, R. & Carrive, D. (1988) Brain Res. 439, 95-106.
Bandler, R. & Depaulis, S. (1988) Neurosci. Lett., 88, 291-296.
Barlow, D.H. (1988) Anxiety and its Disorders, Guilford Press. New York.
Berger, J., Cunningham, C. (1988) Ecology 69, 177-183.
Blanchard, D.C., Blanchard, R.J (1988) Ann. Rev. Psychol. 39, 43-68.
Blanchard, D.C., Blanchard, R.J., Tom, P., Rodgers, R.J.(1990) Psychopharmacology 101, 511-518.
Blanchard, R.J., Blanchard, D.C. Neurosci. Biobehav. Rev. (In press.)
Blanchard, R.J., Blanchard, D.C. (1969a) J. Comp. Physiol. Psychol. 68, 129-135.
Blanchard, R.J., Blanchard, D.C. (1969b) J. Comp. Physiol. Psychol. 67, 370-375.
Blanchard, R.J., Blanchard, D.C. (1989) J. Comp. Psychol. 103, 70-82.
Blanchard, R.J., Blanchard, D.C., Takahashi, L.K. (1978) Behav. Biol. 23, 291-305.
Blanchard, R.J., Blanchard, D.C., Weiss, S.M. (1990b) Alcohol. 7, 375-381.
Blanchard, R.J., Blanchard, D.C., Weiss, S., Agullana, R. (1990) Soc. Neurosci. Abstr., 16, 598.
Blanchard, R.J., Blanchard, D.C., Weiss, S.M., Mayer, S. (1990a) Pharmacol. Biochem. Behav. 35, 775-780.
Blanchard, R.J., Flannelly, K.J., Blanchard, D.C. (1986) J. Comp. Psychol. 100, 101-107.
Brown, J.S., Kalish, H.I., Farber, I.E. (1951) J. Exp. Psychol. 41, 317-328.
Brutus, M., Zuabi, S., Siegel, A. (1988) Brain Res. 473, 147-152.
Brutus, M., Zuabi, S., Siegel, A. (1989) Exp. Neurol. 104, 55-61.
Caine, N.G. (1984) Folia primatol. 43, 59-67.
Caine, N.G., Marra, S.L. (1988) Anim. Behav. 36, 897-904.
Carey, H.V., Moore, P. (1986) Am. Midl. Nat. 116, 267-275.
Crawley, J.N., Goodwin, F.K. (1980) Pharmacol. Biochem. Behav. 13. 167-170.
Curti, M.W. (1935) Psychol. Monogr. 46, 76-98.
Davis, M., Hitchcock, J.M., Rosen, J.B. (1987) In: The Psychology of Learning and Motivation: Advances in research and theory (G.H.Bower, Ed), Academic press, New York, pp 263-305.
Darwin, C. (1872) The Expression of the Emotions in Man and Animals, John Murray, London.
Endler, J.A. (1986) In: Predator Prey Relationship (M.E.Feder and G.V.Lauder, Eds), University of Chicago Press, Chicago, pp 109-134.
Estes, W.K., Skinner, B.F. (1941) J. Exp. Psychol. 29, 390-400.
Fanselow, M.S., Sigmundi, R.A. (1986) J. Exp. Psychol. 12, 301-309.
Fanselow, M.S., Sigmundi, R.A. (1987) Psychol. Rec. 37, 317-334.
File, S.E. (1985) Neurosci. Biobehav. Rev. 13, 55-62.
File, S.E. (1986) Psychopharmacology 89, S4.
File, S.E., Hyde, J.R.G. (1978) Br. J. Pharmacol. 62, 19-24.
File, S.E., Johnston, A.L. (1987) ISI Atlas of Science: Pharmacol. 1, 284-288.
Flynn, J.P. (1972) Nebr. Symp. Motiv. 20, 125-54.
Freud, S. (1936) Inhibitions, symptoms and anxiety (H.0.Butler, trans.), Hogarth Press, London (Original work published 1926).
Fuchs, S.A.G., Siegel, A. (1984) Brain Res. 306, 263-281.
Fuchs, S.A.G., Edinger, H.M., Siegel, A. (1985) Brain Res. 330, 77-92.
Grant, E.C., Mackintosh, J.H. (1963) Behaviour 21, 246-259.
Gray, J.A., Davis, N., Feldon, J., Rawlins, N.P., Owen, S.R. (1981) Progress in Neuropsychopharmacology 5, 143-157.

Handley, S.L., Mithani, S. (1984) Naunyn-Schmiedebergs's Arch. Pharmacol. 109. 1-5.
Hitchock, J.M., Sananes, C.B., Davis, M. (1989) Behav. Neurosci. 108, 509-518.
Holmes, W.G. (1984) Behav. Ecol. Sociobiol. 15, 293-302.
Iwata, J., LeDoux, J.E., Meeley, M., Arneric, S., Reis, D.J. (1986) Brain Res. 383, 105-114.
Jarman, P.J. (1987) Anim. Behav. 35, 1044-1050.
Kaesermann, H.P. (1986) Psychopharmacology 89, 31-37.
Lal, H., Emmett-Oglesby, M.W. (1983) Neuropharmacology 22, 1423-1441.
LeDoux, J.E., Sakaguchi, A., Iwata, J., Reis, D.J. (1986) Neurosci. 17, 615-627.
LeDoux, J.E. (1986) In: Mind and Brain (J.E.LeDoux and W.Hirst, Eds), Mind and Brain, Cambridge Univ. Press, New York, pp 301-354.
Lee, C., Rodgers, R.J. (1990) Psychopharmacology (In press).
Lister, R.G. (1987) Psychopharmacology 92, 180-185.
Lister, R.E. (1990) Pharmac. Therap. 46, 321-340.
Lorenz, K. (1974) Science 229-234.
Marks, I.M. (1987) Fears, Phobias, & Rituals, New York, Oxford University Press.
Mackintosh, N.J. (1974) Academic Press: New York.
Nolan, N.A., Parkes, M.W. (1973) Psychopharmacology 29, 277-288.
Pinel, J.P.J., Mana, M.J. (1989) In: Ethoexperimental Approaches to the Study of Behaviour (R.J.Blanchard, P.F.Brain, D.C.Blanchard, S.Parmigiani, Eds), Kluwer Academic Publishers, Dordrecht, pp 137-151.
Pinel, J.P.J., Mana, M.J., Wilkie, D.M. (1986) Anim. Learn. Behav. 14, 301-304.
Pinel, J.P.J., Treit, D. (1978) J. Comp. Physiol. Psychol. 92, 708-712.
Pinel, J.P.J., Wilkie, D.M. (1983) In: Animal Cognition and Behaviour (R.L.Mellgren, Ed), North-Holland, Amsterdam, pp 285-318.
Rickels, K., Schweizer, E.E. (1987) In: Psychopharmacology: The Third Generation of Progress (H.Y.Meltzer, Ed), Raven press, New York, pp 1193-1203.
Shaikh, M.B., Brutus, M. Siegel, H.E., Siegel, A. (1985) Brain Res. Bull. 14, 217-221.
Siegel, A., Pott, C.B. (1988) Progr. Neurobiol. 31, 261-283.
Stone, C.P. (1932) In: Studies in the dynamics of behaviour (K.S.Lashley, Ed), University of Chicago Press,Chicago, pp 3-55.
Takahashi, L.K., Blanchard, R.J. (1982) Behav. Proc. 49-62.
Underwood, R. (1982) Behaviour 7-9, 8-107.
Wawra, M. (1988) Behaviour 107, 61-1.
Williams, J.L. (1987) Psychol. Rec. 37, 369-380.
Williams, J.L., Scott, D.K. (1989) Anim. Learn. Behav. 17, 383-393.
Yerkes, R.M. (1913) J. Anim. Behav. 3, 286-296.
Ydenberg, R.G., Dill, L.M. (1986) Adv. Stud. Behav. 37, 305-317.

part 2:

DEPRESSION

Animal Models in Psychopharmacology
Advances in Pharmacological Sciences

PHARMACOLOGICAL MODELS OF DEPRESSION

R.D. Porsolt, A. Lenègre and R.A. McArthur

I.T.E.M.-Labo, 93 Avenue de Fontainebleau, 94276 Le Kremlin-Bicêtre Cedex, France

SUMMARY: The present paper reviews some currently used pharmacological models of depression. Pharmacological models of depression are those procedures which have been conceived with the aim of discovering antidepressants in experimental animals. The major criteria for the validity of such models are their sensitivity to known antidepressants (absence of false negatives) and their selectivity (absence of false positives). In addition pharmacological models of depression should be simple, rapid to perform and reliable. The present paper reviews models based on pharmacological interactions (reserpine, apomorphine, yohimbine, clonidine) and those based on induced behavioural change ("behavioural despair", tail suspension, "learned helplessness", olfactory bulbectomy). All these models have a certain utility, apart from clonidine where results are contradictory. Of the models based on a pharmacological interaction, yohimbine potent-iation in the mouse appears to have the greatest sensitivity and is very simple and rapid to perform. Of the behavioural models, "behavioural despair" has been subjected to the most intensive pharmacological validation and is both sensitive to a wide range of antidepressants and is simple and rapid to perform; false negatives include 5-HT uptake inhibitors and beta-adrenergic stimulants. Tail suspension is also simple and rapid but appears to be less sensitive; nonetheless 5-HT uptake inhibitors are active. "Learned helplessness" appears highly sensitive but is more time-consuming. Olfactory bulbectomy, although reasonably sensitive, is extremely time-consuming and of doubtful relevance to clinical depression.

INTRODUCTION

For the purposes of this paper pharmacological models of depression are defined as those procedures which have been conceived specifically with the aim of detecting antidepressants in experimental animals. This definition is deliberately more restrictive than that which would be applied to animal models in general whose aims are much more ambitious, i.e. to attempt to imitate in animals the clinical phenomena of depression. Animal models in the broader sense, also referred to as psychological models (see Willner, this volume), must satisfy at least the four validation criteria defined by McKinney (1977):

1. Similarity of inducing conditions.
2. Similarity of the behavioural state induced.
3. Similarity of the underlying mechanisms.
4. Similarity of clinically effective treatment techniques.

Although it would no doubt be an advantage for a pharmacological model to satisfy all four criteria, the most important criterion is to show a maximum of sensitivity to different kinds of antidepressant drug (absence of false negatives) while at the same time showing a maximum of selectivity (absence of false positives). Absence of false negatives is the most significant requirement in that drugs with potential antidepressant activity should not be missed during first-pass testing. This aspect is difficult to evaluate because potential antidepressant activity can then only be discovered, if at all, by serendipity (e.g. mianserin, levoprotiline) or by pursuit of other drugs discovery strategies (e.g. receptor binding, hypotheses of mechanism of action). To avoid false negatives as far as possible, pharmacological models must be continually reassessed in terms of their response to novel antidepressants, in particular atypical ones. False positives represent less of a problem in that they can mostly be eliminated by further testing.

Unlike psychological models, pharmacological models do not require any assumed similarities with depressive illness or with hypothesized underlying neurobiological substrates; they simply have to show the effectiveness of known drugs. Nonetheless the existence of some behavioural similarities to the clinical target would represent an important advantage in that such models would then be less likely to discover only "me-too" compounds. It seems reasonable to suppose that a purely behavioural model could show a wider pharmacological sensitivity to different kinds of antidepressants than a model based simply on the interaction of a test substance with a pharmacological reagent such as reserpine, a drug whose mechanism is related to the supposed mode of action of a limited class of antidepressants.

Further requirements of pharmacological models are simplicity, rapidity and reliability. The requirement of rapidity and simplicity can be contrasted with that for psychological models which are frequently criticized for oversimplifying complex clinical phenomena. For example, most pharmacological models show acute effects of antidepressants whereas most antidepressants start exerting their clinical effects only after several weeks of treatment. In contrast to psychological models of depression where sensitivity to chronic but not acute treatment is a requirement, insistance on the necessity for chronic vs acute effectiveness represents a major disadvantage for a pharmacological model because of reduced rapidity. A way out of this dilemma is to demonstrate that the pharmacological model, although responsive to acute treatment, shows a more marked effect after repeated treatment. In any case the problem is of minor importance in that the effects of chronic treatment are generally studied in a secondary phase of drug evaluation usually in more psychological models; thus the two kinds of approach complement each other.

The aim of the present paper is to evaluate several currently used pharmacological models of depression in terms of the above criteria.

RESERPINE ANTAGONISM

Rationale: The syndrome induced by reserpine and related compounds, first proposed by Askew (1963), is the model most widely used in experimental pharmacology for the detection of antidepressant drugs. Interest in the psychotropic effects of reserpine, a depletor of central monoamines, arose from clinical observations that some patients treated with reserpine for hypertension developed clear signs of clinical depression. Although the validity of reserpine-induced depression in man has been seriously questioned (Bein, 1978), indirect support for the notion is provided by the observation that rhesus monkeys chronically treated with reserpine show behavioural signs similar to those induced by separation procedures (McKinney et al., 1971).

Pharmacological validation: The main justification for the use of reserpine tests is that most antidepressants in clinical use today antagonize some or all of the symptoms induced by reserpine. This is not a coincidence, however, because most of these compounds were developed in animal pharmacology on the basis of their antireserpine activity. Early false negatives were the atypical compounds iprindole and mianserin. Other false negatives include several 5-HT uptake inhibitors (zimelidine, fluvoxamine, indalpine), the phosphodiesterase inhibitor rolipram and most recently levoprotiline, the pharmacologically inactive isomer of oxaprotiline (Bourin, 1983; Delini Stula et al. 1988; Porsolt, 1985; Wachtel, 1983).

Procedural problems: One of the difficulties in evaluating reserpine antagonism as a pharmacological model is the diversity of different protocols employed. Different results are obtained depending on the parameter measured, the principal ones being ptosis, hypothermia and sedation. Ptosis appears to represent a mainly peripheral effect

Table 1: **RESERPINE : PHARMACOLOGICAL SPECIFICITY**

	HITS	MISSES
TRUE	TRICYCLICS MAOI BETA STIMULANTS	NEUROLEPTICS ANXIOLYTICS ANTICONVULSANTS
FALSE	PSYCHOSTIMULANTS ANTICHOLINERGICS ANALGESICS PERIPHERAL SYMPATHOMIMETICS	MIANSERIN IPRINDOLE ROLIPRAM 5-HT UPTAKE BLOCKERS LEVO-PROTILINE

of reserpine whereas sedation and hypothermia are thought to reflect more central activity. A further effect of reserpine, the induction of PGO spikes in cats, is almost certainly of central origin (Jalfre et al., 1970). Paradoxically, ptosis appears more readily antagonized by classical and atypical antidepressants than sedation or PGO spikes. Different results are also obtained depending on the reserpine-like compound employed (reserpine, tetrabenazine, Ro 4-1284). Reserpine appears to deplete central 5-HT more markedly than central NA whereas the reverse appears to be true with tetrabenazine or Ro 4-1284. The timing is also important; in some protocols the antidepressant is administered before reserpine (preventive treatment) whereas in others afterwards (curative treatment). Clear species differences also exist; tricyclic antidepressants antagonize and even reverse reserpine-induced sedation in rats but have much less marked effects in mice. For further details see Porsolt (1985).

Conclusions: In view of the above it is difficult to draw general conclusions about the reserpine syndrome as a pharmacological model of depression. Ptosis antagonism, although probably a peripheral effect, appears to detect the greatest number of clinically active antidepressants, including mianserin and trazodone (Gouret et al., 1977) which are generally regarded as false negatives in most reserpine procedures. There are, on the other hand, numerous false positives including psychostimulants, anticholinergics, analgesics and peripherally acting sympathomimetics (Colpaert et al., 1975; Bourin et al., 1983). With parameters depending more closely on central effects (sedation, hypothermia, PGO spikes) the number of false positives tends to decrease but there is a corresponding increase in false negatives (Gouret et al., 1977) In general MAOIs show clearer antagonism and reversal of most reserpine parameters than do tricyclic or atypical antidepressants (Colpaert et al., 1975).

APOMORPHINE ANTAGONISM

Rationale: Apomorphine, at low doses (1 mg/kg), is a specific agonist at post-synaptic DA receptors and induces hypothermia, stereotypies and verticalized climbing behaviour in mice. All these effects are antagonized by DA blocking neuroleptics. A similar syndrome is induced by much higher doses of apomorphine (16 mg/kg) but, in contrast to the effects of 1 mg/kg, the hypothermia is not antagonized by neuroleptics. A wide range of antidepressants, on the other hand, clearly antagonize the hypothermia induced by high dose apomorphine without affecting the stereotypies or climbing behaviour. Antagonism of the hypothermia induced by a high dose of apomorphine has thus become a simple and rapid test for detecting antidepressant activity.

Pharmacological validation: Initial studies by Puech et al. (1981) showed that all the tricyclic antidepressants tested (amitriptyline, clomipramine, desipramine, imipramine, nortriptyline) possessed this specific activity and that similar effects were observed with atypical antidepressants such as amoxapine, imafen, nomifensine and viloxazine. Partial hypothermia antagonism was observed with mianserin and TRH. Clear and specific antagonism of apomorphine-induced hypothermia was also observed with the beta-adrenergic agonists (clenbuterol, isoprenaline, salbutamol) and the catecholamine stimulants amphetamine, para-chloro-methyl-amphetamine and L-DOPA whereas diverse compounds such as atropine, clonidine, lithium, morphine, yohimbine and L-5-HTP were without effect. The B-type MAOI pargyline was inactive. Subsequent publications have suggested that the type A MAOI pirlindole (Maj et al., 1986) and the inhibitors of 5-HT uptake, citalopram and fluvoxamine (Pawlowski and Mazela, 1986), are not active either.

Table 2 **APOMORPHINE ANTAGONISM: PHARMACOLOGICAL SPECIFICITY**

	HITS	MISSES
TRUE	**TRICYCLICS** **NA UPTAKE INHIBITORS** **BETA STIMULANTS**	**NEUROLEPTICS** **ANTICHOLINERGICS** **LITHIUM** **ANALGESICS**
FALSE	**AMPHETAMINE** **L-DOPA** **QUIPAZINE**	**MAOI** **MIANSERIN** **IPRINDOLE** **ROLIPRAM** **5-HT UPTAKE INHIBITORS** **(FLUVOXAMINE, CITALOPRAM)**

Conclusions: Taken together, the findings suggest that antidepressants with a marked noradrenergic or dopaminergic component are active against high dose apomorphine-induced hypothermia but not antidepressants acting mainly through serotonergic mechanisms. The procedure could thus represent a useful and rapid test for detecting antidepressants acting primarily on central catecholamine function. On the other hand, the absence of activity in this test could not rule out potential antidepressant activity of other origins.

Table 3 **YOHIMBINE POTENTIATION: PHARMACOLOGICAL SPECIFICITY**

	HITS	MISSES
TRUE	TRICYCLICS MAOI NA AND 5HT UPTAKE INHIBITORS BETA STIMULANTS MIANSERIN IPRINDOLE NOMIFENSINE	NEUROLEPTICS ANXIOLYTICS
FALSE	AMPHETAMINE TRH ANTICHOLINERGICS ANTIHISTAMINES	ROLIPRAM

YOHIMBINE POTENTIATION

Rationale: Yohimbine is a naturally occurring alkaloid which preferentially inhibits central presynaptic alpha-adrenergic receptors, thereby increasing the release of noradrenaline. The compound causes behavioural excitation and increases in heart rate and blood pressure in animals and has been shown to induce panic attacks in man. The potentiation of the behavioural and cardiovascular effects in animals has been claimed to be a selective indicator of potential antidepressant activity (Bourin, 1990; Graaf et al., 1985; ; Malick, 1983; Quinton, 1963; Sanghvi and Gershon, 1977). Yohimbine potentiation is usually studied in the mouse (potentiation of lethality) or the conscious dog (potentiation of behavioural excitation and blood pressure effects).

Pharmacological validationl In the mouse, yohimbine lethality is potentiated by tricyclics, 5-HT uptake inhibitors (fluoxetine, fluvoxamine, femoxetine, zimelidine), beta-adrenergic agonists, MAOIs and atypical antidepressants including bupropion, doxepine, iprindole, mianserin and rolipram but not by neuroleptics or minor tranquillizers; trazodone is not active and thus represents a false negative (Bourin et al., 1990; Graaf et al., 1985; Malick, 1983; Wachtel, 1983). The effects seem to be of central origin because quaternary derivatives of compounds such as imipramine are inactive (Frances and Simon, 1985). False positives in the procedure include psychostimulants, antihistamines and anticholinergics (Graaf et al., 1985; Malick, 1983). The procedure in dogs appears to be more selective, eliminating false positives such as psychostimulants and also several compounds (MJ 1986; EXP 561) which exhibit "antidepressant" profiles in classical pharmacological tests but were subsequently found inactive in the clinic (Sanghvi and Gershon, 1977).

Conclusions: Overall, although the effects of yohimbine in man or in animals are not clearly related to depression or mania, yohimbine potentiation, particularly in mice, appears to represent a simple and rapid test for detecting the antidepressant activity of a wide range of compounds.

CLONIDINE ANTAGONISM

Rationale: Clonidine is an alpha$_2$-adrenergic receptor agonist which acts preferentially at presynaptic receptor sites, thereby reducing the release of noradrenaline. In addition to its marked cardiovascular effects, clonidine causes hypothermia, analgesia and marked sedation. This latter effect, which is probably related to a decrease in central noradrenaline activity, has incited several authors to use clonidine-induced hypoactivity as a test model for antidepressant drugs. Impetus in this direction was undoubtedly given by the demonstration that mianserin, an antagonist of presynaptic alpha-adrenergic receptors (Robson et al., 1978), attenuated clonidine-induced decreases in conditioned avoidance behaviour.

Table 4 **CONTRADICTORY FINDINGS WITH CLONIDINE ANTAGONISM**

PARAMETER	ANTI DEPRESSANT	RESULT	AUTHORS
CONDITIONED AVOIDANCE	MIANSERIN	ANTAGONISM	Robson et al (1978)
HYPOTHERMIA	MIANSERIN	NO ANTAGONISM	Von Voigtlander et al. (1978)
LOCOMOTOR ACTIVITY	MIANSERIN	NO ANTAGONISM	Delini Stula et al. (1979)
LOCOMOTOR ACTIVITY	IMIPRAMINE	ANTAGONISM	Gower & Marriott (1980)
SELF STIMULATION	MIANSERIN	NO ANTAGONISM	Hunt et al. (1981)
LOCOMOTOR ACTIVITY	MIANSERIN	ANTAGONISM	Kostowski & Malatynska (1983)

Pharmacological validation: Similar findings using conditional avoidance behaviour have been reported after chronic treatment with desipramine and imipramine but not with the 5-HT uptake inhibitor zimelidine (Kostowski and Obersztyn, 1988). Decreases in clonidine effects by tricyclics, MAOI, mianserin and trazodone have also been reported using clonidine-induced depression of locomotor activity as a behavioural index (Gower and Marriott, 1980; Kostowski and Malatynska, 1983; O'Connor and Leonard, 1988). Unfortunately, conflicting results have also been reported in which clonidine-induced decreases in locomotor or self-stimulation behaviour could not be antagonized by mianserin (Delini Stula et al., 1979; Hunt et al., 1981) or other antidepressants (von Voigtlander et al., 1978; Passerelli and Scotti de Carolis, 1983) after acute or chronic administration, although von Voigtlander et al (1978) reported that mianserin antagonized clonidine-induced hypothermia after chronic but not acute treatment.

Interactions with "behavioural despair": Another source of conflicting findings are the interactions of clonidine with the "behavioural despair" procedure (see below). Early findings (Porsolt et al., 1979) suggested that low doses of clonidine had antidepressant-like effects in this test. Similar findings of an antidepressant-like effect of clonidine in a forced swimming situation have also been reported when clonidine was injected centrally into the hippocampus of raphe nucleus lesioned rats (Tamborska et al., 1989) or into the nucleus accumbens of rats pre-exposed to electric shocks (Weiss et al., 1986). Furthermore, Bourin et al. (1990) have reported that preinjections of low and behaviourally inactive doses of clonidine potentiate the antidepressant effects of a variety of antidepressants in the behavioural despair test, including compounds such as the 5-HT uptake inhibitors (citalopram, fluvoxamine, paroxetine, indalpine) and trazodone whose activity when given alone is weak or non-existent in this test. In contrast to these findings, Parale and Kulkarni (1986) reported that similar doses of clonidine potentiated the immobility response to forced swimming, an effect which was antagonized by a wide range of antidepressants!

Conclusions: In view of these contradictory findings it is too early to judge whether antagonism of clonidine-induced effects is a useful test for antidepressant activity.

BEHAVIOURAL DESPAIR

Rationale: Rats and mice, when forced to swim in a cylinder from which there is no escape, will rapidly adopt a characteristic immobile posture and make no further attempts to escape apart from the movements necessary to keep their heads above water. It was suggested (Porsolt et al., 1977; 1978) that the immobility reflected a state of lowered mood in which the animals had given up hope of finding an exit and had resigned themselves to the experimental situation. Conceptually, the behaviour is similar to the "learned helplessness" which can be induced in some animals exposed to inescapable aversive stimulation (see below) and parallels could also be drawn with the "protest-despair" reactions described in young primates separated from their mothers or peers (Harlow and Suomi, 1974). Although this point of view has been criticized by numerous authors (Borsini et al., 1986; Hawkins et al., 1978; Nishimura et al., 1988; Thierry et al., 1984) the fact remains that the immobility induced by forced swimming is attenuated by a wide variety of antidepressants.

Table 5 **BEHAVIOURAL DESPAIR: PHARMACOLOGICAL SPECIFICITY**

	HITS	MISSES
TRUE	TRICYCLICS MAOI NA UPTAKE INHIBITORS MIANSERIN IPRINDOLE ROLIPRAM LEVOPROTILINE LITHIUM	NEUROLEPTICS ANXIOLYTICS ANALGESICS
FALSE	PSYCHOSTIMULANTS ANTICHOLINERGICS ATYPICAL NEUROLEPTICS ANTIHISTAMINES	5-HT UPTAKE INHIBITORS BETA STIMULANTS

Pharmacological validation: During the 13 years since the model was first described (Porsolt et al., 1977), the test procedure has been subjected to evaluation in many laboratories. In contrast to the many different protocols used for classical tests such as reserpine (see above), the procedures used with "behavioural despair" experiments have largely followed those originally described in the mouse and the rat (Porsolt et al., 1977; 1978) rendering direct comparisons possible between different publications. The general conclusion from these studies is that most known clinically active antidepressants tested, 87% in the rat and 94 % in the mouse according to Borsini and Meli (1988), particularly after repeated treatment, reduce the duration of immobility induced by forced swimming. The rat version would appear to be more selective (fewer false positives) whereas the mouse version more sensitive (fewer false negatives).

The list of active compounds includes many atypical compounds (mianserin, iprindole, rolipram, idazoxan, levoprotiline), which are not readily detected using classical pharmacological tests. The findings with these compounds are important as they suggest that "behavioural despair", which is not based on any preconceptions concerning the neurochemical substrates of depression, may be capable of detecting antidepressants acting by mechanisms other than those of traditional agents. This possibility has been dramatically demonstrated recently by the finding that levoprotiline, the clinically active but otherwise pharmacologically inactive isomer of oxaprotiline, shows clear activity, particularly after repeated administration in the "behavioural despair" test in the rat (Delini Stula et al., 1988).

Table 6 **CLONIDINE: CONTRADICTORY EFFECTS ON BEHAVIOURAL DESPAIR**

CLONIDINE ALONE	**CLONIDINE + IMIPRAMINE**	**CLONIDINE + AMITRYPTILYNE**	**AUTHORS**
	ANTAGONISM	**ANTAGONISM**	**Parale & Kulkarni (1986)**
	POTENTIATION	**POTENTIATION**	**Bourin et al. (1990)**

In addition to compounds known for their clinical antidepressant activity, many other kinds of compounds have been reported to reduce immobility in the "behavioural despair" test; these include the neuropeptides enkephalin, oxytocin, MIF-1 and Tyr-MIF-1 (Arletti and Bertolini, 1987; Ben Natan et al., 1984; Pulvirenti and Kastin,

1988), the ACE inhibitor captopril (Giardina and Ebert, 1989), the GABA agonists aminoxyacetic acid, muscimol, THIP and fengabine (Borsini et al., 1986; Aley and Kulkarni, 1989), the calcium antagonists nimodipine, nifedipine and nitrendipine (Czyrak et al., 1989), the 5-HT_{1A} agonists 8-OH-DPAT, gepirone and buspirone (Cervo et al., 1988; Robinson et al., 1989), the atypical neuroleptics clozapine, carpipramine, sulpiride and amperozide (Browne, 1979; Gorka and Janus, 1985; Gustafsson and Christensson, 1990), the anti-convulsants diphenylhydantoin, carbamazepine and valproate (Barros and Leite, 1987; Fernandez-Teruel et al., 1988; Poncelet et al., 1984), the benzodiazepine inverse agonists beta-CCE and S-135 (Matsushita et al., 1988; Nishimura et al., 1989) and lithium (Eroglu and Hizal, 1987). Whether these observed effects represent true antidepressant activity or should be considered as false positives remains to be demonstrated. Other compounds which seem more clearly to represent false positives are diverse psychostimulants, anticholinergics and antihistamines (see Porsolt, 1985).

An important false negative is constituted by the 5-HT uptake inhibitors (e.g. indalpine, citalopram, fluvoxamine), which although clinically active, show little or no activity in the "behavioural despair" test, particularly in the rat. Of interest is the report by Bourin et al. (1990) that the antidepressant effects of these compounds can be unmasked by pretreatment with low doses of clonidine (see above). Other potential false negatives are the beta-adrenergic agonists (salbutamol, clenbuterol), although their status as antidepressants is not clearly established.

Conclusions: It is clear from the above that the "behavioural despair" test is widely used for the detection of antidepressant activity. It is sensitive to a large number of atypical antidepressants otherwise inactive in more classical tests. It is simple to perform and rapid, particularly in the mouse. Data obtained in different laboratories worldwide suggest that the results are highly reproducible. The procedure lends itself readily to automation enhancing its usefulness for routine drug testing. Finally, the fact that the model is purely behavioural, without presuppositions concerning the mechanism of action of potential antidepressants, suggests that the method is less prone to producing "me-too" compounds.

TAIL SUSPENSION TEST

Rationale: Stéru and coworkers have proposed a variant of the "behavioural despair" test in which immobility is induced simply by suspending a mouse or a rat by the tail (Chermat et al., 1986; Stéru et al 1985; 1987). Unlike "behavioural despair", tail suspension-induced immobility is not accompanied by marked hypothermia or behavioural changes lasting longer than the test period itself, suggesting that the procedure is less stressful to the experimental animals (Thierry et al., 1986).

Table 7 **TAIL SUSPENSION TEST: PHARMACOLOGICAL SPECIFICITY**

	HITS	MISSES
TRUE	TRICYCLICS MAOI (?) MIANSERIN (?) 5-HT UPTAKE INHIBITORS 8-OH-DPAT IPRINDOLE NOMIFENSINE	NEUROLEPTICS ANXIOLYTICS 5-HT1A AGONISTS (GEPIRONE, BUSPIRONE)
FALSE	PSYCHOSTIMULANTS ANTICHOLINERGICS CLONIDINE	BETA STIMULANTS LEVO-PROTILINE ROLIPRAM

Pharmacological validation: Tail suspension-induced immobility is reduced by a large number of clinically active typical and atypical antidepressants. Included in the latter are viloxazine, bupropion, mianserin and nomifensine; iprindole is active but only after 5 days treatment (Porsolt et al, unpublished findings). One important class of compounds showing clear activity in the tail suspension test are the inhibitors of the uptake of serotonin - clomipramine, paroxetine, indalpine, fluvoxamine, citalopram -

(Ali-Kodja et al., 1986; Stéru et al., 1987; van der Heyden et al., 1987), which differentiates the tail suspension test from the "behavioural despair" test. On the other hand, other classes of compounds such as the MAOIs, which are clearly active in the "behavioural despair" test, show rather weaker effects in the tail suspension test (Porsolt et al., 1987; Stéru et al., 1987) and some atypical compounds found active in the "behavioural despair" test - levoprotiline, rolipram - have not been found active in the tail suspension test (Lenègre et al., unpublished findings). Another difference between the two tests is suggested by the fact that neuroleptics and anxiolytics clearly increase the duration of immobility in the tail suspension test but show little activity in the "behavioural despair" test. In some cases clearly opposite findings are obtained. For example, sulpiride, clozapine, buspirone and gepirone have all been reported to decrease forced swimming-induced immobility (see above), but clearly increase the duration of tail suspension-induced immobility in a similar dose-range (Stéru et al., 1987; Lenègre et al, unpublished findings). Whereas the results in the "behavioural despair" test would suggest atypical antidepressant activity, the effects in the tail suspension test would appear to reflect the neuroleptic or anxiolytic effects of these compounds. Interestingly, the specific 5-HT_{1A} agonist, 8-OH-DPAT, in contrast to buspirone and gepirone, shows antidepressant-like activity in both tests (Cervo et al., 1988; Lenègre et al., unpublished findings).

Conclusions: Despite an obvious conceptual similarity between tail suspension and "behavioural despair", the findings described above suggest subtle differences which require further examination. Nonetheless, the tail suspension test, particularly in its automated form (Porsolt et al., 1987; Stéru et al., 1987) provides a rapid and reliable behavioural screening test for antidepressants and other classes of psychotropics.

LEARNED HELPLESSNESS

Rationale: Another model that has gained wide interest in recent years, particularly for its pharmacological applications, is Seligman's "learned helplessness" (Seligman, 1975). According to this model, animals and even humans, when exposed to aversive events over which they have no control, suffer decreased motivation to respond, cognitive deficits which interfere with new learning, and emotional disruption - first anxiety and then, with prolonged exposure to the precipitating conditions, depression. According to Seligman, the deficit results from the animal's having learned that the situation is uncontrollable. Whatever the theoretical explanation, the "learned helplessness" model, particularly the learning deficits induced in the rat, has become a popular method for evaluating antidepressant activity.

Table 8 **LEARNED HELPLESSNESS: PHARMACOLOGICAL SPECIFICITY**

	HITS	MISSES
TRUE	TRICYCLICS MAOI MIANSERIN IPRINDOLE 5-HT UPTAKE INHIBITORS BETA STIMULANTS NOMIFENSINE	PSYCHOSTIMULANTS ANXIOLYTICS NEUROLEPTICS
FALSE	5-HT RECEPTOR BLOCKERS PCPA PIRACETAM (?) GINKGO BILOBA (?)	

Pharmacological validation: Early pharmacological studies (Leshner et al., 1979; Sherman et al., 1979; 1982) indicated that a wide variety of antidepressants including tricyclics, MAOI and atypical antidepressants such as iprindole and mianserin reversed "learned helplessness" when treatment was given repeatedly after exposure to unavoidable shocks and during the learning of an active avoidance response. Anxiolytics, neuroleptics or psychostimulants, given in the same experimental conditions, were without effect. Interestingly, both classical anxiolytics (benzodiazepines) and atypical compounds such as buspirone or hydroxyzine could prevent the occurrence of "learned helplessness" if administered during the exposure to inescapable shocks (Drugan et al., 1987; Porsolt et al., 1989; Sherman et al., 1979).

Since, a wide range of compounds have been investigated in variants of the "learned helplessness" paradigm with results indicating a high degree of sensitivity to diverse clinically used and potential antidepressants. Therapeutic effects in a "learned helplessness" model have been reported with the atypical antidepressant nomifensine (Kametani et al., 1983) and the neuropeptide oxytocin (Arletti and Bertolini, 1985). Martin and coworkers have shown that "learned helplessness" is reduced by toloxatone and l-deprenyl, specific inhibitors of type A and type B MAO respectively (Martin et al., 1987), by the specific 5-HT uptake blockers citalopram, indalpine, fluvoxamine and zimelidine (Martin et al., 1990), by the 5-HT_{1A} agonists buspirone, gepirone, ipsaperone and 8-OH-DPAT (Giral et al., 1988), by the beta-adrenoceptor stimulants salbutamol and clenbuterol (Martin et al., 1986), by the calcium antagonist nimodipidine (Martin et al., 1989) and by gonadotrophin releasing hormone (Massol et al., 1989). Finally, facilitation of the effects of tricyclic antidepressants in the "learned helplessness" model has been reported with the thyroid hormones triiodothyronine (Brochet et al., 1987) and triiodothyroacetic acid (Massol et al., 1987).

There is some confusion with serotonergic agents in that the uptake blockers are only active at low doses and are not effective when administered immediately before avoidance acquisition sessions (Martin et al., 1990). Furthermore, other authors have reported improvements of "learned helplessness" with agents which decrease serotonin transmission such as the 5-HT receptor blocker methysergide (Brown et al., 1982) or the 5-HT synthesis inhibitor PCPA (Edwards et al., 1986).

Decreases in "learned helplessness" have also been reported with two compounds, piracetam (Cavoy et al., 1988) and the ginkgo biloba extract Tanakan(TM) (Porsolt et al., 1990), neither of which are known for antidepressant activity. On the other hand, both are used for treating cognitive deficits related to aging. It is thus possible that "learned helplessness" reflects not only depressive phenomena but may constitute a model of cognitive deficit as well. This possibility was envisaged even in the earliest publications by Seligman (see Seligman, 1975).

Conclusions: Taken together, the "learned helplessness" model shows clear sensitivity to a wide range of antidepressants although, like the "behavioural despair" test, it provides equivocal results with agents which act on the serotonergic system. Unlike the "behavioural despair" test, the experimental procedure is time-consuming usually requiring several days of test. The procedure would therefore not appear appropriate for primary antidepressant screening but could provide valuable supportive evidence for selected compounds.

Table 9 **OLFACTORY BULBECTOMY: PHARMACOLOGICAL SPECIFICITY**

	HITS	MISSES
TRUE	TRICYCLICS MAOI (BROFAROMINE, MOCLOBEMIDE) 5-HT UPTAKE BLOCKERS MIANSERIN BUPROPION NOMIFENSINE IRPINDOLE GABA AGONISTS	PSYCHOSTIMULANTS NEUROLEPTICS ANXIOLYTICS
FALSE	?	MAOI (TRANYLCYPROMINE) TRAZODONE LEVO-PROTILINE

OLFACTORY BULBECTOMY

Rationale: Bilateral destruction of the olfactory bulbs in rats leads to impairment of passive avoidance acquisition, irritability, mouse killing behaviour and locomotor hyperactivity (Leonard and Tuite, 1981) although decreases in motor activity have also been reported (Stockert et al., 1988). Interest in this syndrome was aroused by the fact that the atypical antidepressant mianserin reversed these effects after repeated but not after acute treatment (Cairncross et al., 1978; van Riezen et al., 1977). These observations were made at a time when the clinical antidepressant activity of mianserin was not apparent in classical tests for antidepressants thereby providing an impetus for any model which could predict these effects in experimental animals.

Pharmacological validation: Subsequent studies have largely confirmed the sensitivity of olfactory bulbectomy as a predictive model. Decreases in one or more of the symptoms induced by bulbectomy have been reported with a wide range of clinically active or potential antidepressants including classical tricyclics (Cairncross et al., 1978; Iwasaki et al., 1986; O'Connor and Leonard, 1988), atypical antidepressants such as bupropion, nomifensine, doxepine, viloxazine, rolipram, iprindole, the 6-aza analog of mianserin (Org 3770) and its (+)-enantiomer (Cairncross et al., 1978; Gorka et al., 1985; Jesberger and Richardson, 1986; Mizokawa et al., 1988; O'Connor and Leonard, 1986; van Riezen et al., 1977), the serotonin uptake inhibitors fluoxetine and sertraline (Broekkamp et al., 1980; Butler et al., 1988) and the GABA agonists progabide and fengabine (Lloyd et al., 1989). Conflicting findings have been reported for the MAOI tranylcypromine (Cairncross et al., 1978; Jesberger and Richardson. 1986) whereas other MAOI - brofaromine, moclobemide appear to be active (Leonard, unpublished findings).

The atypical compounds trazodone and levoprotiline are without effect (Gorka et al., 1985; Leonard, unpublished findings) and therefore constitute false negatives. On the other hand, amphetamine reduces the passive avoidance deficit but increases hyperactivity, whereas both chlorpromazine and chlordiazepoxide reduce hyperactivity but further impair passive avoidance (van Riezen et al., 1977); these compounds therefore do not constitute false positives.

Conclusions: Overall, the olfactory bulbectomy model appears to provide a sensitive procedure for the detection of antidepressant activity in animals. It possesses the major disadvantage, however, of being very time-consuming (2 to 4 weeks experimentation) without this being compensated for by greater predictive power as compared with simpler procedures or by greater face validity in terms of its resemblance to clinical depression.

GENERAL CONCLUSIONS:

The present paper has critically reviewed some of the pharmacological models currently used for the evaluation of antidepressants in animals. Two major classes of model can be distinguished, those based on pharmacological interactions and those based on behaviour.

The former are certainly useful and frequently constitute simple and rapid tests for primary screening. The most broadly sensitive would appear to be the potentiation of yohimbine lethality in the mouse whereas results with clonidine antagonism are contradictory.

Of the behavioural models, "behavioural despair" has been the most widely exploited and remains the behavioural screening test of choice, particularly the mouse version. The conceptually related tail suspension test shows somewhat different pharmacological sensitivity and may thus complement the "behavioural despair" test for primary screening, in particular for revealing the antidepressant activity of 5-HT uptake inhibitors. The "learned helplessness" procedure is highly sensitive to a wide range of antidepressants and is also selective but is probably too time-consuming for screening purposes. A similar criticism can be advanced for olfactory bulbectomy which is even more time-consuming and of doubtful relevance to clinical depression.

In general behavioural models, which are less theory-bound, offer the most promising path towards the detection of new antidepressants with mechanisms of action different from those in current use. The indications of antidepressant activity for compounds such as levoprotiline, idazoxan, 5-HT_{1A} agonists, calcium antagonists, phosphodiesterase inhibitors, anticonvulsants, several atypical neuroleptics and various hormones provide illustrations of this potential.

Acknowledgements: We thank Ms Nathalie de Cerchio for expert secretarial assistance.

REFERENCES

Aley, K.O., Kulkarni, S.K. (1989) Naunyn. Schmied. Arch. Pharmacol. 339, 306-311.
Ali-Kodja, F., Bougard, M., Perrault, G., Zivkovic, B. (1986) Brit. J. Pharmacol. 87, 130P.

Arletti, R., Bertolini, A. (1987) Life Sci. 41, 1725-1730.
Askew, B.M. (1963) Life Sci. 10, 725-730.
Barros, H.M., Leite, J.R. (1987) Psychopharmacology 92, 340-342.
Bein, H.J. (1978) Pharmako-Psychiat. Neuropsycho-pharmakol. 11, 289-293.
Ben Natan, L., Chaillet, P., Lecomte, J.M., Marçais, H., Uchida, G., Costentin, J. (1984) Eur. J. Pharmacol. 97, 301-304.
Borsini, F., Evangelista, S., Meli, A. (1986) Eur. J. Pharmacol. 121, 265-268.
Borsini, F., Meli, A. (1988) Psychopharmacology 94, 147-161.
Borsini, F., Volterra, G., Meli, A. (1986) Physiol. Behav. 38, 385-386.
Bourin, M. (1990) Fund. Clin. Pharmacol. 4, 49-64.
Bourin, M., Colombel, M.C., Malinge, M. (1990) Eur. J. Pharmacol. (In press).
Bourin, M., Poncelet, M., Chermat, R., Simon, P. (1983) Arzneim. Forsch. 33, 1173-1176.
Brochet, D.M., Martin, P., Soubrié, P., Simon, P. (1987) Psychiatry Res. 21, 267-275.
Broekkamp, C.L., Garrigou, D., Lloyd K.G. (1980) Pharmacol. Biochem. Behav. 13, 643-646.
Brown, L., Rosellini, R.A., Samuels, O.B., Riley, E.P. (1982) Pharmacol. Biochem. Behav. 17, 877-883.
Browne, R.G. (1979) Eur. J. Pharmacol. 58, 331-334.
Butler, J., Tannian, M., Leonard, B.E. (1988) Prog. Neuropsychopharmacol. Biol. Psychiatry 12, 585-594.
Cairncross, K.D., Cox, B., Forster, C., Wren, A.F. (1978) J. Pharmacol. Meth. 1, 131-143.
Cervo, L., Grignaschi, G., Samanin, R. (1988) Eur. J. Pharmacol. 158, 53-59.
Chermat, R., Thierry, B., Mico, J.A., Stéru, L., Simon, P. (1986) J. Pharmacol. (Paris) 17, 348-350.
Czyrak, A., Mogilnicka, E., Maj, J. (1989) Neuropharmacology 28, 229-233.
Colpaert, F.C., Lenaerts, F.M., Niemegeers, C.J.E., Janssen, P.A.J. (1975) Arch. Int. Pharmacodyn. Ther. 215, 40-90.
Delini Stula, A., Baumann, P., Buch, O. (1979) Naunyn. Schmied. Arch. Pharmacol. 307, 115-122.
Delini Stula, A., Radeke, E., van Riezen, H. (1988) Neuropharmacology 27, 943-947.
Drugan, R.C., Crawley, J.N., Paul, S.F., Skolnick, P. (1987) Drug Dev. Res. 10, 63-67.
Edwards, E., Johnson, J., Anderson, D., Turano, P., Henn, F.A. (1986) Pharmacol. Biochem. Behav. 25, 415-421.
Eroglu, L., Hizal, A. (1987) Pol. J. Pharmacol. Pharm. 39, 667-673.
Fernandez-Teruel, A., Boix, F., Escorihuela, R.M., Yanez, P., Tobena, A. (1988) Eur. J. Pharmacol. 152, 1-7.
Frances, H., Simon, P., (1985) Neuropharmacology 24, 659-663.
Giardina, W.J., Ebert, D.M. (1989) Biol. Psychiat. 25, 697-702.
Giral, P., Martin, P., Soubrié, P., Simon, P. (1988) Biol. Psychiat. 23, 237-242.
Gorka, Z., Earley, B., Leonard, B.E. (1985) Neuropsychobiology 13, 26-30.
Gorka, Z., Janus, K. (1985) Pharmacol. Biochem. Behav. 23, 203-206.
Gouret, C.J., Mocquet, G., Coston, A., Raynaud, G. (1977) J. Pharmacol. (Paris) 8, 333-350.
Gower, A.J., Marriott, A.S. (1980) Brit. J. Pharmacol. 69, 278P-288P.
Graaf, J.S. de, van Riezen, H., Berendsen, H.G., van Delft, A.M.L. (1985) Drug Dev. Res. 5, 291-301.
Gustafsson, B., Christensson, E. (1990) Pharmacol. Toxicol. 66 (Suppl. 1), 34-39.
Harlow, H.F., Suomi, S.J. (1974) Behav. Biol. 12, 213-220.
Hawkins, J., Hicks, R.A., Phillips, N., Moor, J.D. (1978) Nature 274, 512.
Hunt, G.E., Atrens, D.M., Johnson, G.F.S. (1981) Eur. J. Pharmacol. 70, 59-63.

Iwasaki, K., Fujiwara, M., Shibata, S., Ueki, S. (1986) Pharmacol. Biochem. Behav. 24, 1715-1719.
Jalfre, M., Monachon, M.A., Haefely, W. (1970) Experienta 26, 691.
Jesberger, J.A., Richardson, J.S. (1986) Gen. Pharmacol. 17, 293-307.
Kametani, H., Nomura, S., Shimizu, J. (1983) Psychopharmacology 80, 206-208.
Kostowski, W., Malatynska, E. (1983) Psychopharmacology 79, 203-208.
Kostowski, W., Obersztyn (1988) Pol. J. Pharmacol. Pharm. 40, 341-349.
Leonard, B.E., Tuite, M.C. (1981) Int. Rev. Neurobiol. 22, 251-286.
Leshner, A.I., Remler, H., Biegon, A., Samuel, D. (1979) Psychopharmacology 66, 207-208.
Lloyd, K.G., Zivkovic, B., Scatton, B., Morselli, P., Bartholini, G. (1989) Prog. Neuropsychopharmacol. Biol. Psychiatry 13, 341-351.
Maj., J., Michaluk, J., Rawlow, A., Rogoz, Z., Skuza, G. (1986) Arzneim. Forsch. 36, 1198-1201.
Malick, J.B. (1983) Drug Dev. Res. 3, 357-363.
Martin, P., Laurent, S., Massol, J., Childs, M., Puech, A.J. (1989) Eur. J. Pharmacol. 162, 185-188.
Martin, P., Soubrié, P., Puech, A.J. (1990) Psychopharmacology. (In press).
Martin, P., Soubrié, P., Simon, P. (1986) Pharmacol. Biochem. Behav. 24, 177-181.
Martin, P., Soubrié, P., Simon, P. (1987) Prog. Neuropsychopharmacol. Biol. Psychiatry 11, 1-7.
Massol, J., Martin, P., Adessi, G., Puech, A.J. (1989) Eur. J. Pharmacol. 160, 395-399.
Massol, J., Martin, P., Soubrié, P., Simon, P. (1987) Eur. J. Pharmacol. 134, 345-348.
Matsushita, A., Kawasaki, K., Matsubara, K., Eigyo, M., Shindo, H., Takada, S. (1988) Prog. Neuropsychopharmacol. Biol. Psychiatry 12, 951-966.
McKinney, W.T. (1977) In: Animal Models in Psychiatry (E.Usdin and I.Hanin, Eds), Pergamon, Oxford, pp 117-126.
McKinney, W.T., Eising, R.G., Moran, E.C., Suomi, S.J., Harlow, S.F. (1971) Dis. Nerv. Sys. 32, 735-741.
Mizokawa, T., Kimura, K., Ikoma, Y., Hara, K., Oshino, N., Yamamoto, T., Ueki, S. (1988) Jap. J. Pharmacol. 48, 357-364.
Nishimura, H., Tsuda, A., Oguchi, M., Yoshishige, I., Tanaka, M. (1988) Physiol. Behav. 42, 93-95.
Nishimura, H., Yoshishige I., Tsuda, A., Tanaka, M. (1989) Pharmacol. Biochem. Behav. 33, 227-231.
O'Connor, W.T., Leonard, B.E. (1986) Neuropharmacology 25, 267-270.
O'Connor, W.T., Leonard, B.E. (1988) Prog. Neuropsychopharmacol. Biol. Psychiatry 12, 41-51.
Parale, M.P., Kulkarni, S.K. (1986) Psychopharmacology 89, 171-174.
Passerelli, F., Scotti de Carolis, A. (1983) Neuropharmacology 22, 875-789.
Pawlowski, L., Mazela, H. (1986) Psychopharmacology 88, 240-246.
Poncelet, M., Hakkou, F., Simon, P. (1984) Prog. Neuropsychopharmacol. Biol. Psychiat. 8, 373-378.
Porsolt, R.D. (1985) In: Pharmacotherapy of Affective Disorders (W.G.Dewhurst and G.B.Baker, Eds), Croom Helm, New-York, pp 108-150.
Porsolt, R.D., Anton, G., Blavet, N., Jalfre, M. (1978) Eur. J. Pharmacol. 47, 379-391.
Porsolt, R.D., Bertin, A., Blavet, N., Deniel, M., Jalfre, M. (1979) Eur. J. Pharmacol. 57, 201-210.
Porsolt, R.D., Bertin, A., Jalfre, M. (1977) Arch. Int. Pharmacodyn. Ther. 229, 327-336.
Porsolt, R.D., Chermat, R., Lenègre, A., Avril, I., Janvier, S., Stéru, L. (1987) Arch. Int. Pharmacodyn. Ther. 288, 11-30.

Porsolt, R.D., Martin, P., Lenègre, A., Fromage, S., Giurgea, C. (1989) Prevention of "learned helplessness" in the rat by hydroxyzine. Drug Dev. Res. 17, 227-236.
Porsolt, R.D., Martin, P., Lenègre, A., Fromage, S., Drieu, K. (1990) Pharmacol. Biochem. Behav. (In press).
Puech, A.J., Chermat, R., Poncelet, M., Doaré, L., Simon, P. (1981) Psychopharmacology 75, 84-91.
Pulvirenti, L., Kastin, A.J. (1988) Eur. J. Pharmacol. 151, 289-292.
Quinton, R.M. (1963) Brit. J. Pharmacol. 21, 51-66
Robinson, D.S., Alms, D.R., Shrotriya, R.C., Messina, M., Wickramaratne, P. (1989) Psychopathology 22 (Suppl. 1), 27-36.
Robson, R.D., Antonaccio, M.J., Saelens, J.K., Liebman, J. (1978) Eur. J. Pharmacol. 47, 431-442.
Sanghvi, I.S., Gershon, S. (1977) In: Animal Models in Psychiatry and Neurology (I.Hanin and E.Usdin, Eds), Pergamon, Oxford, pp 157-171.
Seligman, M.E.P. (1975) Helplessness: on Depression, Development and Death, Freeman, San Francisco.
Sherman, A.D., Allers, G.L., Petty, F., Henn, F.A. (1979) Neuropharmacology 18, 891-893.
Sherman, A.D., Saquitne, J.L., Petty, F. (1982) Pharmacol. Biochem. Behav. 16, 449-454.
Stéru, L., Chermat, R., Thierry, B., Mico, J.A., Lenègre, A., Stéru, M., Simon, P., Porsolt, R.D. (1987) Prog. Neuropsychopharmacol. Biol. Psychiatry 11, 659-671.
Stéru, L., Chermat, R., Thierry, B., Simon, P. (1985) Psychopharmacology 85, 367-370.
Stockert, M., Serra, J., de Robertis, E. (1988) Pharmacol. Biochem. Behav. 29, 681-686.
Tamborska, E., Plaznik, A., Kostowski, W., Hauptmann, M., Bidzinski, A. (1989) Pharmacol. Toxicol. 64, 233-236.
Thierry, B., Stéru, L., Chermat, R., Simon, P. (1984) Behav. Neur. Biol. 41, 180-189.
Thierry, B., Stéru, L., Simon, P., Porsolt, R.D. (1986) Psychopharmacology 90, 284-285.
van der Heyden, J.A.M., Molewijk, E., Olivier, B. (1987) Psychopharmacology 92, 127-130.
van Riezen, H., Schieden, H.G., Wren, H.F. (1977) Brit. J. Pharmacol. 60, 521-528.
von Voigtlander, P.F., von Triezenberg, H.G., Losey, E.G. (1978) Neuropharmacology 17, 375-381.
Weiss, J.M., Simson, P.G. (1986) Ciba Foundation Symposium 123, 191-215.
Wachtel, H. (1983) Neuropharmacology 22, 267-272.

Animal Models in Psychopharmacology
Advances in Pharmacological Sciences

SIMPLE ETHOLOGICAL MODELS OF DEPRESSION: SOCIAL-ISOLATION INDUCED "DESPAIR" IN CHICKS AND MICE

Jaak Panksepp, Gordon Yates, Satoshi Ikemoto and Eric Nelson

Department of Psychology, Bowling Green State University, Bowling Green, OHIO 43403, US

SUMMARY: Social isolation is widely considered to be one of the major vectors which contributes to the onset of depression. Although an isolation-induced "despair" model of depression is available for primates, neuroscience work would be facilitated by a simplified model that could be used routinely. In the following work we analyzed the ability of social isolation to facilitate immobility in the mouse forced swimming test and to decrease separation induced distress vocalizations (DVs) in young chicks. Isolation increased immobility only in very young mice (17-21 days of age) but not in older juveniles or adults. Chronic isolation reduced DVs in chicks and made them more susceptible to the DV reducing effects of reserpine. The chronically isolated young chick appears to be an excellent and inexpensive model for the analysis of despair/depression processes in brain.

INTRODUCTION

Depression probably reflects the failure of regulatory neurochemical controls in emotional circuits of the brain. Such imbalance can probably occur spontaneously because of constitutional weaknesses in the underlying neural circuits which control emotions, but more commonly the dysregulations are precipitated by stressful life events sustained beyond the ability of emotional circuits to maintain synaptic homeostasis. Whether all forms of stress can ultimately provoke depression is unknown. Indeed, it remains uncertain how stress should be conceptualized and taxonomized in the nervous system (Puglisi-Alegra & Oliverio, 1990). It is possible that there are subtypes of stress which are related to the natural taxonomies of primary emotional and arousal circuits of the brain (Panksepp, 1982; 1990). From that perspective, probably the most commonly involved emotional systems which lead to clinical depression are those which provoke organisms to establish social dependencies with each other. Perhaps the most common form of depression is the extreme form of melancholy arising from what used to be called "lovesickness" (Ferrand, 1623) -- from the unfullfillment or severance of attachment bonds. Long-term separation produces an irresistable demoralization on the psyche, and clinical depression results when organisms are pulled into its bleak center of gravity. Although we do not yet know how social dependencies are created in the brain (Rajecki et al., 1978), one possible mechanism is an addictive one mediated by brain opioid effects on primary process separation-distress mechanisms of the brain (Panksepp, 1981; Panksepp et al., 1985). Indeed, it is possible that there are common neurochemical denominators in the brain which mediate social attachments and which create attachments to inanimate objects. For instance, place attachments (or preferences) are easy to generate via the pairing of opiates or psychomotor stimulants with specific environments (Carr et al., 1988). Indeed, brain mechanisms for social attachments may have emerged evolutionarily from pre-existing brain mechanisms which mediate object attachments. Attachments, whether social or territorial (including possessions), provide individuals with a sense of power and purpose. When such attachments are severed, the potential to undergo depression is dramatically amplified.

The loss of attachment objects leads to a cascade of prolonged and powerful emotional states - protest, despair, and in some species detachment - which are not easily alleviated except by the return of the object (or subject) of attachment, or via the administration of powerful neuroactive agents such as opiates, clonidine or posterior pituitary peptides such as oxytocin and vasotocin. Although quite variable from species to species, and experiment to experiment, the despair and detachment phases which follow protest are widely considered to represent a brain condition which is especially closely related to the underlying dynamics of depression. Harlow et al. (1965; 1986) most vividly demonstrated the initial emotional distress and eventual despair that resulted from chronic social isolation in young rhesus monkeys. A similar syndrome of despair/depression resulting from inadequate social contact had already been noted in institutionalized human infants (Bowlby, 1982; Spitz & Wolf, 1946). Accordingly, the most important avenue of depression research, at least from an ethological perspective, may be the analysis of behavioural and neurochemical changes resulting from the severance of attachment bonds.

The clinical literature is fairly clear on the point that social loss is one of the most powerful antecedents to clinical depression. Prolonged social isolation produces various neurochemical changes in the brain as well as powerful bodily changes such as compromised immunocompetence (see Calabrese et al., 1987; Melnechuk, 1988; and chapters in Reite & Fields, 1985). The most important role of animal models for understanding depression is the characterization of the underlying brain changes that cause despair, and thereby the discovery of medical and behavioural solutions to halt the cascade of events that terminate in depression. We remain far from that goal, but credible and efficient animal models of spontaneous depression will be essential for the effective analysis of the causal issues. The best models that presently exist are the primate social-isolation models (Rosenblum & Paully, 1987;Suomi, 1983), but it seems unlikely, especially within the present research climate, that primate models can ever be effectively deployed for routine laboratory analyses. Clearly, more simplified animal models which can be related to separation distress mechanisms of the brain are needed for substantial future progress to be made. That approach will be the focus of the present paper. Other approaches such as learned-helplessness and chronic imposition of other stressors (e.g., Katz,1981) will not be emphasized here. There have been several excellent reviews contrasting the various other animal models which are presently available (e.g., Jesberger & Richardson, 1985; Willner, 1984; 1985; Whybrow et al., 1984). The focus here will be on the role of social separation processes in the genesis of depression.

Although the objective behavioural changes which characterize social-loss induced despair are visually evident in some species -- especially primates -- only a few objective studies have formally attempted to analyze the postural and acoustic changes that characterize the despair syndrome (e.g., Suomi et al., 1970; Mason & Capitanio, 1988). This is because the despair symptoms are so "self-evident". It almost seems that we have affective feature-detectors which allow us to resonate with the plight of the distressed animals. The picture generated is one of dejection and chronic saddness. The despondent creatures are quite immobile. Their heads droop, their facial muscles seem slack, and their bodies are commonly hunched over in a collapsed, defeated posture. There has been some recent interest in using changes in expressive behaviours as indicators of clinical depression in humans (e.g., Maser, 1987), but the development of such ethological approaches remains in its infancy. Work with animal models where postures are documented in a quantified way have been few and far between not only because of the difficulty in adequately quantifying such expressive changes but also because there is considerable within- and between-species variability in the incidence of despair. For instance, in several prolonged separation studies in primates no despair-like symptoms have been observed (e.g., Levine et al., 1984; Rosenblum & Kaufman, 1968).

Perhaps one of the most promising objective indicators of despair is the vocal behaviour of socially isolated animals. During prolonged periods of separation, animals exhibit a gradual diminution in the number of distress vocalizations (DVs), and we use these decrements as indicators of the despair phase. Although there are obvious problems with this strategy, including the fact that social-comfort also reduces DVs as do processes related to habituation and fatigue, this should not dissuade us from considering the vocal channel as a potentially excellent indicator of the status of underlying central states. The problem that the same objective response (DVs in the present case) may reflect diametrically opposing processes (comfort vs. despair) is not an uncommon one in psychobiology (for instance, both satiety and sickness reduce food intake). Careful additional work will be needed to discriminate between such distinct central state controls. Although this has yet to be achieved in the vocal domain, it is generally agreed that during the despair phase of social-isolation there is little vocal activity (Hinde & Davies, 1972; Mineka & Suomi,1978). It now needs to be determined whether manipulations which promote and/or alleviate clinical depression also have the predicted reciprocal effects on DVs. Recent work indicates that tricyclic antidepressants are quite effective in sustaining the vocal behaviour of chicks during prolonged (several hour) periods of social-isolation (Lehr, 1989). This pattern of results suggests that antidepressants keep young chicks in the protest phase for a longer period and presumably retard the latency to enter the despair phase. It should be kept in mind that imipramine has little or no effect on the immediate post-isolation DVs of chicks (Panksepp et al., 1980), whereas in several "higher" species, such as dogs and primates, tricyclics can be highly effective in reducing such isolation-induced DVs (Scott, 1974; Suomi et al., 1978). The cause of these diametrically different results obtained in different species remains unresolved at the present time.

In the present paper we will explore the utility of the "social-separation model" in newborn chicks. Also, because of the widespread use of the rodent "swimming despair" model in pharmacological research (Porsolt, 1981), we have also re-evaluated the role of social isolation on the "despair" generated in that system. The basic hypotheses of this work were as follows: 1) Social isolation should increase indices of despair -- facilitate immobility in the mouse swimming test, and decrease DVs in chicks during prolonged separation tests. 2) Pharmacological manipulations which can promote depression, such as reserpine, should prolong immobility and should also reduce DVs, especially in animals maintained chronically in isolation. 3) Antidepressants should inhibit the decline in DVs produced by depressive manipulations, whether environmental or pharmacological. Of course, the swimming test was originally validated by the ability of antidepressants to counteract immobility, but in the present paper we will highlight the fact that the traditionally employed 5 minute tests in small chambers may not be a valid way to measure unconditional immobility in the forced swimming test. In general the results to be presented support the conclusion that the chronically isolated domestic chick may be an especially useful ethological model for depression research.

THE PORSOLT SWIMMING-IMMOBILITY (DESPAIR) TEST

Animals placed into a pool of water from which they cannot escape initially exhibit a "panic" response of rapid agitated swimming with seemingly desperate grasping at the sides of the chamber, followed by a less agitated methodical swimming around the chamber, and eventually the animals exhibit bouts of immobility reminiscent of hopelessness and despair. Although this test proved to be an effective and rapid screening procedure for anti-depressant drugs (Porsolt et al., 1977), the utility of the model for understanding depression has never been clear and has recently become quite controversial. First, performance of rats on learned-helplessness tasks does not predict helpless behaviour on the swimming test (Drugan, et al., 1989). More importantly, prior social isolation, which should presumably evoke some despair, does not appear to amplify the immobility (Hilakivi et al., 1989). Also, one of the best (albeit controversial) pharmacological models of depression -- the behavioural sluggishness that follows reserpine administration -- does not yield heightened immobility in the forced swimming test (Porsolt et al., 1979).

Several recent studies have raised the possibility that the immobility generated during the swimming test may largely reflect some type of learning (adaptation) rather than emotional despair (Borsini et al., 1986; Depablo et al., 1989; Nishimura et al., 1988a). There is now some evidence that the immobility (or lack thereof) may be partially due to situational variables such as the size of the chamber, the depth of the water or the presentation of possible escape routes (DePablo et al.,1989), which permit animals to use various operant strategies to prevent sinking (Nishimura et al., 1988b). Therefore, the immobility that has typically been observed in short observation periods may largely be an instrumental response generated by characteristics of the test situation rather than a conditional emotional response to the threat of survival that is inherent in this frightening situation.

Because of these issues, we decided to re-evaluate the ability of the immobility test to generate an emotional response (helplessness) that might be related to the underlying etiology of depression. We were especially interested in determining whether social isolation could promote the immobility response. Although Hilakivi et al. (1989) found no amplification of immobility by prior social isolation, we thought that their use of adult animals might have precluded observation of synergism between isolation and immobility. Since the species which are typically used in the immobility test (i.e., rats and mice) are relatively socially independent creatures that do not exhibit strong separation responses except for a short period during early juvenile development (Hofer & Shair, 1987), we decided that it would be most appropriate to evaluate the effects of isolation in much younger animals than have been used in past work.

In the first experiment, we evaluated the ability of 24 hrs of prior isolation to promote immobility in 18 day old Swiss-Webster mice during a 5 minute observation test in a 7.5 cm diameter chamber, with 9 cm of 25°C water (all animals had been habituated to the chamber for 15 min on the previous day). All animals were food deprived during the 24 hr isolation period so as to minimize differential weight changes between socially and isolate-housed animals. There was absolutely no facilitation of immobility by prior social isolation during the standard 5 minute test: The social animals exhibited an average of 183.3 sec of immobility while the isolates exhibited 173.6 sec. However, animals tested in a larger 14 cm diameter chamber (12 cm water depth) exhibited essentially no immobility at all during the 5 minute test. Accordingly, we concluded that the immobility observed in the other groups must have been due to the small size of the test chambers. Apparently if animals are able to get three limbs (including tail) touching the sides of the chamber, they became immobile in an instrumental attempt to stabilize themselves.

Accordingly, in the next experiment, we simply used the larger 14 cm test chamber and extended the observation period to 15 minutes. In this testing situation, a non-instrumental form of immobility emerged. Separate groups of animals were tested at 18, 21 and 26-30 days of age. As summarized in Figure 1, prior social isolation clearly increased immobility in the youngest animals, and the effect, albeit attenuated, was also present in the 20 day old animals. The effect was completely gone in the oldest juvenile animals, as well as in other older adult groups we have tested.

However, body weights indicated that the isolated animals lost slightly more weight than the socially housed animals at the youngest age. This may have reflected the fact that huddling in socially housed animals provides substantial thermoregulatory / metabolic savings. Although the older animals exhibited similar weight differentials (with no immobility differentials), we ran several control experiments to determine whether weight changes might be contributing to the isolation effect observed in the younger animals. In the first control study, we kept 17 day old isolated animals at a warmer temperature (36°C) while socials were kept at room temperature. This equilibrated weight loss between the two groups at 18 days of age, and there was still a reliable isolation effect. In a second series, we compared socially housed animals kept on ad lib food vs those deprived of food overnight. Food deprivation had absolutely no effect on immobility time. Thus, we feel confident that social isolation was the factor that promoted the immobility observed in Figure 1. However, it should be emphasized that this effect was observed only in quite young animals and only when relatively long test sessions were used. Presumably the reason Hilakivi et al. (1989) did not observe such an effect was due to their use of adult animals observed in too brief a test session (8 min), and perhaps also because of their failure to use an adaptation training session 24 hr prior to formal testing.

Since much of the immobility that has been reported in the past using brief observation periods in relatively small test chambers may not reflect the type of immobility we observed in the longer test sessions, we decided to re-evaluate the effects of reserpine in isolated and socially-housed mice starting at 19 days of age with daily injections continuing untill 29 days of age (with testing at 19, 22 and 29 days of age). Briefly, 0.25 mg/kg of reserpine given 4 hrs before testing yielded no clear prolongation of immobility. However, the youngest animals did exhibit a slight trend in the predicted direction, but it was only apparent during the third 5 minute period of testing (only in the socially isolated animals). When tested again at the older ages, the reserpine groups (whether social or isolate) exhibited reliably less immobility than the controls (albeit their swimming appeared to be less agitated than that of controls). In sum, we were not able to obtain clear evidence that reserpine amplifies the isolation effect, even though such a modest trend was apparent in the youngest animals. Also, we must note that using the present procedures, we were unable to observe a clear reduction of immobility with imipramine (10 mg/kg), suggesting that the instrumental-response inducing characteristics of the traditional 5 minute test situation may be essential for the effect.

In sum, it appears that immobility in the forced swimming test can be promoted by prior social isolation but only at quite young ages. This suggests that the presumed despair arising from social loss may add to the despair which emerges from stressful environmental situations in which animals feel helpless, which may indicate that both stressors are operating on a common underlying dimension. Although there is no good predictive relationship between immobillity and other measures of depression such as that evoked by "learned helplessness" paradigms (Drugan et al., 1989), it should be kept in mind that most of the literature using the forced swimming test has observed animals in small diameter chambers with shallow depths of water during short 5-minute observation periods. As summarized herein, there is reason to believe that that type of immobility is different from the type observed here in prolonged 15 minute test sessions in young animals. As a consequence, we feel that many of the findings in this rather extensive literature are now in need of re-evaluation.

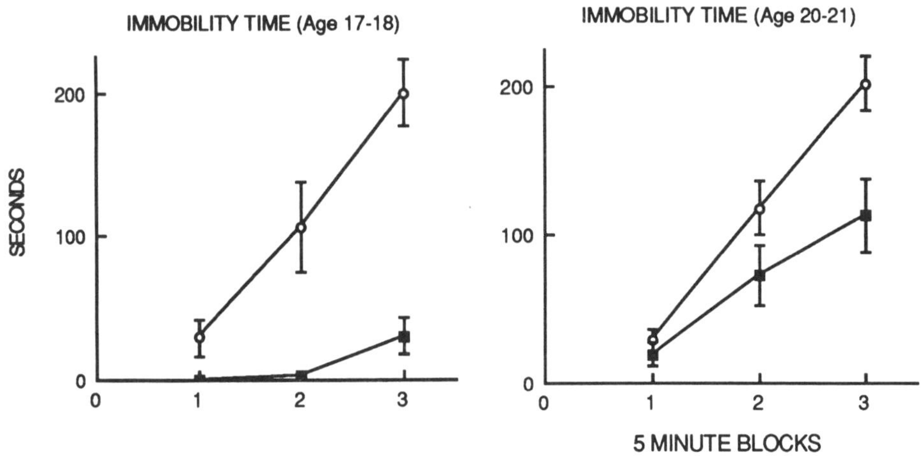

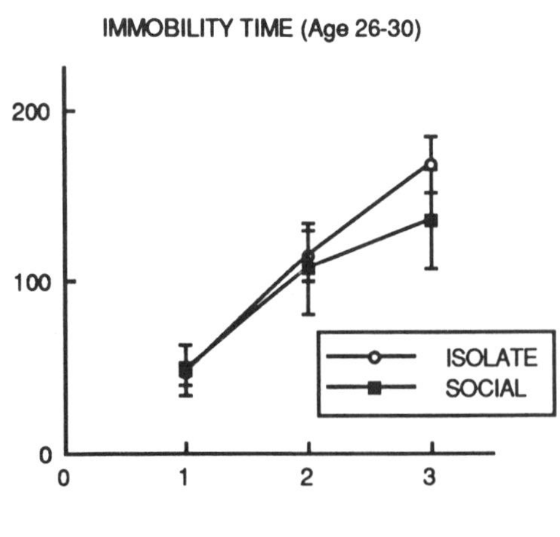

Fig. 1 Immobility times in the forced swimming test as a function of age in Swiss-Webster mice during a 15 minute test session. Animals had been habituated to the situation 24 hr before and were maintained food deprived until the next day either in social housing (with a surrogate non-lactating "aunt") or individually at room temperature of about 20°C. (Error bars indicate SEMS).

THE NEWBORN-CHICK ISOLATION MODEL

In the following work we investigated whether chronic isolation of the newborn chick would lead to a pattern of despair-like responses resembling primate depression. There have been several previous reports which indicate that chronic isolation can lead to adverse effects in young chicks. Several early studies briefly mentioned that social isolation leads to a variety of abberant behaviours in newborn chicks (Hess et al., 1959), and it has occassionally been noted that some isolated chicks fail to thrive and die after several days of social isolation (e.g., Schrenk et al., 1963). Early social isolation can also have other depressive-like effects which last beyond the isolation period, such as the disruption of aggregative behaviour up to four weeks later (Baron & Kish, 1960). Eating and drinking are also affected (Tolman,1964), with isolated chicks exhibiting markedly decreased levels of imitative pecking relative to socially housed animals (May & Dorr, 1968). Hence, isolated chicks exhibit higher mortality levels than socially housed ones (Schrenk, et al., 1963). Rajecki et al. (1976a) were the first to systematically follow a group of chicks that were isolated for a prolonged period (i.e., from the time of hatching until three weeks of age). Relative to a social group, the isolates initially exhibited higher levels of emotional behaviours (DVs and escape behaviour) and lower levels of exploratory and consummatory behaviours. Over time the distress behaviours diminished and the amount of time spent inactive increased dramatically, which may have represented a shift from protest to despair.

After noticing a substantial amount of variation in isolation-induced distress responses, Rajecki et al., (1976a) conducted an internal analysis on a group of chicks which were separated following two weeks of social housing. The response to separation fell into three patterns: Nine out of 20 birds responded by becoming inactive and lethargic; six out of 20 responded with a sustained distress-disturbance reaction; and five out of 20 exhibited little observable change. Over time the distress sub-group appeared to move from a period of protest to a period of inactivity, which was deemed to resemble anaclitic depression.

Accordingly, the young chick may serve as a useful model for isolation-induced depressive reactions. This would be particularly useful because more psychobiological knowledge is available about the separation response of the domestic chick than any other species (Panksepp et al., 1985; 1988). If the protest phase which immediately follows separation establishes the neurochemical conditions from which isolation-induced despair ultimately emerges, then we may be able to analyze the cascade of events that lead from the acute emotional reactions of separation to eventual depression.

Accordingly, in the following work we evaluated the utility of the young chick as an effective model for isolation-induced depression research. In preliminary work (not reported here in detail), birds were isolated at 5 days of age and behaviour was compared to animals kept in social groups. Also, the consequences of thrice daily injections of corticotrophin releasing factor (CRF, 0.5 μg) into the fourth ventricle were evaluated. Individual housing at this age had no untoward effect on growth (all animals seemed to thrive), and animals continued to exhibit high levels of DVs in our isolation chambers, albeit reliably lower than social animals. The CRF injections made animals less sensitive to social stimuli (i.e., reduced the comforting effect of mirrors). At thirteen days of age, the DVs of isolated animals were more reduced by 1.25 mg/kg of reserpine than in socially housed animals, suggesting that prior isolation had magnified the despair-like response (i.e., reduced DVs) to reserpine. Overall, a single injection of reserpine inhibited DVs for approximately 2-9 hrs after injection. The aim of the work to be detailed here, was to determine whether such depressive-like symptoms could be amplified if isolation was started at an earlier age.

EFFECTS OF ISOLATING CHICKS BETWEEN ONE AND TWO DAYS OF AGE

Young male domestic chicks (Hubbard White Mountain Cornish Rock Broilers) were shipped (under socially housed conditions) from a local hatchery and arrived when animals were 1-2 days of age (designated as 1 day old animals). The group housing during shipping should have given animals sufficient time to become imprinted upon each other. Immediately upon arrival, animals were divided into two housing conditions--individually housed animals (n=16) were kept with free access to food and water in 30.4 x 18 x 12.8 cm high opaque boxes, and socially housed animals were kept with free access to food and water in a group of 16 (animals kept in separate rooms at approximately 20°C). Although the isolated animals could not see each other, they remained in auditory and olfactory contact since they were all kept in the same room. Isolation conditions, with no testing, were maintained until 6 days of age, during which time weight gain was monitored. While all the socially housed animals thrived, 6 out of 16 isolated animals failed to thrive, started to lose weight and died with in the first 4 days following isolation. These animals exhibited bodily postures indicative of despair/debilitation: They were immobile, with drooping heads, and eventually splayed wings. At six days of age, the isolation-induced DVs of the surviving animals were initially tested for 20 minutes in the presence and absence of mirrors (5 mins plainboxes, followed by 5 mins of mirrors, replicated a second time). Although their no-mirror vocalization levels were reliably reduced, they still exhibited normal DV inhibition in response to mirror chambers. At seven days of age, animals were given a prolonged 5 hr isolation test in the plain boxes (with DVs recorded for each of 10 successive 30 minute periods). An additional group of ten same-aged animals which had only been isolate-housed for the preceeding 2 days were also included. The highest levels of vocalization were emitted by the socially housed animals (approximately 1700 DVs/30 minutes) followed by 2-day isolates (~80 % of socials) followed by the chronic 7-day isolates (~65 % of the socials).

Although only the chronic isolates were reliably different from controls, it is equally noteworthy just how much vocal activity was still emitted by the chronically isolated chicks simply by being moved to the test chamber (the birds were emitting very few DVs in their home cages). Whether this home-isolation induced vocalization was due to the residual effects of auditory or olfactory social bonding, or to the fact that they had become attached to their home environments remains unclear at the present time. However, it is worth noting that even when the isolate-housed chicks were placed into the isolation chambers in their home cages, there was a marked amplification of DVs suggesting that there had been some bonding to the local housing environment. However, since these DV levels were not as high as those when the chick was removed from its home cage suggests that part of the DV response was due to a "place-separation" phenomenon.

Two days after the 5 hr isolation test, the responsivity of all three groups to a 0.25 mg/kg intraperitoneal reserpine challenge was evaluated at 2, 5, 9 and 24 hrs following drug injection (for 10 minute test periods). As summarized in Figure 2, at this low dose, there was only a marginal reduction of DVs in controls, while the reductions were larger in the 4 day isolates, and even larger in the chronic isolates, and these differential effects were still present when data were calculated as percent of baseline responding.

In a separate group of chicks, the ability of 5 mg/kg of imipramine to reverse the reduction of DVs produced by 0.5 mg/kg of reserpine was evaluated. Animals were treated with reserpine 6 hrs prior to testing, and injected with imipramine 10 minutes prior to the first 10 minute isolation test (two 5 minute recording epochs). Twenty minutes later (during which time animals were housed individually) another 10 minute DV test was administered. After a 2 hr interval, during which animals were kept socially housed, the two segment test cycle was repeated. As summarized in Figure 3, the depression of DVs produced by reserpine was reliably reversed with imipramine.

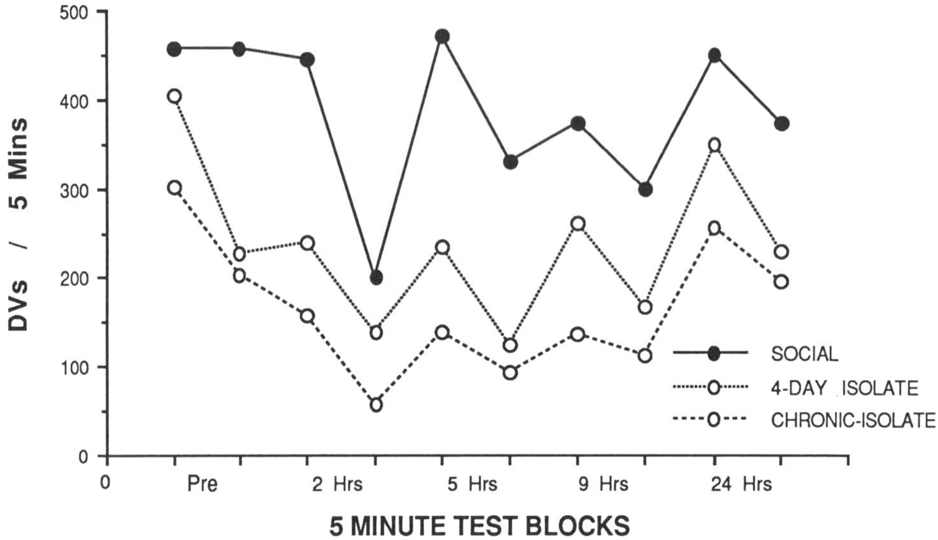

Fig. 2 DVs as a function of time following i.p. administration 0.25 mg/kg of reserpine in 3 groups of subjects. At each time point, animals were tested for 2 consecutive 5 minute periods. The "pre" test period occurred just prior to reserpine injection.

Also, as reported by Lehr (1989), we found that the decline in DVs that is normally observed during a prolonged isolation session could be markedly attenuated by prior administration of 5-10 mg/kg of imipramine. Although more work is essential to clarify the underlying neurochemical and psychological underpinnings of this result, this data suggests that imipramine can keep the chicks in the active protest phase of separation distress for a prolonged time (i.e., the despair phase is delayed).

This pattern of results was replicated in a subsequent experiment which included 1) sex as a variable, 2) social housing in groups of 4 as opposed to the larger groups of the previous experiment (with both isolates and socially-housed birds kept in the same room), 3) daily body temperature recordings, and 4) 2 doses of reserpine (0.5 and 1.0 mg) and 5) an attempt to rehabilitate animals which exhibited the despair-like failure to thrive. In brief, the results yielded a pattern of results similar to that described in the preceeding experiment. It is noteworthy however, that females exhibited a greater rate of failure to thrive (45%) than the males (25%). Using the previous experiment as an indicator that such animals would die without rehabilitation, we determined whether we could help the isolates recover by re-establishment of social housing. Three quarters of the animals were restored to seemingly normal health through this maneuver, but at later ages, they remained especially sensitive to reserpine-induced "despair" as measured by reductions in DVs.

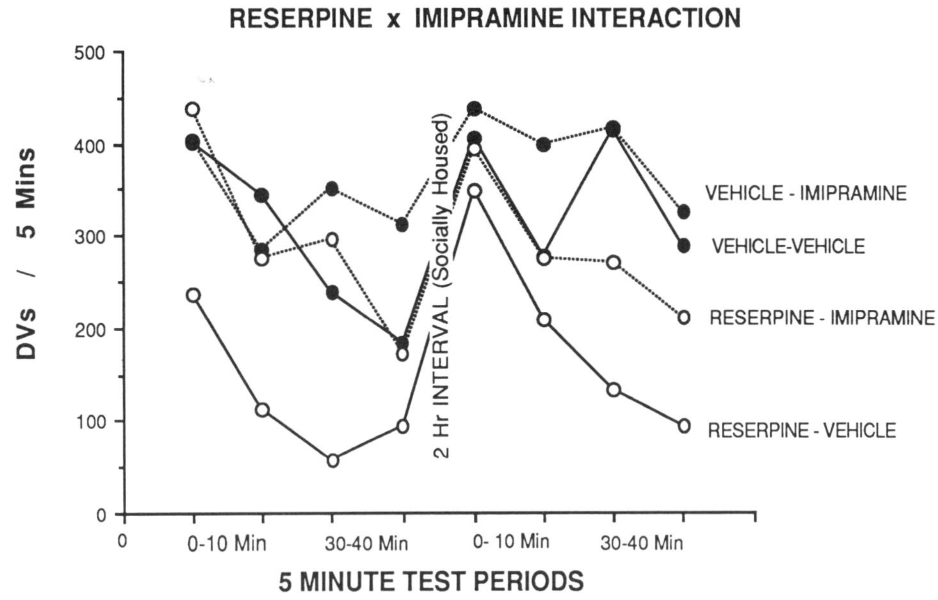

Fig. 3 Effects of imipramine (5 mg/kg) in reserpine (0.5 mg/kg) or vehicle pre-treated birds. Although imipramine marginally increased DVs in control animals (during later parts of the test session), imipramine was effective during all periods in the reserpine pre-treated animals.

In sum, the present data suggest that early social isolation of the domestic chick may be a useful model for depression induced by social loss. The advantages of this model for analyzing the underlying neurochemical systems are obvious. The chick is inexpensive, easy to run in large numbers, and it is a precocious animal which, with respect to acquired knowledge, can be considered a "blank slate" at birth but quite capable of sophisticated forms of learning. Also, a subset of animals displays a strong tendency to exhibit a severe failure to thrive syndrome in a manner reminiscent of indvidual-differences that have been apparent in primate research (Mendoza & Mason, 1989; Suomi, 1985). This severe form of "despair" was readily reversed by providing social contact. This leads to the possibility that one can analyze the constitutional characteristics of an animal which lead to the most severe form of depression and the endogenous neurochemical changes that lead to its alleviation by social contact. We have conducted some preliminary neurochemical work along these lines which will be briefly summarized next, and a brief synopsis of what is known about the bodily changes in other species that result from social isolation is provided in appendix fashion at the end of this paper.

ISOLATION-INDUCED NEUROCHEMICAL CHANGES IN CHICKS

The chick model of depressive despair provides a useful model of "depression" which may be useful for pursuing basic neuroscience issues that cannot be easily implemented in the available primate models. Although the biogenic amine theories of depression have guided thinking about underlying issues for 30 years, with highly promising insights (Willner, 1985; Whybrow et al., 1984), it is likely that the neuropeptides will provide totally unexpected insights for understanding depressive as well as other psychiatric disorders (Nemmeroff, 1988). In preliminary work with Drs. Alois Saria (Innsbruck) and Gunther Bernatzky (Salzburg) brain levels of Calcitonin Gene Related Peptide (CGRP), Substance P (SP) and Neurokinin-A(NA) like immunoreactivity have been measured in domestic chicks that have been socially isolated for 24 hrs from the first to second days after hatching. While there was no change in NA activity, and a modest increase in SP (from 17 to 21 picomoles/brain, $p < .07$), CGRP was reduced from 14 to 10 picomoles/brain, $p < .05$). The SP/CGRP ratio was twice as large in isolated animals (ratio = 2.58) than in socially housed animals (ratio = 1.25, $t(38) = 3.60$, $p < .001$). Also, while the CGRP and SP in brains of socially housed animals was highly correlated ($r = .76$), that was not the case in isolated animals ($r = .05$), suggesting that some type of neurochemical coupling was changed in isolated animals. Because of these results, we evaluated the ability of intraventricularly administered CGRP (1 μg) to modulate chick DVs and found no immediate effect on DVs but a substantial 50% inhibition of DVs when testing was conducted 30 minutes following peptide administration.

Although these brain changes are difficult to interpret within the context of the existing knowledge about the functions of these neuropeptides, the chick model affords a useful way to analyze the brain changes wrought by social isolation in a thoroughgoing "Baconian" (i.e., non-theoretical) way. When enough basic data is collected about changes in the behavioural functions of neuropeptides, we may be in a position to theorize credibly about the underlying neurodynamics of depression.

At present, the only system that allows some coherent theoretical statement is the CRF system of the brain. As noted elsewhere, that system may be an executive neuromodulator for separation distress in the avian (Panksepp et al., 1988; Panksepp, 1989) and mammalian brain (Kalin, 1985), and it is now well established that this system has strong excitatory connections with brain norepinephrine systems (Brown & Fisher, 1985; Siggins et al., 1985; Valentino et al.,1983). One can easily envision that chronic arousal of the CRF system could promote depletion of noradrenergic resources in the brain, which would be compatible with a depressive response style. Hopefully, future work will be able to yield a more comprehensive picture of the underlying neural dynamics through the use of the chick animal model of despair.

CONCLUSIONS

Although social separation does not always induce depressive symptoms and is not the only causal antecedent of depression, the relationship between the two appears to be quite intimate. The initial period of protest which follows social isolation is followed in some species by a despair phase characterized by decreased vocal activity, disruptions of physiological systems, and psychomotor retardation. Such symptoms commonly accompany endogenous depression in humans. Ethological and other animal behavioural models are especially useful in studying such reactive forms of depression, and the aim of the present paper was to review such models and to provide some new data concerning two potential simplified models--namely the Porsolt's (1979; 1981) forced swimming test and the isolation-induced "despair" model for young chicks. The obvious pragmatic advantages of the chickmodel over other available models invites investigators to attempt to utilize this system to analyze the cascade of neurochemical and psychobiological events that generate depressive symptoms as a result of social loss. The young chick model may be an especially useful one for facilitating work on the underlying brain changes that lead from protest to despair since primate models are so expensive and ethically troublesome.

Whether the depressive response following social loss has some type of adaptive function (e.g., preventing needless dissipation of energy) remains unclear. At present, the common viewpoint is that the protest phase of the separation response clearly has the potential adapative effect of an increased probability of being reconnected with social resources, while a comparable adaptive function is not obvious for the ensuing period of despair. Thus, whether depression should be viewed as the pathological end-result of insurmountable emotional stressors (an exhaustion response) or as a spontaneously occurring psychobiological strategy that can help promote the re-establishment of brain and social homeostasis (i.e., it is a regulatory response) deserves further consideration in future research.

APPENDIX: A BRIEF REVIEW OF THE BODILY CONSEQUENCES OF SOCIAL SEPARATION

Non-primate work: Social separation can have remarkably powerful physiological effects, confirming that social affect is mediated by robust regulatory systems of the brain which can have wide-ranging influences over bodily functions. In rats, the effects of maternal separation have been studied fairly extensively (see Hofer, 1983, 1987 for review). behavioural and physiological changes in offspring include decreased orthonine decarboxylase activity in the brain (Butler et al., 978), altered norepinephrine (NE) and dopamine (DA) levels (Stone et al., 1976), learning deficiencies (Navcikovi, 1966a,b), altered sleep patterns (Hofer, 1976), and higher susceptibility to gastric erosion following stress (Ackerman et al., 1975). These symptoms of maternal deprivation highlight the significance of the mother for normal infant development.

When infants are socially isolated, they suffer not only from maternal deprivation but also experience the absence of any social, affective interactions with peers. In rats, peer isolation also leads to marked physiological and behavioural changes. Only 24-hrs of total social isolation was enough to increase serotonin [5-hydroxytryptamine (5-HT)] levels in the brains of 4- and 10-day-old rats (Spear & Scalro, 1985). Vocalizations of 14-day-old rats were markedly amplified as soon as pups were left alone (Hofer & Shair, 1978, 1980, 1987). Moreover, only 5-min of social isolation was necessary to modulate pain responses in 10-day-old pups (Kehoe & Blass, 1986a,b). Relatively longer periods of social deprivation have been used in post-weanling rats. behavioural changes include modified pain responses (e.g., Kostowski et al., 1977; Panksepp, 1980), increased locomotor activity (e.g., Gentsch et al., 1982; 1982), disrupted learning (e.g., Juraska et al., 1984; Watson & Livesey,1982), increased aggression (e.g., Day et al., 1982), disrupted sexual behaviour (Gerall et al., 1967), and increased social proximity-maintenance (Latané et al., 1970; 1972). These behavioural changes may be attributed partially, if not mostly, to changes in the CNS. In fact, biochemical changes in the brain have been reported after sustained social isolation. These changes include altered 5-HT turnover (Valzelli & Garattini, 1972; Parker & Morinan, 1986), decreased NE and DA turnover in some areas of the brain (Thoa et al., 1977; Weinstock et al., 1978), and altered NE and DA concentrations (Thoa et al., 1977). Such effects have typically been seen in young animals, while in mature animals isolation may actually produce increased NE turnover with no change in 5-HT (Stolk et al., 1974). Isolation can also increase α_1-noradrenergic receptor binding in the striatum (de Ceballos et al., 1983), DA receptor binding in the striatum (Guisado et al., 1980), decrease opiate receptor binding in the whole brain analysis (Schenk et al., 1982), and alter 5-HT, DA, NE, and acetylcholine (ACh) sensitivity at various sites of the brain (Oehler et al., 1987).

Effects in Primates: The mere act of social interaction can modify physiological responses inprimates (Mendoza & Mason, 1989). Powerful behavioural consequences after prolonged social isolation have also been reported (see Harlow et al., 1965; Mitchell, 1970). Prolonged early social isolation diminished subsequent social outputs, including play and sexual behaviours. That is, young animals subjected to prolonged social isolation are not able to deal effectively on a social level with other members of their species. In primates, many behavioural and physiological changes have also been reported after social separation (for excellent reviews see Reite & Fields, 1985). These changes include altered plasma cortisol levels (Smotherman et al., 1979), modified immune response (Laudenslager et al., 1982), and changes in heart rate, body temperature, sleep patterns, locomotor activities, and play (Hinde & Spencer-Booth, 1971; Reite et al., 1974). With respect to neurochemical vectors, it has been found that NE depletion can lead to behavioural withdrawal resembling depression (Kraemer & McKinney, 1979). McKinney (1985) has also reported that separation-induced despair is associated with lowered cerebrospinal (CSF) NE and that amelioration of despair is associated with increased CSF NE. Also the level of CSF NE is a stable trait that was reciprocally related to huddling behaviour, and hence was speculated to be predictive of separation-induced depression. Functional evidence for such a possibility has been provided by the observation that long-term social deprivation in rhesus monkeys leads to hypersensitivity to the indirect catecholamine stimulating agent d-amphetamine (Kraemer et al., 1984). A similar hypersensitivity has been observed in dopamine systems of rhesus monkeys, but no difference was apparent in isolated and socially housed animals in CSF levels of DA, NE and 5-HT metabolites (Lewis et al., 1990). To our knowledge, comparable neurochemical analyses have yet to be done in isolated chicks.

In general, the effects of social isolation on brain biogenic amine systems appear to be compatible with classic catecholamine hypotheses of depression (Willner, 1987).

Whether they are also compatible with more current revisions of the catecholamine hypotheses such as upregulation of norepinephrine receptor systems (e.g., Mobley & Sulser, 1981; Stone, 1983) remains to be seen.

REFERENCES

Ackerman, S.H., Hofer, M.A., Weiner, H. (1975) Psychosom. Med. 37, 180-184.

Baron, A., Kish, G.B. (1960) J. Comp. Physiol. Psychol. 53, 459-463.

Bowlby, J. (1982) Attachment: Attachment and loss (2nd ed., vol.1). New York: Basic Books, Inc.

Borsini, F., Volterra, G., Meli, A. (1986) Physiol. Behav., 38, 385-386.

Brown, M.R., Fisher, L.A. (1985) Fed. Proc. 44, 243-248.

Butler, S.R., Suskind, M.R., Schanberg, S.M. (1978) Science 199, 445-447.

Calabrese, J.R., Mling, M.A., Gold, P.W. (1987) Am. J. Psychiatry 144, 1123-1134.

Carr, G.D., Fibiger, H.C., Phillips, A.G. (1988) In: The Neuropharmacological Basis of Reward (J.M.Liebman and S.J.Cooper, Eds), Oxford, Clarendon Press.

Day, H.D., Seay, B.M., Hale, P., Hendricks, D. (1982) Devl. Psychobiol. 15, 47-59.

De Ceballos, M.L., Guisado, E., Sanchez-Blazquez, P., Garzon, J., del Rio, J. (1983) Neurosci. Letters 39, 217-222.

De Pablo, J.M., Parra, A., Segovia, S., Guillamón, A. (1989) Physiol. Behav. 46, 229-237.

Drugan, R.C., Skolnick, P., Paul, S.M., Crawley, J.N. (1989) Pharmacol. Biochem. Behav. 33, 649-654.

Ferrand, J. (1623) A Treatise on Lovesickness. (Translated and Edited by D.A. Beeher & M. Ciavolella) 1990 Syracuse, NY, Syracuse University Press.

Gentsch, C., Lichtsteiner, M., Frischknecht, H.-R., Feer, H., Siegfried, B. (1988) Physiol. Behav. 43, 13-16.

Gentsch, C., Lichtsteiner, M., Kraeuchi, K., Feer, H. (1982) Behav. Brain Res. 4, 45-54.

Gerall, D.H., Ward, I L., Gerall, A.A. (1967) Anim. Behav. 15, 54-58.

Guisado, E., Fernandez-Tome, P., Garzon, J., del Rio, J. (1980) Eur. J. Pharmacol. 65, 463-464.

Harlow, H.F. (1986) From Learning to Love: The Selected Papers of H.F.Harlow, C.M. Harlow (Ed), New York, Praeger.

Harlow, H.F., Dodsworth, R.O., Harlow, M.K. (1965) Proc. Natl. Acad. Sci. 54, 90-96.

Hess, E.H., Polt, J.M., Goodwin, E. (1959) In: The pharmacology and Clinical Usefulness of Carisoprodol (J.G.Miller, Ed), Detroit, Wayne State Univ. Press. pp 51-65.

Hilakivi, L.A., Ota, M., Lister, R.G. (1989) Pharmacol. Biochem. Behav. 33, 371-374.

Hinde, R.A. (Ed.) (1983) Primate social relationships. Sunderland, MA: Sinauer Associates, Inc., Publishers.

Hinde, R.A., Davies, L.(1972) J. Child Psychol. Psychiat. 13, 227-237.

Hinde, R.A., Spencer-Booth, Y. (1971) Science 173, 111-173.

Hofer, M.A. (1976) Devl. Psychobiol. 9, 189-203.

Hofer, M.A. (1983) In: Emotion (R.Plutchik, Ed), New York: Academic Press pp 199-219.

Hofer, M.A. (1987) In: Perinatal development, Orland (N.A.Krasnegar, E.M.Blass, M.A.Hofer and W.P.Smotherman, Eds), FL: Academic Press, pp 251-274.

Hofer, M.A., Shair, H.N. (1978) Devl. Psychobiol. 11, 495-504.

Hofer, M.A., Shair, H.N. (1980) J. Comp. Physiol. Psychol. 24, 271-179.

Hofer, M.A., Shair, H.N. (1987) Devl. Psychobiol. 20, 465-476.
Jesberger, J.A., Richardson, J.S. (1985) Biol. Psychiatry 22, 764-784.
Juraska, J.M., Henderson, C., Muller, J. (1984) Devl. Psychobiol. 17, 209-215.
Kalin, N.H. (1985) Fed. Proc. 44, 249-253.
Katz, R.J. (1981) Neurosci. Biobehav. Rev. 5, 231-246.
Kehoe, P., Blass, E.M. (1986a) Devl. Psychobiol. 19, 385-398.
Kehoe, P., Blass, E.M. (1986b) Behav. Neurosci. 100, 624-630.
Kostowski, W., Czlonkowski, A., Rewerski, W., Piechocki, T. (1977) Psychopharmacology 53, 191-193.
Kraemer, G.W., McKinney, W.T. (1979) J. Aff. Dis. 1, 33-54.
Kraemer, G.W., Ebert, M.H., Lake, C.R., McKinney, W.T. (1984) Psychopharmacology 82, 266-271.
Latané, B., Cappell, H., Joy, V. (1970) J. Comp. Physiol. Psychol. 70, 221-227.
Latané, B., Nesbitt, P., Eckman, J., Rodin, J. (1972) J. Comp. Physiol. Psychol. 81, 69-75.
Laudenslager, M.L., Reite, M., Harbeck, R.J. (1982) Behav. Neural Biol. 36, 40-48.
Lehr, E. (1989) Psychopharmacology 97, 145-146.
Levine, S., Franklin, D., Gonzalez, C.A. (1984) Child Devl. 55, 1386-1393.
Lewis, M.H., Gluck, J.P, Beauchamp, A.J., Keresztury, M.F., Mailman, R.B. (1990) Brain Res. 513, 67-73.
Maser, J.D. (Ed) (1987) Depression and Expressive behaviour. Hillsdale, New Jersey, Lawrence Erlbaum.
Mason, W.A., Capitanio, J.P. (1988) Devl. Psychobiol. 21, 401-430.
May, J.G., Dorr, D. (1968) Psychon. Sci. 11, 175-176.
McKinney, W.T. (1985) In: The psychobiology of Attachment and Separation (M.Reite and T.Fields, Eds), New York, Plenum Press, pp 201-222.
Melnechuk, T. (1988) Emotions, brain, immunity, and health: a review. In: Emotions and Psychopathology (M.Clynes and J.Panksepp, Eds), New York, Plenum Press, pp 181-247.
Mendoza, S.P., Mason, W.A. (1989) In: Perspectives in Primate Biology (P.K.Seth and S.Seth, Eds), Vol. 2. New Delhi, India, Today & Tomorrow's Printers and Publishers, pp 129-143.
Mineka, S., Suomi, S.J. (1978) Psychol. Bull. 85, 1376-1400.
Mitchell, G. (1970) In: Primate behaviour: Developments in Field and Laboratory Research (L.A.Rosenblum, Ed), Vol. 1. New York, Academic Press.
Mobley, P.L., Suiser, F. (1981) In: Antidepressants: Neurochemical, behavioural, and Clinical Perspectives (S.J.Enna et al., Eds), New York, Raven Press, pp 31-51.
Naváková, V. (1966a) Physiol. Behav. 1, 219-221.
Naváková, V. (1966b) Science 151, 475-476.
Nemeroff, C.B. (Ed) (1988) Neuropeptides in Psychiatric and Neurological Disorders. Baltimore, The Johns Hopkins University Press.
Nishimura, H., Tsuda, A., Ida, Y., Tanaka, M. (1988a) Physiol. Behav. 43, 665-668.
Nishimura, H., Tsuda, A., Oguchi, M., Ida, Y., Tanaka, M. (1988b) Physiol. Behav. 42, 93-95.
Oehler, J., Jahkel, M., Schmidt, J. (1987) Physiol. Behav. 41, 187-191.
Panksepp, J. (1980) Psychopharmacology 72, 111-112.
Panksepp, J. (1981) In: Progress in Theory in Psychopharmacology (S.Cooper, Ed), Vol. 1. London, Academic Press, pp 149-175.
Panksepp, J. (1982) Behav. Brain Sci. 5, 407-465.
Panksepp, J. (1989) In: Handbook of Social Psychophysiology (H.Wagner and T.Manstead, Eds), Chichester, England, John Wiley & Sons, pp 5-26.
Panksepp, J. (1990) In: Psychobiology of Stress (S.Puglisi-Allegra and A.Oliverio, Eds), Dordrecht, the Netherlands, Kluwer, In Press.
Panksepp, J., Meeker, R., Bean, N.J. (1980) Pharmacol. Biochem. Behav. 12, 437-443.

Panksepp, J., Normansell, L., Herman, B., Bishop, P., Crepeau, L. (1988) In: The Physiological Control of Mammalian Vocalizations (J.D.Newman, Ed), New York, Plenum Press, pp 263-299.
Panksepp, J., Siviy, S.M., Normansell, L.A. (1985) In: The Psychobiology of Attachment and Separation (M.Reite and T.Fields, Eds), New York, Academic Press, pp 3-49.
Parker, V., Morinan, A. (1986) Neuropharmacology 25, 663-664.
Porsolt, R.D. (1979) Biomedicine 30, 139-140.
Porsolt, R.D. (1981) In: Antidepressants: Neurochemical, behavioural, and clinical perspectives (S.J.Enna et al., Ed), New York: Raven Press, pp 121-139.
Porsolt, R.D., LePichon, M., Jalfre, M. (1977) Nature 20, 730-732.
Porsolt, R.D., Bertin, A., Blavet, N., Deniel, M., Jalfre, M. (1979) Eur. J. Pharmacol. 57, 201-210.
Puglisi-Allegra, S., Oliverio, A. (Eds) (1990) Psychobiology of Stress, Dordrecht, The Netherlands, Kluwer Academis Publishers. (In Press).
Rajecki, D.W., Lamb, M.E., Obmascher, P. (1978) Behav. Brain Sci. 3, 417-436.
Rajecki, D.W., Suomi, S.J., Scott, E.A., Campbell, B. (1976a) Devl. Psychol. 13, 143-155.
Rajecki, D.W., Wilder, D.A., Kidd, R.F., Jaeger, J. (1976b) Anim. Learn. Behav. 4, 30-32.
Reite, M., Fields, T. (Eds) (1985) The psychobiology of Attachment and Separation, New York, Plenum Press.
Reite, M., Kaufman, I.C., Pauley, J.D., Stynes, A.J. (1974) Psychosom. Med. 36, 363-367.
Rosenblum, L.A., Kaufman, I.C. (1968) Amer. J. Orthopsychiatry 38, 418-426.
Rosenblum, L. A., Paully, G.S. (1987) Psychiatric Clin. N. Am. 10, 437-447.
Schenk, S., Britt, M.D., Atalay, J., Charleson, S. (1982) Pharmacol. Biochem. Behav. 16, 841-842.
Scott, J.P. (1974) Proc. IX CINP. Exc. Med. Int. Congr. Series No. 357, pp 735-745.
Shreck, P.K., Sterritt, G.M., Smith, M.P., Stilson, D.W. (1963) Anim. Behav. 11, 306-309.
Siggins, G.R., Gruol, D., Aldenhoff, J., Pittman, O. (1985) Fed. Proc. 44, 237-242.
Smotherman, W.P., Hunt, L.E., McGinnis, L.M., Levine, S. (1979) Devl. Psychobiol. 12, 211-217.
Spear, L. P. & Scalzo, F. M. (1985) Devl. Brain Res. 18, 143-157.
Spitz, R.A., Wolf, K.M. (1946) In: The Psychoanalytic Study of the Child, V.2. New York, International University Press, pp 313-342.
Stolk, J.M., Conner, R.L., Barchas, J.D. (1974) J. Comp. Physiol. Psychol. 87, 203-207.
Stone, E.A. (1983) Behav. Brain Sci. 6, 535-578.
Stone, E.A., Bonnet, K.A., Hofer, M.A. (1976) Psychosom. Med. 38, 242-249.
Suomi, S.J. (1983) Psychol. Med. 13, 465-468.
Suomi, S.J. (1985) In: Comprehensive Textbook of Psychiatry/IV (H.I.Kaplan and B.J.Sadock, Eds), Baltimore, Williams & Wilkins, pp 226-237.
Suomi, S.J., Harlow, H.F., Domek, C.J. (1970) J. Abn. Psychol. 76, 161-172.
Suomi, S.J., Seaman, S.F., Lewis, J.K. et al. (1978) Arch. Gen. Psychiatry 35, 321-325.
Thoa, N.B., Tizabi, Y., Jacobowitz, D.M. (1977) Brain Res. 131, 259-269.
Tolman, C.W. (1965) An. Behav. 13, 493-496.
Tolman, C.W. (1967) Behaviour 30, 275-286.
Valentino, R.J., Foote, S.L., Aston-Jones, G. (1983) Brain Res. 270, 363-367.
Valzelli, L., Garattini, S. (1972) Neuropharmacology 11, 17-22.
Watson, J.S., Livesey, P.J. (1982) Behav. Neural Biol. 36, 98-101.
Weinstock, M., Speiser, Z., Ashkenazi, R. (1978) Psychopharmacology 56, 205-209.
Willner, P. (1984) Psychopharmacology 23, 1-16.

Willner, P. (1985) Depression: A Psychobiological Synthesis, New York, John Wiley & Sons.
Willner, P. (1987) Rev. Neurosci. 1, 77-99.
Whybrow, P.C., Akiskal, H.S., McKinney Jr., W.T. (1984) Mood Disorders: Toward a New Psychobiology, New York, Plenum Press.

Animal Models in Psychopharmacology
Advances in Pharmacological Sciences

ANIMAL MODELS FOR INVESTIGATING THE SYMPTOMS OF DEPRESSION AND THE MECHANISMS OF ACTION OF ANTIDEPRESSANT DRUGS

Paul Willner and Richard Muscat

Department of Psychology, City of London Polytechnic, Old Castle St, London E1 7NT, United Kingdom

SUMMARY: The list of putative animal models of depression is now quite extensive. The prominent behavioural features of animal models of depression are: decreased or increased locomotor activity; decreased motivation; decreased social contact; impulsivity; and decreased sensitivity to rewards (anhedonia). Of these symptoms, only anhedonia realistically models a core symptom of depression. Most models feature only a single behavioural symptom. The two psychological procedures most frequently used to generate animal models of depression are stress and social isolation. In rodents, stress (typically inescapable footshock) causes a variety of antidepressant-reversible behavioural impairments, including anhedonia. Models of stress-induced anhedonia are reviewed, and evidence is presented that anhedonia may be induced by chronic exposure to very mild stressors. This procedure causes a decrease in the consumption of rewarding sucrose solutions and the failure of a variety of rewards to support conditioned place preferences. These impairments reflect a dysfunction of dopamine transmission in the nucleus accumbens; antidepressants act in this model by potentiating transmission at dopamine synapses. The effects of social isolation have usually been studied ethologically. However, social isolation also impairs performance in a test of cooperative social behaviour. This impairment, and its reversal by antidepressants, appears to involve serotonergic rather than dopaminergic mechanisms. This dissociation may have important clinical implications.

ANIMAL MODELS OF DEPRESSION

From its early beginnings in the reserpine reversal, amphetamine potentiation and muricide (mouse killing) tests, the list of animal models of depression has grown to include upwards of twenty experimental procedures in current use (see Table 2, and Willner, 1984, 1990a). Their common feature is that behavioural abnormalities are reversed by antidepressant drugs. In other respects, animal models of depression vary widely in the means of inducing abnormal behaviour, in the aspects of behaviour chosen for study, and in the time course of antidepressant action.

Animal models of depression are used for two distinct purposes: as screening tests in the context of antidepressant discovery and development programmes, and as simulations within which to study aspects of depression (see Willner, 1990b). It is important to recognize that these two uses have different and to some extent conflicting requirements, and therefore that a model developed for one purpose may be unsuitable for the other. The most obvious area of conflict is the time course of antidepressant action. Screening tests are subject to a number of logistical imperatives, such as ease of use, high capacity, and speed. For all of these reasons, it is highly desirable that a screening test responds acutely to antidepressant treatment, with a week of treatment representing an effective upper limit. However, a simulation aims to mimic aspects of the clinical situation, and from this perspective, an early response to antidepressant treatment is viewed with suspicion, particularly when low doses are used. Thus, a particular time course of antidepressant action may be a desirable or an undesirable feature of a particular model, depending upon the purpose for which it is being used.

Table 1 Diagnosis of major depression (DSM-III, 1980)

Requires the presence
- (a) for at least two weeks of
- (b) one core symptom and
- (c) four subsidiary symptoms (or three with both core symptoms)

Core symptoms:
1. Dysphoric mood
2. Loss of interest* or pleasure*

Subsidiary symptoms:
3. Psychomotor retardation* or agitation*
4. Feelings of worthlessness, self reproach or excessive or inappropriate guilt
5. Suicidal ideation
6. Decreased ability to concentrate*
7. Loss of energy*
8. Decreased sex drive*
9. Sleep disturbance*
10. Appetite disturbance*

* = **Could in principle be modelled in animals**

The present paper examines selected aspects of the use of animal models to investigate depression and the mechanisms of therapeutic action of antidepressant drugs. For this purpose, the response to antidepressant drugs is insufficient to define an animal model of depression, since antidepressants are active in at least some animal models of anxiety, following chronic treatment (Bodnoff et al., 1988; Fontana et al., 1989). An important additional requirement, therefore, is that the behavioural features of the model bear a realistic resemblance to the symptoms of depression (see Abramson & Seligman, 1978; Willner, 1990a,b).

SYMPTOMS OF DEPRESSION

Depression is a complex psychological disorder, which in the DSM-III diagnostic system (American Psychiatric Association, 1980) is defined by the presence of at least one core symptom combined with a number of additional subsidiary symptoms (see Table 1). It is clear that certain of the symptoms of depression could in principle be modelled in animals, though equally clearly, others could not. Table 2 provides an impressionistic summary of the major behavioural features of animal models of depression. The list of symptoms surveyed is certainly not exhaustive: biological symptoms such as neuroendocrine changes are omitted, and some of the models involve more complex behavioural changes than are included in the Table (for example, deficits of associative learning and/or attention in the learned helplessness model: see Minor et al., 1988; or postural changes in isolated primates: see Suomi, 1976). Nevertheless, despite this qualification, Table 2 serves to illustrate that most current models provide at best a superficial reflection of clinical symptomatology.

The behaviour most frequently studied in animal models is a decrease in locomotor activity. However, while it is true that psychomotor changes are a central feature of severe depression (Nelson & Charney, 1981), it can not be simply assumed that retardation and agitation are equivalent to decreases or increases in locomotor activity. Slowness of thinking is at least as important a feature of psychomotor retardation as a decrease in physical activity, and psychomotor agitation involves such features as hand-wringing and outbursts of complaining, in addition to a restless increase in locomotion. With the exception of the primate separation model, the degree of resemblance between the changes in locomotor activity observed in animal models of depression and the psychomotor features of depression is questionable.

In three of the models (see Table 2), the model could be described as demonstrating a decrease in motivation or persistence, rather than, or in addition to, a decrease in locomotion. This kind of behaviour may have some relevance in relation to the symptom of 'loss of interest'. However, in only one of these models, learned helplessness, have behavioural studies been carried out that support this interpretation (Maier & Seligman, 1976).

Table 2 Some animal models of depression (see Willner, 1990a for references)

Major symptom(s) (defined below):	**D-LA**	**I-LA**	**D-MP**	**IMP**	**D-SC**	**ANH**
Stress models:						
Learned helplessness	+		+			+
'Behavioural despair' and variants (eg. tail suspension)			+			
Failure to adapt to stress	+					
Chronic unpredictable stress	+					+
Chronic unpredictable mild stress						+
Amphetamine withdrawal	+					+
Separation models:						
Primate separation models (infant and adolescent)	+				+	
Distress calling in isolated chicks			+			
Separation in pair-bonded hamsters	+				+	
Social isolation in rats		+			+	
Pharmacological models:						
Reserpine reversal	+					
Amphetamine potentiation	+					
5-HTP-induced behavioural depression	+					
Miscellaneous:						
Olfactory bulbectomy		+		?		
Muricide				?		
Waiting behaviour				+		
Differential reinforcement of low rate				+		
Circadian rhythm models	+	+				

D-LA = decreased locomotor activity; **I-LA** = increased locomotor activity; **D-MP** = decreased motivation/persistence; **IMP** = impulsivity; **D-SC** = decreased social contact; **ANH** = anhedonia (decreased response to rewards)

Two further behaviours, characterized in Table 2 as impulsivity and decreased social contact, do not feature prominently among the diagnostic criteria for depression, perhaps because they have low specificity for depression as compared to a range of other psychiatric disorders. A variety of social factors have been described that powerfully influence vulnerability to, and recovery from, depression. Two of these, the lack of social support (Brown & Harris, 1978) and social skill (Lewinohn, 1974) could be said to involve a decrease in social contact. However, decreased social contact appears among the DSM-III criteria for depression as an aspect of 'loss of interest', rather than as a symptom in its own right. Similarly with impulsivity: This concept is used increasingly by basic neuroscientists to organize the behavioural literature on the functions of the forebrain serotonin (5-HT) projections (Soubrié, 1986), and may lead to a reconceptualization of the relationship between depression and other disorders in which 5-HT is implicated (see Willner, 1989a); at present, however, viewed as a symptom of depression, impulsivity is at best a minor aspect of psychomotor agitation.

The final symptom shown in Table 2, anhedonia, is somewhat different. Anhedonia, the decreased ability to experience pleasure (Fawcett et al., 1983), is one of the core symptoms of depression, and defines the endogenous subtype (melancholia). Anhedonia has been conceptualized as a decrease in the sensitivity of brain reward mechanisms (Klein, 1974), and as such may realistically be modelled in animals as a decrease in sensitivity to rewards. In four of the models surveyed, animals display impairments of rewarded behaviour, following exposure to stressor of various kinds; in three of the four cases, there is evidence that a decreased sensitivity to reward is responsible for these impairments (see Willner et al., 1990a). These models merit examination in some detail, since they may provide useful information about the clinical condition.

MODELS OF ANHEDONIA

Withdrawal from psychomotor stimulants: Until recently, and since its discovery by Olds and Milner in 1954, intracranial self stimulation (ICSS) has been the major paradigm used for the investigation of reward processes. In this procedure, electrical stimulation of the brain is used to reinforce behaviour, typically, lever pressing. ICSS may be elicited from many parts of the brain, including most regions of the limbic system. Although early studies suggested that ICSS had unusual properties compared to natural rewards (for example, rapid extinction), it was later recognized that these properties derive from differences in the experimental procedures typically employed (such as the delay of reinforcement); when such extraneous factors are equated. ICSS appears very comparable to a high incentive natural reward presented under conditions of low drive (eg. Gibson et al., 1965).

These parallels, together with the observation that responding for brain stimulation reward is influenced by many of the factors that control responding for natural rewards, have justified the assumption that the ICSS electrode stimulates directly the neural substrates that are activated indirectly by natural rewards (Hoebel, 1976). A degree of caution is required, since although people implanted with self-stimulation electrodes report a variety of pleasurable sensations, they also report other reasons for stimulating, such as curiosity (Atrens, 1984; Valenstein, 1973). Nevertheless, the commonality of anatomical substrate between ICSS and other types of reward supports the use of this procedure as an animal model of hedonic behaviour (see Koob, 1989; Wise, 1989).

A number of studies have reported that responding for ICSS was reduced in the days following withdrawal from chronic amphetamine treatment. In these studies, amphetamine was administered to rats for between 4 and 10 days, typically using several administrations each day, at increasing doses (Barrett & White, 1980; Kokkinidis & Zacharko, 1980; Leith & Barrett, 1976; Simpson & Annau, 1977). The threshold for ICSS was elevated following amphetamine withdrawal, confirming that the rate reduction reflects a subsensitivity to brain stimulation reward rather than a depression of motor activity (Cassens et al., 1981; Leith & Barrett, 1980). In a single pharmacological study, this effect was reversed by chronic treatment with imipramine or amitriptyline treatment (Kokkinidis et al., 1980).

In a variant of this procedure, animals self-administer cocaine, rather than having amphetamine administered to them (Koob, 1989). In these experiments, the threshold for brain stimulation reward, administered through electrodes in the posterior lateral hypothalamus, was obtained using a discrete trial procedure (Kornetsky & Esposito, 1981), which is sensitive to changes in reward value, but not to changes in motor performance (Markou et al., 1989). Following 24h of cocaine self-administration, ICSS thresholds were elevated for several hours (Koob, 1989). Acute administration of the dopamine (DA) agonist bromocryptine restored ICSS thresholds to normal (Markou & Koob, 1989). Conventional antidepressants have not been tested in this paradigm.

Uncontrollable shock: Responding for brain stimulation reward is also impaired, in mice, following exposure to inescapable electric footshock; this effect is not seen following exposure to comparable escapable shock. These studies have mainly used the rate of ICSS as the dependent variable, which is susceptible to a variety of non-specific influences. However, a specific impairment of sensitivity to reward is indicated by the observation that the effects of inescapable shock are anatomically specific. ICSS elicited from the ventral tegmental area (the origin of the mesolimbic dopamine (DA) projection), or from the nucleus accumbens or frontal cortex (two of its terminal fields) was suppressed by inescapable shock; however, ICSS elicited from the substantia nigra (the origin of the nigrostriatal DA projection) was unaffected. Furthermore, within the ventral tegmental area, ICSS from certain regions was affected, while comparable rates of ICSS elicited from other regions were not (Zacharko et al., 1983, 1987; Zacharko & Anisman, 1990). Normal sensitivity to brain stimulation reward was restored by chronic, but not acute, treatment with tricyclic antidepressants or nomifensine, administered prophylactically or following exposure to inescapable shock (Zacharko et al., 1984; Zacharko & Anisman, 1990).

The decrease in ICSS following inescapable shock was long-lasting, and showed little recovery during the course of a week. However, this long-term impairment was only found if the animals were tested in the immediate aftermath of stress; otherwise the effect dissipated rapidly (Zacharko et al, 1983). This effect could potentially be of importance for understanding how the risk of depression is elevated for several months in the aftermath of a stressful life event. Another important observation in this model is that uncontrollable electric shock has variable behavioural effects (most of which are antidepressant-reversible) in different inbred mouse strains. To take an extreme example, in the C57BL/67 strain, exposure to inescapable shock severely impaired subsequent learning to escape shock, but had no effect on responding for brain stimulation reward, while the DBA/2J strain showed exactly the opposite pattern of deficits (Shanks & Anisman, 1988; Zacharko et al., 1987). These studies may provide a starting point for investigation of the physiological mechanisms underlying individual differences in responses to stress.

Chronic mild unpredictable stress: While the stimulant withdrawal model clearly has considerable face validity in relation to stimulant withdrawal depressions (Watson et al., 1972), its relevance to other types of depression is debatable; a similar question may be raised in relation to the use of severe electric shocks. We have developed a third stress-anhedonia paradigm, which involves chronic exposure to a mild stress regime that arguably provides a relatively realistic analogue of the stress of daily living. Furthermore, this model uses natural rewards rather than ICSS, which, again, provides a more realistic parallel to clinical anhedonia.

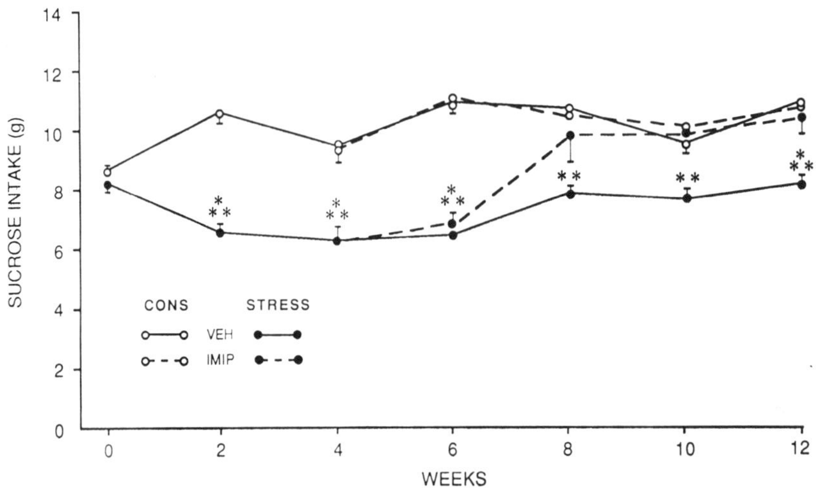

Fig. 1 Reduction of sucrose consumption by chronic mild stress and its restoration by imipramine. Stress (closed circles) was administered for a total of 12 weeks. Imipramine treatment (5 mg/kg daily) commenced after 4 weeks of stress (dotted lines) Imipramine restored normal behaviour, after 4 weeks of treatment, in the stressed animals, but had no effect in non-stressed controls. $^{**}p<0.01$, $^{***}p<0.001$ (From Willner et al., 1990a).

These studies involve the sequential presentation of a variety of low grade, mild stressors: typical elements include overnight illumination, tilting of the cage, periods of food or water deprivation, or changes of cage mates. Rats subjected chronically to this regime reduce their consumption of, and preference for, highly rewarding weak solutions of sucrose and saccharin, measured periodically in brief tests. The impairment of sucrose consumption in stressed animals is restored to normal following several weeks of treatment with the tricyclic antidepressants DMI, imipramine or amitriptyline. Induction of the deficit requires between one and four weeks of exposure to stress, and reversal of the deficit by antidepressants requires between two and five weeks of treatment (Fig. 1). Following the cessation of stress, behaviour remains abnormal, in untreated animals, for at least two weeks (Muscat et al., 1987, 1990; Sampson et al., 1990; Willner et al., 1987, 1990a).

In order to investigate whether the reduction of sucrose consumption by chronic mild stress reflects a decrease in sensitivity to rewards, we have examined the effects of chronic mild stress in the place conditioning paradigm, which has been used extensively to study the rewarding properties of self-administered drugs (Carr et al., 1989), and is also sensitive to natural reinforcers such as food or sucrose (Spyraki et al., 1982; White & Carr, 1985). In this paradigm, rewards are administered in one of two distinctive environments (in our laboratory, a white chamber with a smooth plastic floor and a black chamber with a wire mesh floor). The animal is first allowed to explore both chambers, and its initial preferences are assessed; under our conditions, animals display a strong preference for the black side. Then, on alternate trials, animals are either administered the reward and confined in one chamber, or not administered the reward and confined in the alternative chamber. Following a series of conditioning trials (in our procedure, four on each side), the strength of the reward is assessed by measuring the increase in the amount of time the animal chooses to spend on the reward-associated side, in the absence of the primary reward. Following 2-7 weeks of chronic mild stress, rats were subsensitive, in this paradigm, to the rewarding properties of 0.7% sucrose, 34% sucrose, food pellets, and amphetamine (0.5 or 1.0mg/kg) (Fig. 2).

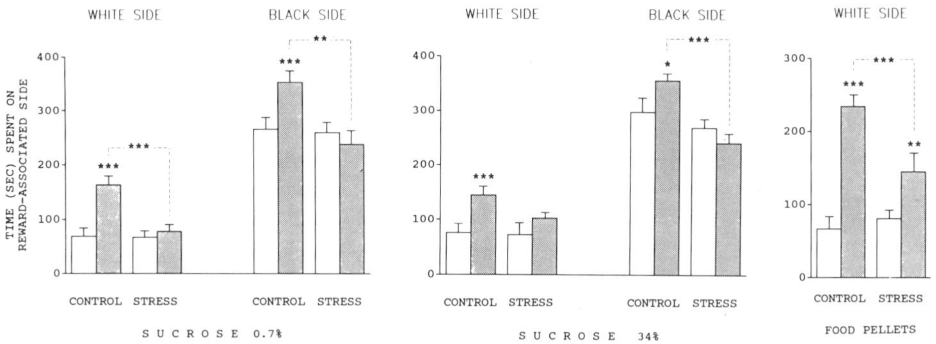

Fig. 2 Place preference conditioning following 2-4 weeks of unpredictable mild stress. On each of 4 days, food-deprived animals were confined on one side of the chamber (white or black), where they were allowed to consume sucrose solutions (0.7 or 34% w/v) or food pellets; on alternate days they were simply confined on the other side of the chamber. White bars = time spent in reward-associated side prior to conditioning; hatched bars = time spent in reward-associated side following conditioning. Place conditioning was abolished (sucrose) or substantially attenuated (food) in stressed animals. ** $p<0.01$; *** $p<0.001$. (Willner, Papp & Muscat, submitted).

The effects of chronic mild stress on sensitivity to reward closely resemble those of DA antagonists (see Willner et al., 1990a), and this parallel also extends more widely: for example, both stress and the DA antagonist pimozide increased the consumption of a very sweet wet mash, while decreasing the rate of food consumption (Muscat et al.,

1989). In light of the extensive evidence implicating the mesolimbic DA system in reward (see Liebman & Cooper, 1989; Willner & Scheel-Kruger, 1990), the pronounced similarities between the effects of chronic mild stress and those of DA antagonists suggest strongly that chronic mild stress causes a functional antagonism of DA transmission in the nucleus accumbens. We have recently confirmed that administration of a DA antagonist to the nucleus accumbens has effects on sucrose consumption comparable to those seen in chronically stressed animals (Phillips et al., 1990).

Somewhat surprisingly, the results of neurochemical studies present a rather different picture. In two experiments, we found that the levels of DA and its three major metabolites in the nucleus accumbens were substantially increased following 3 or 7 weeks of exposure to chronic mild stress. These changes were not seen in the caudate nucleus or septal area (Willner et al., 1990a,b). In subsequent experiments, in collaboration with Drs. J.Stamford and Z.Kruk, we have confirmed using fast cyclic voltammetry that chronic mild stress increases both the release of DA in response to electrical stimulation of the medial forebrain bundle, and the size of the releasable pool of DA; again these effects were seen in the nucleus accumbens but not in the caudate nucleus (in preparation). There are a number of ways in which these data may potentially be reconciled with the functional picture of DA impairment following chronic mild stress (see Willner et al., 1990a), but the precise mechanism remains to be established. Nevertheless, these studies confirm the anatomical specificity of the neurochemical abnormalities, which were observed only within the nucleus accumbens.

PATTERNS OF DEPRESSIVE SYMPTOMATOLOGY

The availability of a range of models of different kinds affords the opportunity to examine patterns of depressive symptoms across models, and to ask whether different types of model relate to different aspects of depression. The usefulness of this approach is to some extent limited by the reservations expressed earlier concerning the extent to which the behaviours typically observed in animal models correspond to the clinical symptoms of depression. However, it is important not to overinterpret these limitations. As with any other type of model, the purpose of constructing an animal model of psychopathology is to develop testable hypotheses about the condition modelled. If the validity of the model is relatively low, then hypotheses deriving from the model should be treated with greater scepticism, but they may, nonetheless, turn out to be correct (see Willner, 1990b).

Despite their common response to antidepressants, a survey of animal models of depression reveals at least two distinct patterns of symptoms. Animals subjected to uncontrollable stress appear to resemble retarded endogenous depressives in their symptomatology, which includes, in addition to anhedonia, hypoactivity, decreased food intake and weight loss, and decreased aggression. However, certain manipulations, such as olfactory bulbectomy or social isolation, increase aggression and locomotor activity (see Table 2, and Willner, 1989a). Social isolation has also been found to cause a deterioration in cooperative social behaviour in rats. This observation derives from a paradigm originally described by Berger and Schuster (1982), in which pairs of rats are trained to run in a shuttle box for food or water rewards, which they receive only if they run in close physical proximity, but not if they fail to coordinate their performance. This behaviour is readily learned by socially-housed animals, but not by singly-housed animals. We have recently confirmed that cooperative performance deteriorates if, following training, the animals are housed singly for four to six weeks; this impairment could be reversed by chronic treatment with imipramine or fluoxetine (Willner et al., 1989). While anhedonia is a core symptom in depression (see Table 1), social dysfunction is found in a wide range of psychopathologies: as, indeed, are the sequelae of social deprivation, which this paradigm may model (see Willner, 1989a).

There is evidence that stress-induced anhedonia and the disruption of social cooperation by isolated housing involve different neurobiological substrates. Thus, the effect of stress on consumption of a sweet reward is mimicked by DA antagonists such as pimozide (Towell et al., 1987) but not by the 5-HT antagonist metergoline (Muscat et al., 1990). However, the opposite is true of cooperative social behaviour, which is disrupted by metergoline but not by pimozide (Willner et al., 1989). This suggests that whereas a dysfunction of DA may mediate the anhedonia that defines melancholia, the deleterious effects of social deprivation may involve a dysfunction of 5-HT. These simple deductions give rise to a number of readily testable clinical hypotheses: for example, that social deprivation may be implicated in the range of disorders in which the clinical evidence suggests impairment of serotonergic neurotransmission (Willner, 1989a).

MECHANISMS OF ANTIDEPRESSANT ACTION

Although animal models of depression may in principle provide valuable insights into the neurobiology and psychopathology of affective disorders, they are currently making their greatest contribution as tools for investigating the mechanisms of action of antidepressant drugs.

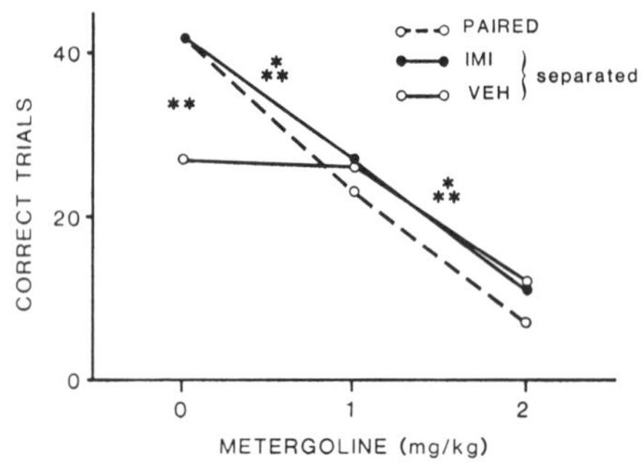

Fig. 3 Effects of metergoline on performance in the paired running task of pair-housed animals, and of singly-housed animals treated prophylactically with imipramine (IMI) or vehicle (VEH) for 7 weeks. 1 mg/kg metergoline reduced performance in group-housed and imipramine-treated singly-housed animals to the level of untreated singly-housed animals. A higher dose further reduced performance in all groups. ** $p<0.01$, *** $p<0.001$ (From Willner et al., 1989).

Following chronic administration, antidepressants cause multiple changes in the functioning of a variety of neuronal systems (reviewed by Willner, 1989b), but the functional significance of these effects is uncertain. Indeed, a prior question is which if any of the changes described in normal animals we might expect to occur in the depressed brain, bearing in mind the failure of antidepressants to elevate mood significantly in non-depressed individuals. This type of problem can only be addressed using animal models of depression that maintain abnormal behaviour over the long periods of time necessary for chronic drug treatment. There are few such models, and those that exist are not widely used. Consequently, there is as yet very little information about the biochemical consequences of chronic antidepressant administration in conditions that simulate the conditions under which the drugs are used clinically.

Initial studies using chronic and functionally relevant models indicate that the mechanisms by which antidepressants normalize behaviour depend upon the behavioural model used. In our experiments, metergoline reversed the improvement by chronic imipramine of cooperative social behaviour in singly-housed animals (Fig. 3: Willner et al., 1989). However, using the same dose of imipramine (5mg/kg) and a similar period of administration, metergoline did not antagonize the effect of imipramine on responsiveness to rewards in animals subjected chronically to mild stress. However, DA antagonists did reverse this effect of imipramine, and also of other tricyclics (Fig. 4: Muscat et al., 1990; Sampson et al., 1990).

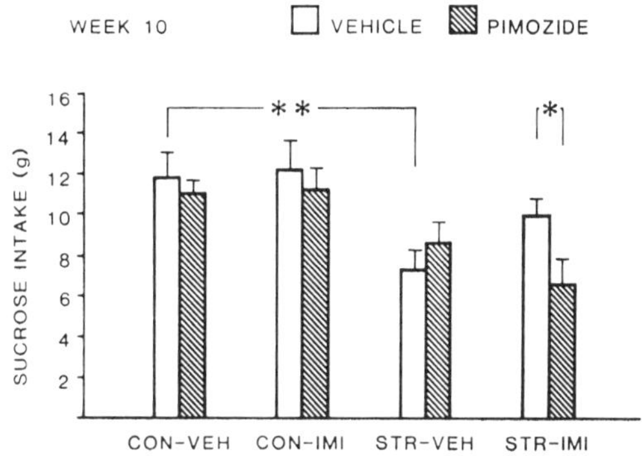

Fig. 4 Effect of pimozide on sucrose consumption in stressed (STR) and non-stressed (CON) animals treated chronically with imipramine (IMI) or vehicle (VEH). Stress was administered for a total of 10 weeks, and imipramine from week 5 onwards. Pimozide specifically reduced sucrose consumption in the imipramine-treated stressed animals. * $p<0.05$; ** $p<0.01$. (From Muscat et al., 1990).

Thus, the effects of chronic imipramine appear to be mediated by an increase in DA receptor sensitivity in the chronic stress/anhedonia model, but by an increase in 5-HT receptor sensitivity in the isolation/social dysfunction model. These conclusions are consistent with our understanding of the neuropharmacological basis of these two behaviours (see above).

There are indications of this differentiation in earlier studies. For example, following one week of treatment, the effect of tricyclic antidepressants in the forced swim test (Porsolt, 1981) may be blocked by catecholamine antagonists (Borsini & Meli, 1988), while their effect in olfactory bulbectomized animals (Cairncross et al., 1979) was reversed by metergoline (Broekkamp et al., 1980). A preferential role of 5-HT in the olfactory bulbectomy model is also indicated by the observation that 5-HT uptake inhibitors work acutely in this model, while chronic treatment is needed with tricyclics (Broekkamp et al., 1980); conversely, in the forced swim test, tricyclics work subacutely but 5-HT uptake inhibitors are ineffective (Borsini & Meli, 1988).

Very little progress has been made in demonstrating clinically that antidepressants with a specific pharmacological profile act selectively in particular patient groups; and it has been widely assumed that this lack of clinical selectivity arises from the fact that after chronic treatment, pharmacological specificity is also lost. It is now becoming apparent that the different pharmacological actions of antidepressants that are present following chronic treatment have different functional consequences. This finding may have far-reaching implications.

Acknowledgements: The studies reported in this chapter were partly supported by the Medical Research council of Great Britain.

REFERENCES

Abramson, L.Y., Seligman, M.E.P., Teasdale, J.D. (1978) J. Abnorm. Psychol. 87, 49-74.

American Psychiatric Association (1980) DSM-III: Diagnostic and Statistical Manual of Psychiatric Disorders. APA, Washington.

Atrens, D.M. (1984) In: Animal Models in Psychopathology (N.S.Bond, Ed), Academic, Sydney, pp 227-256.

Barrett, R.J., White, D.K. (1980) Pharmacol. Biochem. Behav. 13, 555-559.

Berger, B.D., Schuster, R. (1982) In: Behavioural Models and the Analysis of Drug Action (M.Y.Spiegelstein and A.Levy, Eds), Elsevier, Amsterdam. pp 415-428.

Bodnoff, S.R., Suranyi-Codotte, B., Aitken, D.H., Quirion, R., Meaney, M.Y. (1988) Psychopharmacology 95, 298-302.

Borsini, F., Meli, A. (1988) Psychopharmacology 94, 147-160.

Brown, G.W., Harris, T. (1978) Social Origins of Depression. Tavistock, London.

Broekkamp, C.L., Garrigou, D., Lloyd, K.G. (1980) Pharmacol. Bioch. Behav. 13, 643-646.

Cairncross, K.D., Cox, B., Forster, C., Wren, A.F. (1979) Psychoneuroendocrinology 4, 253-272.

Carr, G.D., Fibiger, H.C., Phillips, A.G. (1989) In: The Neuropharmacological Basis of Reward (J.M.Liebman and S.J.Cooper, Eds), Oxford University Press, Oxford, pp 264-319.

Cassens, G.P., Actor, C., Kling, M., Schildkraut, J.J. (1981) Psychopharmacology 73, 318-322.

Fawcett, J., Clark, D.C., Scheftner, W.A. and Gibbons, R.D. (1983) Arch. Gen. Psychiatr. 40, 79-84.

Fontana, D.J., Carbary, T.J., Commisaris, R.L. (1989) Psychopharmacology 98, 157-162.

Gibson, W.E., Reid, L.D., Sakai, M., Porter, P.B. (1965) Science 148, 1357-1358.

Hoebel, B.G. (1976) In: Brain Stimulation Reward (A.Wauquier and E.Rolls, Eds), American Elsevier, New York, pp 331-372.

Klein, D.F. (1974) Arch. Gen. Psychiatr. 31, 447-454.

Kokkinidis, L., Zacharko, R.M. (1980) Psychopharmacology 68, 73-76.

Kokkinidis, L., Zacharko, R.M., Predy, P.A. (1980) Pharmacol. Biochem. Behav. 13, 379-383.

Koob, G.F. (1989) In: Animal Models of Depression (G.F.Koob, C.L.Ehlers and D.J.Kupfer, Eds), Birkhauser, Boston, pp 162-183.

Kornetsky, C., Esposito, R.U. (1981) Brain Res. 209, 496-500.
Leith, N.J., Barrett, R.J. (1976) Psychopharmacology 46, 19-25.
Lewinsohn, P.M. (1974) In: The Psychology of Depression: Contemporary Theory and Research (R.J.Friedman and M.M.Katz, Eds), Winston/Wiley, New York, pp 157-185.
Liebman, J.M., Cooper, S.J. (1989) The Neuropharmacological Basis of Reward. Oxford University Press, Oxford.
Maier, S.F., Seligman, M.E.P. (1976) J. Exp. Psychol. Gen. 1, 3-46.
Markou, A., Koob, G.F. (1989) American College of Neuropsychopharmacology, Abstracts, 157.
Minor, T.R., Pelleymounter, M.A., Maier, S.F. (1988) Psychobiology 16, 135-145.
Muscat, R., Towell, A., Willner, P. (1987) Psychopharmacology 94, 545-550.
Muscat, R., Sampson, D., Phillips, G., Willner, P. (1989) Behav. Pharmacol. 1, Suppl.1, 55.
Muscat, R., Sampson, D., Willner, P. (1990) Biol. Psychiatr., 28, 223-230.
Nelson, J.C., Charney, D.S. (1981) Arch. Gen. Psychiatr. 138, 1-13.
Olds, J., Milner, P. (1954) J. Comp. Physiol. Psychol. 47, 419-427.
Phillips, G., Willner, P., Muscat, R. (1990) Psychopharmacology 101, Suppl., S43.
Porsolt, R.D. (1981) In: Antidepressants: Neurochemical, Behavioural and Clinical Perspectives (S.J.Enna, J.B.Malick and E.Richelson, Eds), Raven, New York, pp 121-139.
Sampson, D., Muscat, R., Willner, P. (1990) Psychopharmacology. (In press).
Shanks, N., Anisman, H. (1988) Behav. Neurosci. 102, 894-905.
Simpson, D.M. and Annau, Z. (1977) Pharmacol. Biochem. Behav. 7, 59-64.
Soubrie, P. (1986) Behav. Brain Sci. 9, 319-364.
Spyraki, C., Fibiger, H.C., Phillips, A.G. (1982) Psychopharmacology 77, 379-382.
Suomi, S.J. (1976) In: Animal Models in Human Psychobiology (G.Serban and A.Kling, Eds), Plenum, New York, pp 9-26.
Towell, A., Muscat, R., Willner, P. (1987) Psychopharmacology 92, 262-264.
Valenstein, E.S. (1973) Brain Control: A Critical Examination of Brain Stimulation and Psychosurgery. Wiley, New York.
Watson, R., Hartman, E., Schildkraut, J.J. (1972) Am. J. Psychiatr. 129, 263-269.
White, N.M., Carr, G.D. (1985) Pharmacol. Biochem. Behav. 23, 37-42.
Willner, P. (1984) Psychopharmacology 83, 1-16.
Willner, P. (1989a) In: Behavioural Pharmacology of 5-HT (P.Bevan, A.R.Cools and T.Archer, Eds), Erlbaum, New York, pp 157-178.
Willner, P. (1989b) In: Psychoactive Drugs: Tolerance and Sensitization (M.W.Emmett-Oglesby and A.J.Goudie, Eds), Humana Press, Clifton, N.J., pp 407-459.
Willner, P. (1990a) Pharmacol. Ther. 45, 425-455
Willner, P. (1990b) In: Behavioural Models in Psychopharmacology: Theoretical, Industrial and Clinical Perspectives (P.Willner, Ed), Cambridge University Press. (In press).
Willner, P., Scheel-Kruger, J. (Eds), (1990) The Mesolimbic Dopamine System: From Motivation to Action. Cambridge University Press, Cambridge.
Willner, P., Towell, A., Sampson, D., Muscat, R., Sophokleous, S. (1987) Psychopharmacology 93, 358-364.
Willner, P., Sampson, D., Phillips, G., Fichera, R., Foxlow, P., Muscat, R. (1989) Behav. Pharmacol. 1, 85-90.
Willner, P., Sampson, D., Papp, P., Phillips, G., Muscat, R. (1990a) In: Animal Models of Psychiatric Disorders, Vol.3 (P.Simon, P.Soubrie and D.Widlocher, Eds), Karger, Basel, pp 71-91.
Willner, P., Golembiowska, K., Klimek, V., Muscat, R. (1990b) Psychobiology. (In press).

Wise, R.A. (1989) In: The Neuropharmacological Basis of Reward (J.M.Liebman and S.J.Cooper, Eds), Oxford University Press, Oxford, pp 377-424.
Zacharko, R.M., Anisman, H. (1990) In: The Mesolimbic Dopamine System: From Motivation to Action (P.Willner, J.Scheel-Kruger, Eds), Wiley, Chichester, pp 411-442.
Zacharko, R.M., Bowers, W.J., Kokkinidis, L., Anisman, H. (1983) Behav. Brain Res. 9, 129-141.
Zacharko, R.M., Lalonde, G.T., Kasian, M., Anisman, H. (1987) Brain Res. 426, 164-168.

Animal Models in Psychopharmacology
Advances in Pharmacological Sciences

NEUROENDOCRINE EVIDENCE FOR HYPERSENSITIVITY IN SEROTONERGIC NEURONAL SYSTEM AFTER PSYCHOSOCIAL STRESS OF DEFEAT

S.M. Korte, J. Smit, G.A.H. Bouws, J.M. Koolhaas and B. Bohus

Department of Animal Physiology, University of Groningen, Biological Centre, 9750 AA Haren, The Netherlands

SUMMARY: The effect of the 5-HT_{1A} agonist ipsapirone on behaviour and plasma corticosterone levels was studied in the same male Wistar rats before and after psychosocial stress of defeat. Ipsapirone in a dose of 5 mg/kg i.p. failed to affect corticosterone levels before social stress whereas after defeat the hormone response was much higher than in drug-free controls. The exaggerated stress response and particularly the presence of high post-defeat corticosterone levels in ipsapirone treated rats already before the exposure to the dominant rat lead to the hypothesis that postsynaptic 5-HT_{1A} receptor hypersensitivity develops after stress of defeat.

INTRODUCTION

Anxiolytic or anxiogenic actions of drugs are usually deduced from changes of the behaviour in the various test situations. Studying components of social behavioural patterns (e.g. postdefeat) even under seminatural (laboratory) conditions in the rat may serve as an additional model of anxiety syndrome in man living under social conditions. We selected a psycho-social stress paradigm in which the experimental rat was re-exposed to a caged dominant male rat one day after defeat. Anxiety or reduction of anxiety are usually accompanied by physiological and neuroendocrine changes. Being interested in behavioural and neuroendocrine interactions in relation to stress and anxiety, stress or anxiety reducing interventions we used the novel anxiolytic 5-HT_{1A} agonist ipsapirone in experiments in a psychosocial stress paradigm. The dose of the drug used (5 mg/kg given i.p.) appeared to be anxiolytic in a number of former studies (e.g. Dourish, 1987; Korte and Bohus, 1990). Plasma corticosterone levels were determined and used as neuroendocrine measure.

MATERIALS AND METHODS

Animals: Male Wistar rats weighing 290-340 g at the beginning of the experiments were used. They were housed individually in clear plexiglas cages (25 x 25 x 30 cm) with a reversed light-dark cycle (light on between 21.30 - 09.30 h). The experiments were carried out in the first half of the dark period. All animals were provided with a silicon heart catheter for blood sampling (Steffens, 1969). In order to study the consequences of psychosocial stress, male Tryon Maze Dull S3 rats were used because they display high rate of aggression and dominance in a neutral environment.

Procedure: The post-defeat psychosocial stress model (Koolhaas et al., 1986) was used to investigate stress related behavioural and hormonal responses. Shortly, a neutral observation cage (85 x 60 x50 cm) with a front glass wall was used. The S3 rats enclosed in a small wire mesh cage was placed in the neutral observation cage which prevented direct physical interactions. Three 5 min exposures of the experimental rats to the neutral cage with caged dominant on day 1 and 2 served to habituate. On day 3, the rats were tested with an encaged S3 rat. Injections of ipsapirone or saline were given i.p. 30 min prior to testing the rats. Ipsapirone (TVX Q 7821, Troponwerke, Köln, FRG) was dissolved in saline in a dose of 5 mg/kg and a constant volume of 2.0 ml/kg. Blood samples were taken during confinement in the waiting cage, during 5 min social-confrontation with the encaged dominant and afterwards in the home-cage. After a wash out period of six days all animals were exposed to the home cage of the same S3 rat for 10 minutes (day 9). The experimental rats were invariably threatened and/or attacked by the S3 rat and displayed submissive postures and flight, thereby considered as being defeated. The following day (day 10) the rats were divided into two equally sized groups with similar pre-defeat treatment history. The groups were allocated to ipsapirone or saline treatment 30 minutes before testing them in the experimental room. At this post-defeat test, the psychosocial stress situation, the dominant rat was enclosed in a small wire mesh cage with the same blood sampling schedule as before the defeat. Plasma corticosterone was measured by means of reversed phase high performance liquid chromatography, as described earlier (Dawson et al., 1984). The behaviour of the experimental animals was continuously recorded by distinguishing immobility, rearing, exploring, sniffing cage and grooming as typical elements.

RESULTS

Behaviour: Ipsapirone has no influence upon the pre-defeat behaviour of the experimental rat. On the day after defeat saline treated rats spent more time on rearing and exploring and less on sniffing the cage of the dominant. Post-defeat administration of ipsapirone significantly increased mean time spent on immobility behaviour (17% vs. 5% in controls; $p<0.05$), whereas both time spent on sniffing the cage and rearing was diminished (29% vs. 45% and 8% vs. 16%; $p<0.05$, respectively).

Corticosterone: Ipsapirone in the dose of 5 mg/kg failed to have a significant effect on hormone levels before defeat (Fig. 1). On the day after defeat higher elevations of corticosterone were seen in the ipsapirone treated rats as compared to the defeated saline treated controls both in the home cage 30 min after injection and in the observation cage during exposure to the encaged dominant. The hormone levels remained elevated even after return to the home cage.

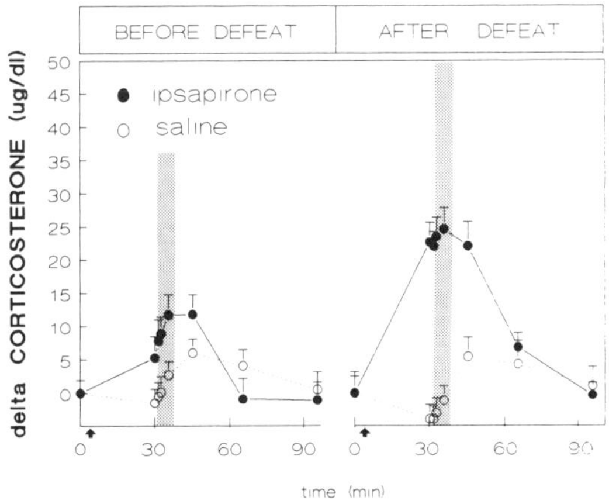

Fig. 1 The effects of ipsapirone (5 mg/kg,i.p.) on plasma levels of corticosterone in the home and experimental environments before and after defeat. Each point is the mean ± SEM of 11 to 12 rats. (↑:i.p.injection; o:saline; o:ipsapirone). MANOVA after defeat resulted in a significant effect of drug treatment, $F(1,20)=34.54$, $p<0.001$ and treatment by sampling time interaction, $F(7,14)=7.92$, $p=0.001$. Dunnett's test showed significant changes in hormone levels at t=30 (home cage), 31.5, 32.5 and 36.5 min (experimental room) and in the home cage at t=45 min.

DISCUSSION

Observations on the behaviour of the rats do not allow a simple conclusion with respect to anxiolytic-like properties of the given dose of ipsapirone. Increased immobility and decreased amount of time spend on sniffing the cage of the dominant by the drug treated rat following defeat suggest an anxiogenic action of ipsapirone whereas decreased rearing behaviour assumes an anxiolytic one. Ipsapirone in a dose of 5 mg/kg, i.p. has practically no effect on the plasma corticosterone level in rats before defeat, but leads to an exaggerated increase of corticosterone levels after defeat in the home cage and during the confrontation with the encaged dominant. This difference in the neuroendocrine measure may be a consequence of hypersensitivity of the serotonergic neuronal system. One may postulate that postsynaptic 5-HT_{1A} hypersensitivity develops after stress of defeat. This postsynaptic phenomenon may 'mask' the anxiolytic effect the drug. Ipsapirone and the related compounds buspirone and gepirone are supposed to reduce anxiety by inhibiting serotonergic impulse flow, an action thought to be mediated through agonistic activity at somato-dendritic 5-HT_{1A} autoreceptors on dorsal raphe neurons (Dourish et al., 1986). Postsynaptic stimulation on the other hand would facilitate 5-HT neuro-transmission and accounts for the release of ACTH (Gilbert et al., 1988) and consequent elevation of plasma corticosterone (Koenig et al., 1988). Therefore the exaggerated corticosterone response may be the consequence of a functional hypersensitivity in serotonergic neuronal systems. Changes in receptor density, a shift from low to high affinity state of the receptor or an increased functional coupling of 5-HT receptors to their second messengers are suggested as possible underlying mechanisms. This hypersensitivity may narrow the therapeutic dose-range of the serotonergic agonists. The occurrence of this hypersensitivity might be of particular interest in view of the underlying mechanisms in the development of anxiety and/or depression, wherein abnormal 5-HT function may have a causal or predisposing role (Kahn et al., 1988; Van de Kar, 1989).

REFERENCES

Dourish, C.T., Hutson, P.M., Curson, G. (1986) Trends Pharmacol. Sci. 7, 212-214.

Dourish, C.T. (1987) In: Brain 5-HT_{1A} receptors (C.T.Dourish, S.Åhlenius and P.H.Hutson, Eds), Ellis Harwood Ltd., Chichester, England, pp 261-277.

Gilbert, F., Dourish C.T., Brazell, C., McClue, S., Stahl, S.M. (1988) Psychoneuroendocrinology 13, 471-478.

Kahn, R.S., van Praag, H.M., Wetzler, S., Asnis, G.M., Barr, G. (1988) Biol. Psychiatry 23, 189-208.

Koenig, J.I., Meltzer, H.Y., Gudelsky, G.A. (1988) Pharmacol. Biochem. Behav. 29, 711-715.
Koolhaas, J.M., Fokkema, D.S., Bohus, B., Oortmerssen van, G.A. (1986) In: Biological and psychological factors in cardiovascular disease (T.H.Schmidt, T.M.Dembroski and G.Blümchen, Eds), Springer-Verlag Berlin Heidelberg, pp 517-526.
Korte, S.M., Bohus, B. (1990) Eur. J. Pharmacol. 18, 307-310.
Steffens, A.B. (1969) Physiol. Behav. 4, 883-836.
Van de Kar, L.D. (1989) Neurosc. Biobehav. Rev. 13, 237-246.

in the experiments presented here the effect of different serotonin agonists and antagonists were investigated using LH.

MATERIAL AND METHODS

Groups of at least 8 rats (200 g) were used for each dose level. The control groups consisted of 176 animals. The procedure was as described by Geoffroy et al. (1990): a 30 min. session of inescapable footshock, an interval of 4 days during which drugs were administered, and finally a shuttlebox test which consisted of 10 FR1 trials followed by 15 FR2 non-signalled two-way escape trials. The groups were injected with vehicle (control groups) or the compound 2 hours after exposure to a session of inescapable shock (IS+), or a no-shock control session (IS-). Vehicle or drug treatment continued for five days. The last day either the vehicle or the drug were given 2 hours before the escape test in the shuttlebox.

During the FR1 schedule the rats had to change side only once, and during the FR2 schedule the rats had to change side twice in order to terminate shock.

Statistical analyses of the behavioural data were carried out by the means of the Crunch program, using the non-parametric test Kruskal-Wallis followed by Mann-Whitney U-tests.

RESULTS

As shown in Fig. 1 imipramine has a dose-dependent effect on both FR1 and FR2 escape failures. Ipsapirone has no effect after intraperitoneal (i.p.) and subcutaneous (s.c.) administration. Buspirone increases rather than decreases the effect on escape behaviour whereas citalopram show a slight but significant effect but only at one dose (10mg). Ritanserin and GR 38032F (Fig. 2) show a small but not dose dependent effect.

During the year vehicle groups were pooled in "winter", "spring", "summer" and "autumn" (Fig. 3). The results show a significant differentiation between the four fractions. Within the fractions no difference was seen during the four seasons between IS+ FR1, IS+ FR2 and IS-FR1, whereas only the IS- FR2 results from spring was significantly different from the results in winter. For that reason we pooled IS+ control groups (n = 176).

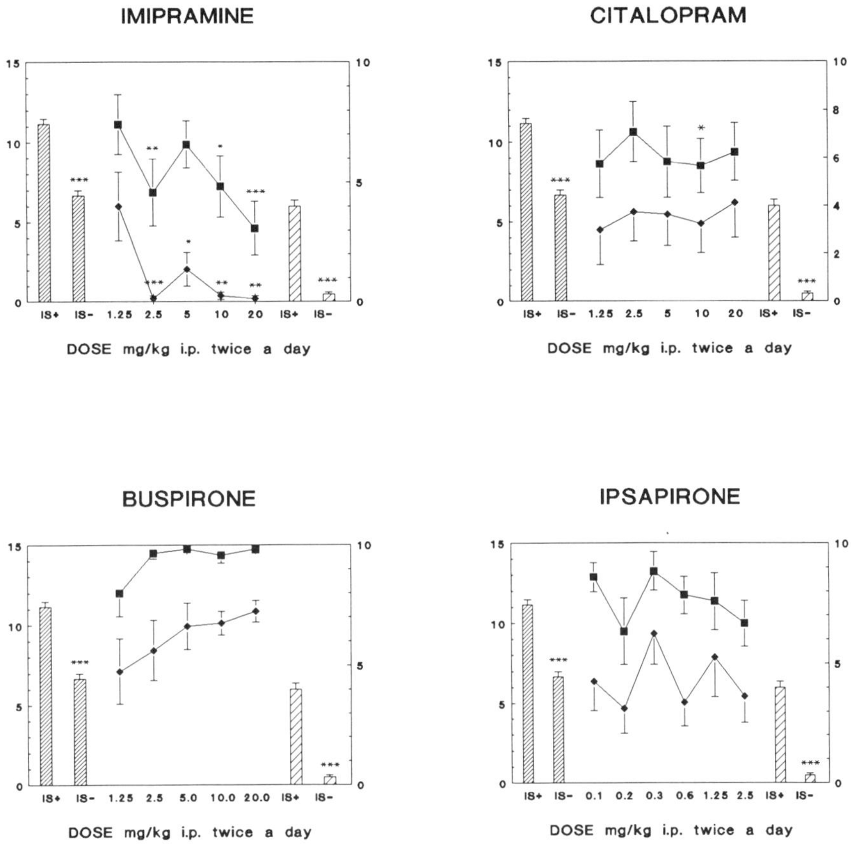

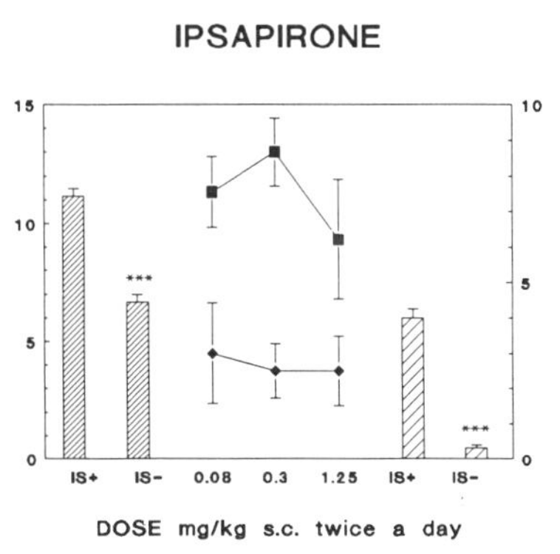

Fig. 1 Escape failures after treatment (4 days). The right ordinate indicates FR1 and the left one, FR2 escape failures. IS+ = plus inescapable shock. IS- = no inescapable shock.

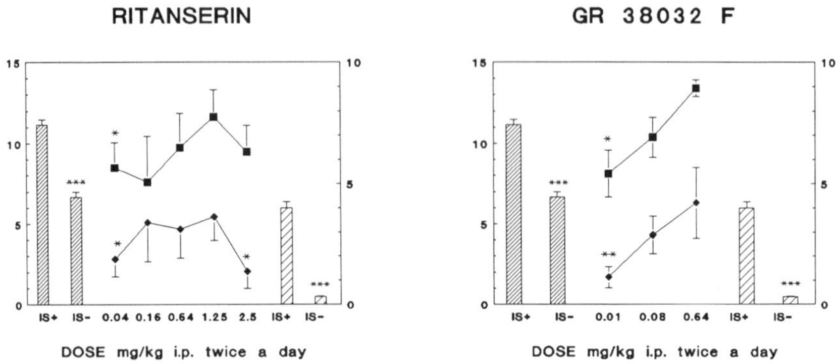

Fig. 2 Escape failures after ritanserin (5-HT_2 antagonist) and GR 38032F (5-HT_3 antagonist). Right ordinate indicates FR1 and the left one, FR2 escape failures. IS+ = plus inescapable shock. IS- = no inescapable shock.

DISCUSSION

As previously shown (Sherman et al., 1982) imipramine shows a dose-dependent effect on escape behaviour in LH. The 5-HT uptake-inhibiting compound citalopram, the 5-HT_{1A} agonists ipsapirone and buspirone, the 5-HT_2 antagonist ritanserin and the 5-HT_3 antagonist GR 38032F show only slight or no effect on escape failures compared to imipramine.

Soubrié et al. (1986) have shown that lesion of serotonergic neurons did not prevent imipramine from reversing the escape deficit induced by inescapable shock. This questions the role of serotonergic systems in LH and could account for the lack of a dose-dependent effect with 5-HT compounds in this model.

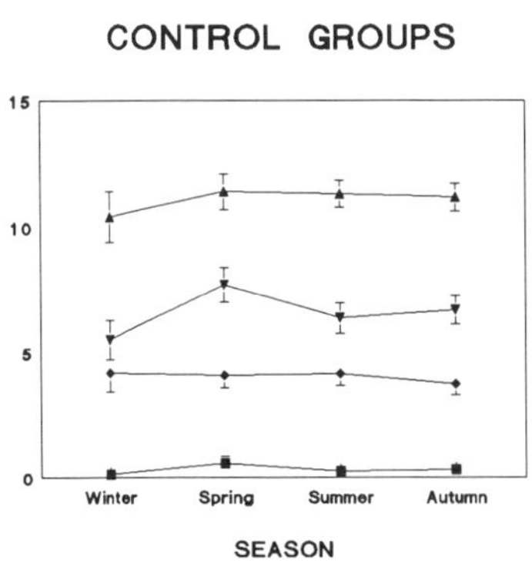

Fig. 3 The number of escape failures (ordinates) as a function of the season in control groups.
IS+ = plus inescapable shock.
IS- = no inescapable shock.

—▲— IS+ FR2
—◆— IS+ FR1
—▼— IS- FR2
—■— IS- FR1

CONCLUSION

As far as data derived from an animal model of depression can be relevant in predicting drug-effects in humans, the present results cast doubt on the usefullness of the LH model as a screening model for new antidepressants drugs.

Acknowledgements: This work was supported by the Lundbeck Foundation. The authors whishes to thank Mrs. Berit Marx for preparing the manuscript and Mr. Jørgen Kring for tecnical assistance.

REFERENCES

Brett, C.W., Pavlik, W.B., Burling, T.A. (1981) An. Learn. Behav. 9, 38-44.
Geoffroy, M., Scheel-Kruger, J., Christensen, A.V. (1990) Psychopharmacology 101, 3, 371-375.
Giral, P., Martin, P., Soubrié, P., Simon, P. (1988) Biol. Psychiatry 23, 237-242.
Maier, S.F., Seligman, M.E.P. (1976) J. Exp. Psychol. Gen. 105, 3-46.
Martin, P., Soubrie, P., Simon, P. (1987) Prog. Neuro-Psychopharmacol. Biol. Psychiat. 11, 1-7.
Montgomery S.A. (1985) J. Clin. Psychiatry 46 (3 sec 2), 2-6.
Sherman, A.D., Sacquitne, J.L., Petty, F. (1982) Pharmacol. Biochem. Behav. 16, 449-454.
Soubrié, P., Martin, P., Mestikawy, S., Thiebot, M.H., Simon, P., Hamon, M. (1986) Pharmacol. Biochem. Behav. 14, 823-826.
Traber, J., Davies, M.A., Dompest, W.U., Glaser, T., Schuurman, T., Siedel, P.R. (1984) Brain Res. Bull. 12, 741-744.
Weiss, J.M., Goodman, P.A., Losito, B.G., Corrigan, S., Charry, J.M., Bailey, W.M. (1981) Brain Res. Bull. 3, 167-205.

Animal Models in Psychopharmacology
Advances in Pharmacological Sciences

THE BEHAVIOURAL DESPAIR MODEL AS A PREDICTOR OF ANTIDEPRESSANT ACTIVITY: EFFECTS OF SEROTONERGIC DRUGS

J.A.M. van der Heyden, B. Olivier and T.J.J. Zethof

Department of Pharmacology, Duphar B.V., P.O.Box 900, 1380 DA Weesp, The Netherlands

SUMMARY: The forced swimming test in rats can discriminate clinically effective antidepressants from other psychoactive drugs under the conditions described here (pre-exposure to stress and repeated drug administration). Drugs stimulating central β-adrenoceptors also reduced immobility. Drugs stimulating the 5-HT_{1A} receptor also showed a clearcut antidepressant-like effect in this model. Of the various 5-HT antagonists tested, only those sharing a high affinity for the 5-HT_{1C} receptor were active in the forced swim model.

Traditionally antidepressant drugs have been identified on basis of their inhibition of NA and/or 5-HT re-uptake. More recently, numerous models have been proposed to predict antidepressant activity (for review see Willner, 1984). The forced swimming test, introduced by Porsolt et al. (1978) is one of the behavioural despair tests which are highly predictive for antidepressant activity in man. Thus, all types of clinically used antidepressant drugs (tricyclics, atypical and MAO-inhibitors) and electroconvulsive shocks tend to reduce the acquired immobility in this model (Duncan et al., 1985; Porsolt et al., 1978). A role for the 5-HT_{1A} receptor has recently been suggested in the action of antidepressants (Goodwin, 1989). Moreover, it has been found that 5-HT_{1A} agonists like 8-OH-DPAT, reduce the immobility in the forced swimming model (Cervo and Samanin, 1987).

Here we describe the effects of a number of drugs affecting various neurotransmitter systems and we also tested a variety of drugs affecting the different subtypes of 5-HT receptors to study their contribution to an antidepressant-like effect in the forced-swimming model, and hence for possible antidepressant activity in man.

MATERIALS AND METHODS

The procedure used is adapted from that published by Porsolt et al. (1978). Male rats (weighing approximately 200g) are placed in a cylinder (diameter 20cm) containing water of 25°C (height 15cm) for 15 minutes. Immediately hereafter the animals are treated subcutaneously with a drug. The following day the animals are again treated with the same drug (23.5 hours after the first treatment) and 30 minutes later the rats are placed in the same water-containing cylinders for 6 minutes during which activity was recorded. Data analysis was performed on the mean duration of inactivity over the first 3 minutes using the Student T-test. Activity is presented as an ED value for 8 sec reduction in immobility which is based on 20% of mean historic control value of 42 sec.

RESULTS

We confirmed the previously reported findings of Porsolt et al. (1978) in that all clinically effective antidepressants were active in our procedure. Thus, tricyclic antidepressants like imipramine and amitriptyline, a MAO inhibitor like clorgyline and the atypical antidepressants iprindol, mianserine, nomifensine, trazodone and fluvoxamine all reduced the acquired immobility (data not shown).The results obtained with the drugs affecting various neurotransmitter systems are presented in Table 1. The centrally acting β-adrenoceptor agonists isoxsuprine and clenbuterol reduced immobility, whereas salbutamol and ritodrine, which do not penetrate the blood-brain-barrier and the β-adrenoceptor antagonists propranolol and tiprenolol were without effect. Drugs either stimulating or blocking the α_2-adrenoceptors (clonidine and idazoxan, respectively) were inactive. Of the drugs affecting the DA system, both the uptake inhibitor nomifensine and the releaser amphetamine reduced immobility. Both drugs, however, also strongly stimulate locomotor behaviour.

Clozapine, in contrast to haloperidol and chlorpromazine unexpectedly reduced immobility (see also Table 2). Of the drugs affecting the benzodiazepine receptor, diazepam showed an antidepressant-like effect that was not observed after administration of chlordiazepoxide or the antagonist Ro 15-1788. Neither an agonist (morphine) nor antagonist (naloxone) of the opiate receptors affected the duration of immobility. The anticholinergic drug scopolamine reduced immobility, but again this drug also strongly enhanced locomotor activity (data not shown).

Table 1 Drugs affecting various neurotransmitter systems.

AGONIST	$ED_{8''}$ (mg/kg sc)	ANTAGONIST	$ED_{8''}$ (mg/kg sc)
		β-NORADRENALINE	
isoxsuprine	0.4	propranolol	>10
clenbuterol	11	tiprenolol	>10
salbutamol	>10		
ritodrine	>10		
		α-NORADRENALINE	
clonidine	>0.1	idazoxan	>10
		DOPAMINE	
d-amphetamine	0.1*	haloperidol	>0.3
nomifensine	<1*	chlorpromazine	>10
		clozapine	1.5
		BENZODIAZEPINE	
diazepam	0.2	Ro 15-1788	>3
chlordiazepoxide	>1		
		OPIATE	
morphine	>1	naloxone	>10
		ACETYLCHOLINE	
		scopolamine	0.01*

*: locomotor stimulation present

The effects of drugs affecting the various subtypes of 5-HT receptors are presented in Table 2.

Drugs stimulating all 5-HT receptors through inhibition of the reuptake (fluvoxamine and trazodone) reduced immobility. Of the various agonists tested, drugs selectively stimulating the 5-HT_{1A} receptor (8-OH-DPAT, flesinoxan, ipsapirone and buspirone) were clearly active in the swimming model. The agonists stimulating more than one subtype of 5-HT receptor only showed an effect if a prominent 5-HT_{1A} component was present (N,N-DOMT, eltoprazine). Those lacking a pronounced 5-HT_{1A} activity were without effect (TFMPP, DOI). The immobility in the swimming model was reduced by those antagonists blocking both 5-HT_{1C} and 5-HT_2 receptors (propranolol, methiotepine, ritanserine, mianserine and clozapine). Drugs either selectively blocking the 5-HT_2 receptor (ketanserine) or 5-HT_3 receptor (ondansetron) were inactive, as was the non-selective 5-HT antagonist methysergide.

Table 2 Drugs affecting the serotonin system

DRUG	NATURE OF INTERACTION	$ED_{8''}$ (mg/kg sc)
fluvoxamine	reuptake inhibition	5
trazodone	,,	0.4
8-OH-DPAT	1A agonist	<0.03
flesinoxan	,,	0.2
ipsapirone	,,	6.8
buspirone	,,	29
N,N-DOMT	1A/1C agonist	1
TFMPP	1B/1C agonist	>3
eltoprazine	1A/1B agonist	1.5
DOI	1C/2 agonist	>1
propranolol	1A/1B antagonist	>10
methiotepine	1C/2 antagonist	8.3
ritanserin	,,	11.5
mianserine	,,	1.7
clozapine	,,	1.5
methysergide	1/2 antagonist	>3
ketanserine	2 antagonist	>10
ondansetron	3 antagonist	>10

DISCUSSION

Under the presently described conditions (pre-exposure to stress and repeated drug administration) reversal of acquired immobility in the rat swimming model is a useful predictor for antidepressant activity in man. It appeared that drugs stimulating central β-adrenoceptors also show an antidepressant-like profile. Of the drugs interacting with the 5-HT system, those either stimulating the 5-HT_{1A} receptor and those blocking the 5-HT_{1C} receptor reduced the duration of immobility.

CONCLUSION:

Drugs stimulating β-adrenoceptors, without their peripheral side effects may be clinically useful antidepressants. Drugs stimulating the 5-HT_{1A} and/or blocking the 5-HT_{1C} receptors may also possess antidepressant activity in man.

REFERENCES

Cervo, L., Samanin, R. (1987) Eur. J. Pharmacol. 144, 223-229.
Duncan, G.E., Paul, I.A., Harden, T.K., Mueller, R.A., Stumpf, W.E., Breese, G.R. (1985) J. Pharmacol. Exp. Ther. 234, 409-414.
Goodwin, G.M. (1989) Prog. Neuro-Psychopharmacol. & Biol. Psychiat. 13, 445-451.
Porsolt, R.D., Anton, G., Blavet, N., Jalfre, M. (1978) Eur. J. Pharmacol. 47, 379-391.
Willner, P. (1984) Psychopharmacology 83, 1-16.

RELATIONSHIP BETWEEN EMOTIVITY LEVEL AND SUSCEPTIBILITY TO THE LEARNED HELPLESSNESS MODEL OF DEPRESSION IN THE RAT

P. Tejedor del Real, J. Gibert-Rahola, I. Leonsegui and J.A. Micó

Departments of Neurosciences and Toxicology, Neuropsychopharmacology Unit, University of Cádiz, Spain

INTRODUCTION

Animal modeling research has been widely studied in the last years and particularly animal models of depression. Animal models are basically experimental preparations developed in some species in order to study specific aspects of human illness.

Shock escape deficits evident in animals previously exposed to inescapable, uncontrollable and unsignaled electric shocks, have been postulated as a model of depression known as "Learned Helplessness" (LH). This model proposes that exposure to uncontrollable stress provides the basis, in animals and in humans, for learning that stress is uncontrollable (helplessness). Such learning could have a number of debilitating consequences, possibly including depression (Seligman et al., 1975), which makes this model very useful for testing of putative antidepressive drugs.

The set of validity criteria has been discussed in detail for this model. Learned Helplessness in rats is one of the models with a very high overall validity, along with models like primate separation and intracranial self-stimulation. LH has good predictive validity as it responds to a wide range of clinically effective treatments. Chronic administration of tricyclic antidepressants and MAOI, which are effective as antidepressants in humans, may also be effective in eliminating some of the behavioural disturbances associated with uncontrollable stressors in animals (Kornetsky, 1989; Martin et al.,1990; Weiss and Goodman-Simson, 1985; Zacharko et al., 1984). This is the most important validity criteria for using this model to test new antidepressive drugs.

LH has considerable face validity; it shows a variety of other behavioural changes and there is a large number of symptoms induced by inescapable shocks with obvious similarities to the symptomatology of depression such as decreased locomotor activity (Wagner et al., 1977), poor performance in motivated tasks (Andersen et al., 1968), loss of appetite and weight (Weiss, 1967), decreased aggression (Maier et al., 1972) and others.

The construct validity of LH is based on three assumptions: 1) animals exposed to uncontrollable aversive events become helpless, 2) a similar state is induced in man by uncontrollability, and 3) helplessness in people is the central symptom of depression (Willner, 1984, 1986).

Uncontrollable stress influences the levels and activity of central neurotransmitters (norepinephrine, dopamine and serotonin) in animals, which have been implicated as mechanisms subserving clinical depression in humans (Jimerson, 1987; Meltzer and Lowy, 1987; Siever, 1987).

It has been shown that the opioid system is implicated as well in this mechanism. The administration of enkephalinase inhibitors (Roques et al., 1986) to rats is able to induce a reversal of escape deficits caused by inescapable shock pretreatment (Gibert-Rahola et al., 1990). It may be that depression results from a deficiency in the release of enkephalins in the brain, minimizing the rewarding and euphorogenic effects produced by the release of these opioid peptides.

When we were working on this model of depression in our laboratory, we observed that only some of the tested animals had the same behavioural disturbances after inescapable shocks, and that some strains of rats evidenced lower susceptibility to LH training. Wieland et al. (1986) found that wide differences exist in susceptibility to LH training between and within strains after exposure to inescapable shocks. A genetic component for susceptibility to helplessness in rats has been suggested (Henn et al., 1985).

Therefore, the aim of this study was to look for a factor able to discriminate rats with either low or high susceptibility to LH training. Since LH includes cognitive, motivational and emotional changes, we thought that an emotional factor might exist so that we could use it to discriminate rats.

The limbic system is important for affect and emotion. In line with this, there is increasing evidence that peptide-monoamine interactions are fundamental in the control of behaviour and many limbic regions, rich in monoaminergic innervations, contain high levels of several neuropeptides (Simon, 1982). Many studies have been made in this huge field. Thus, the reversal of "Learned Helplessness" can be achieved by injection of desipramine or 5-HT into the neocortex. Moreover the activity of serotonin in the septum appears to be implicated in the effect of tricyclics on LH, because there is a decrease in septal 5-HT release in helpless animals. Sherman and Petty (1980) suggest that tricyclics activate a cortico-hipocampal-septal circuit depending respectively on cortical-serotoninergic, hippocampal-GABAergic and septal-serotoninergic neurotransmission. The amygdala might represent an important site of action for tricyclics antidepressant drugs, as the direct administration of tricyclic into the central-medial amygdala has identical effects to the ones observed after systemic administration (Watanabe et al., 1979).

Learned Helplessness and muricide behaviour have been the only test models in neuropharmacology for testing antidepressants used for studying the effect of intralimbic injections of antidepressants and lesions to limbic structures.

On the other hand, the opioid peptides are also implicated in analgesia. In fact, moderate to high opiate binding is displayed by all limbic regions that have been implicated in pain perception. In order to know whether the emotivity level had any relationship with the susceptibility to pain or not, nociceptive thresholds were also recorded in all animals using vocalization after electric discharge as emotional response to pain.

MATERIALS AND METHODS

Animals

Twenty male Wistar rats, weighing 200-250g at the beginning of the experiment, were individually housed under a constant cycle of 12h light and 12h darkness, temperature 21 $\pm 1°C$ and free acces to water and food.

Open-field test

Emotivity was tested at the beginning of the experiment in a round open field, 1m diameter and 50 cm high walls. The open field was situated in a dark and soundproof room. The apparatus was iluminated by a 100W bulb suspended at a height of 60 cm above the floor of the open field. A white noise provided a deep and uniform sonorous background. The test was carried out individually in three daily sessions of 5 min. The number of boluses excreted were recorded during each period. Defecation in a strange situation is a valid measure of emotivity (Broadhurst, 1960). Two groups of rats were selected according to their average rate of defecation: a high (n=8) and low emotive group (n=10). Rats with medium values were eliminated from the experiment. After inescapable shock treatment, the open field test was repeated in the same conditions as before the treatment.

Learned Helplessness induction

Seventy-two hours after the last open field session, all the animals received an inescapable shock session in order to induce helplessness. Electric foot shocks were delivered in 20x12x10 cm chambers with plexiglas walls and cover, the floor was made of stainless-steel grids (1.5 cm mesh). A constant current shocker was used to deliver 60 scrambled, randomized and inescapable shocks (duration 15s, intensity 1mA, intershock interval 60±20s) to the grid floor. The inescapable shock session was performed in the morning, starting at 9.00 am. Each animal was weighed before and after the session.

Conditioned avoidance training

In order to evaluate escape and avoidance performance, avoidance training was initiated 48h after the inescapable shock session in an automated two-way shuttle-box, divided into two equal sized chambers by a stainless-steel partition with a gate that provides access to the adjacent compartment through a 7x7 cm space. The animals were singly placed in the shuttle-box and subjected to 30 avoidance trials.

During the first 3s of each trial a light signal was presented (CS). The animals were allowed to avoid shock during this period. If an avoidance response did not occur, a 1mA shock, 3s, was delivered. If no escape response occurred within this last period, shock and light were terminated. The response required of the rat during each trial was to pass through the gate once into the other compartment of the shuttle-box (FR 1). Intertrials interval was 30s. The very first seconds following shock onset seem to be critical for detecting interference effects in animals pre-exposed to inescapable shocks, despite the fact that escape failure is defined as failure to escape within a 30-50s period in many procedures used for helplessness assessment. In this way, the animals are not exposed to the aversive stimulus for a long time, avoiding damage.

Avoidance sessions were performed for 3 consecutive days in the morning. The number of escape failures were recorded during shock delivery.

Electric stimulation of the tail

Twenty-one days after the inescapable shock session, nociceptive thresholds were measured in all the animals. The test employed was partially modified from Carroll and Lim (1960). This method permits, depending upon the intensity of the painful stimulation, to study different pain responses integrated at different levels of the pain transmission. The day before the test started, two stainless-steel electrodes, 2-3 cm long were inserted subcutaneously into the middle section of the tail in each animal. All the animals were deprived of food but not water 4h before the test was carried out. Nociception was induced by electrical stimulation of the rat tail, via electrodes, with high frecuency square waves, of increasing voltage, from a Hugo Sachs Electronic stimulator type IIMZ (using a frecuency of 125Hz, a pulsewidth of 1.6ms and a train duration of 0.5s). The first voltage applied was 0.25V, it was increased in steps of 0.25V until 1V; 0.5V until 5V; 1V until 10V and 5V until 70V wich was considered the cut-off voltage.

RESULTS

Susceptibility to LH training

When the animals were tested in the shuttle-box (conditioned avoidance training), the high emotive group of rats showed an average higher escape failure than did the low emotive group (see Table 1).

Statistical analysis showed that the difference between the averages of both escape failures were not significant but the difference is quite evident, in fact the percentage of rats becoming helpless was 50% in the high emotive group while it was 20% in the low emotive one (see Table 2), so that, the level of the emotivity factor could be important enough to discriminate rats susceptible to helplessness.

Table 1 Shows the means (±s.e.m.) of escape failures in both groups of rats during the sessions in the shuttle-box.

	1st session	2nd session	3rd session
High emotive rats (n=8)	12.3 ± 3.7	12.3 ± 4.3	11.3 ± 4.4
Low emotive rats (n=10)	8.6 ± 1.2	6.1 ± 2.5	6.0 ± 2.9

Table 2 Shows the percentage of rats in each group with more than 30% of escape failures in the shuttle-box.

	1st session	2nd session	3rd session
High emotive rats (n=8)	50%	50%	50%
Low emotive rats (n=10)	30%	20%	20%

Effect of LH training on emotivity level measured by the rate of defecation in the open field.

After LH training the rate of defecation in the open-field increased in the low emotive group when compared with the defecation rate before inescapable shock pretreatment. On the contrary, the rate of defecation in the high emotive rats decreased after LH training (Table 3).

Statistical analysis revealed a significant decrease in the emotivity level after LH training in high emotive animals (Wilcoxon $T=0$; $p<0.01$); on the other hand, low emotive animals showed an increase in emotivity when compared with their inescapable pretreatment rates (Wilcoxon $T=2$; $p<0.01$).

Table 3 Shows the average (±s.e.m.) number of boluses excreted by both high and low emotive rats before and after LH training

	Pre LH training	Post LH training
High emotive rats (n=8)	7.18 ± 0.38	5.12 ± 0.54
Low emotive rats (n=10)	3.30 ± 0.28	5.35 ± 0.63

Electrical stimulation of the tail: correlationship between emotivity level and susceptibility to pain

As we have explained in the method, 21 days after inescapable shocks pretreatment, nociceptive thresholds were measured in all the animals in accordance to three parameters: withdrawal of the tail or hind quarters movement, vocalization at the same time as electric discharge, and vocalization remaining after electric discharge as high emotional response to pain.

We did not find any important differences in the first and second parameters between both groups, but however, a significant difference was found in the third parameter between both high and low emotive groups (see Table 4).

Table 4 Shows the means (±S.E.M.) of the three nonciceptive thresholds (volts).

	1st session	2nd session	3rd session
High emotive group (n=8)	0.78 ± 0.65	1.14 ± 0.13	4.78 ± 0.54
Low emotive group (n=10)	0.61 ± 0.66	1.61 ± 0.29	7.27 ± 1.25

Statistical analysis revealed that there was a significant negative relationship between the defecation rate and the third nociceptive threshold parameter ($r=-0.4923$; $p<0.05$).

DISCUSSION AND CONCLUSION

Firstly, and from the data presented here, we would like to point out that animals with a high emotional level are more susceptible to acquire the LH paradigm and that this susceptibility depends on the initial level of emotivity. This could be a very important factor to discriminate animals at the beginning of the experiments when the LH paradigm is going to be employed for testing drugs with an hypothetical antidepressive effect.

We have observed as well that the emotivity level changes after the LH training, thus, this level is increased in low emotive animals, meanwhile, it decreases in high emotive rats.

It has been shown that after inescapable shock pretreatment, that some disturbances in the level of enkephalins appeared. It has also been found that the limbic system is a place with a high level of enkephalins and a large number of opiate receptors. Since the limbic system is the place where emotion occurs, it is possible to think about any common pathway for both procedures.

Moreover, we have found a relationship as well between emotivity and nociceptive threshold using vocalization remaining after discharge as a high emotional response. Any complex system could be implied.

REFERENCES

Anderson, D.C., Cole, J., McVaugh, W. (1968) J. Comp. Phisiol. Psychol. 65, (supl): 1-17.

Broadhurst, P.L. (1960) In: Experiments in Personality, vol. 1, (H.J.Eisenck, Eds), Praeger, New York.

Carroll, M.N., Lim, R.K.S. (1960) Arch. Int. Pharmacodyn. 75, 383-403.

Gibert-Rahola, J., Tejedor del Real, P., Chover, A.J., Puyana, M., Rodriguez, M.M., Leonsegui, I., Mellado, M., Micó, J.A., Maldonado, R., Roques, B.P. (1990) Eur. J. Pharmacol. 183, 2317.

Henn, F.A., Jhonson, J., Edwards, E., Anderson, D. (1985) Neurobiol. Pharmacol. Psychopharma. Bull. 21, 443-446.

Jimerson, D.C. (1987) In: Psychopharmacology: the third generation of progress. Raven Press, New York, pp 505-511.

Kornetsky, C. (1989) In: Animal Models of Depression (G.F.Koob, C.L.Ehler and D.J.Kupfer, Eds), Birkhauser, Boston-Basel-Berlin.

Maier, S.F., Anderson, C., Lieberman, D.A. (1972) J. Comp. Physiol. Psychol. 81, 94-100.
Martin, P., Soubrié, P., Puech, A.J. (1990) Psych. Psychobiol. 5, 123-128.
Meltzer, H.Y., Lowy, M.T. (1987) In: Psychopharmacology: the third generation of progress. Raven Press, New York, pp 513-526.
Roques, B.P., Daugé, V., Gacel, G., Fournié-Zaluski, M.C. (1986) In: Biological Psychiatry, 1985. Developments in Psychiatry, 7. (C.Shagass, R.C.Josiassen, W.H.Bridger, K.J.Weiss, D.Stoff and G.M.Simpson, Eds), Elsevier sciences Publishing Co, New York, pp 287-289.
Seligman, M.E.P., Beagly, G. (1975) J. Comp. Physiol. Psychol. 88, 534.
Sherman, A.D., Petty, F. (1980) Behav. Neurol. Biol. 30, 119-134.
Siever, L.J. (1987) In: Psychopharmacology: the third generation of progress. Raven Press, New York, pp 493-504.
Simon, E.J. (1982) Theories, clinical observations and treatment possibilities (K.Verebey, Ed), Published by the New York Academic of Sciences.
Wagner, H.R., Hall, T.L., Coté, I.L. (1977) J. Gen. Psychol. 96, 313-318.
Watanabe et al. (1979) J.Pharmacol. 29, 493-496.
Weiss, J.M., Goodman-Simson, P.A. (1985) In: Stress and coping, vol. 1. Lawrence, Erlbaum, New York, pp 93-116.
Weiss, J.M. (1968) J. Comp. Physiol. Psychol. 65, 251-260.
Wieland, S., Boren, J.L., Consroe, P.F., Martin, A. (1986). Life Sciences, 39, 937-944.
Willner, P. (1984) Psychopharmacology 83, 1-16.
Willner, P. (1986) Prog. Neuro-Psychopharmacol. Biol. Psychiat. 10, 677-690.
Zacharko, R.M., Bowers, W.J.M.S., Anisman, M. (1984a) Prog. Neuropsychopharmacol. Biol. Psychiat. 8:, 601-606.
Zacharko, R.M., Boers, W.J., Kelley, M.S., Anisman, M. (1984 b) Brain. Res. 321, 175-179.

Animal Models in Psychopharmacology
Advances in Pharmacological Sciences

LONG TERM ISOLATION AS AN ANIMAL MODEL OF DEPRESSION

N. Enginar and L. Eroglu

Department of Pharmacology, Istanbul Medical Faculty, University of Istanbul, Istanbul, Turkey

SUMMARY: Isolation induced exploratory and motor activity changes in rats were determined after an isolation period of 11 months. Isolated rats showed more interest in the environment. However, no motor hyperactivity in isolated animals was found. Neither amitriptyline (15 mg/kg) nor imipramine (30 mg/kg) showed a selective effect on isolated rats and decreased the motor and exploratory activities in both groups.

INTRODUCTION

Various studies have been performed in the search for an appropriate animal model of depression (Katz, 1981). One of these models is the chronically isolated rat presented by Garzon and co-workers (1979, 1981). However this behavioural model of depression is less studied and requires more extensive testing in order to gain wider use. The present study was, therefore, designed to evaluate the isolation induced changes in motor and exploratory activities and to evaluate the validity of long term isolation, 11 months, as an animal model of depression.

MATERIALS AND METHODS

Male Wistar albino rats were placed individually in cages of 13x37x17 cm at the time of calcification of the incisors, i.e. at approximately 18 days of age. Their male litter mates were housed in groups of four in cages of 34x47x17 cm. Each group consisted of 24 rats. Both isolated and group-housed rats were kept in a temperature controlled room (22 ± 1°C) under a 12 hr light-dark cycle and provided with food and water ad.lib. Handling of the animals was restricted to the cleaning of the cages -once a month- and to the time of the experiments. All behavioural experiments were carried out between 10.00 a.m. and 3.00 p.m. Exploratory activity was carried out in a hole-board and motor activity was measured in an activity cage.

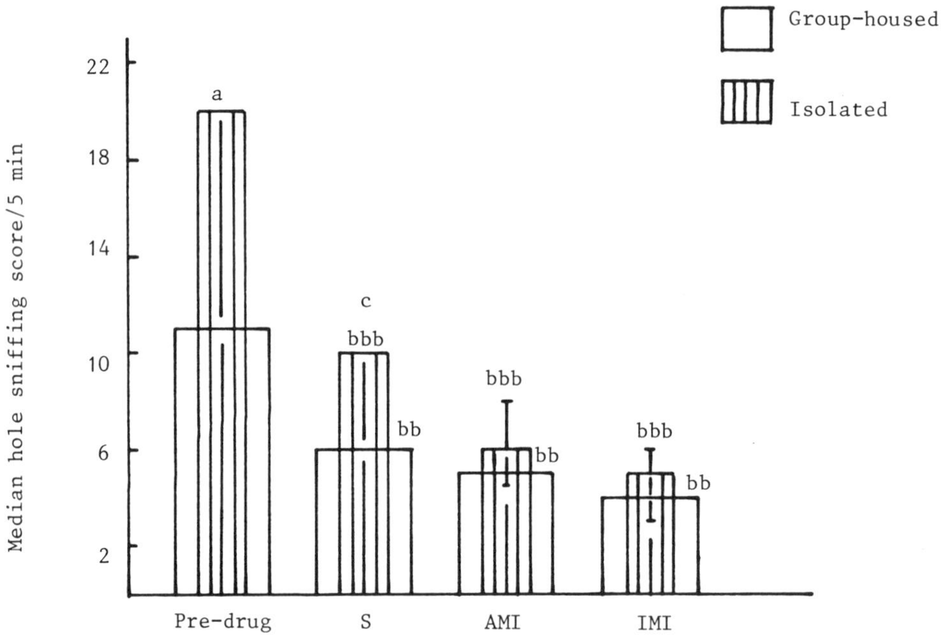

Fig. 1 Pre-drug (baseline) and post-drug median exploratory activity scores of rats. Pre-drug activity was determined 1 day before drug treatment. Rats were placed single 1 hour after saline [S], amitriptyline HCl (15 mg/kg) [AMI], or imipramine HCl (30 mg/kg) [IMI] treatment [number of subjects = 24 (pre-drug) and 8 (post-drug). pre-drug difference between group-housed and isolated rats: $^{a}p<.002$ difference between pre-drug and post-drug scores in each group: $^{b}p<.05$; $^{bb}p<.02$; $^{bbb}p<.002$, difference between S and AMI or IMI treatments: $^{c}p<.05$.

After an isolation period of 11 months, pre-drug (baseline) exploratory and motor activity values of rats in each group were determined. The following day, both the isolated and group-housed rats were divided into 3 subgroups, 8 rats in each, according to drug treatments (i.p.): Saline (1 ml/kg)-treated; amitriptyline HCl (15 mg/kg)-treated; and imipramine HCl (30 mg/kg)-treated. Post-drug exploratory and motor activities were measured 1 hr after drug treatments.

Results were analyzed using Mann-Whitney's 2 tailed U test.

RESULTS

As seen in Fig. 1, there were significant differences in pre-drug exploratory activity scores between isolated and group-housed rats. The exploratory activity was markedly higher in rats which were isolated. The motor activity, on the other hand, was not significant altered by long term isolation (Fig. 2). As for the post-drug results, saline, as well as amitriptyline and imipramine, reduced the exploratory activity scores in both isolated and group-housed rats.

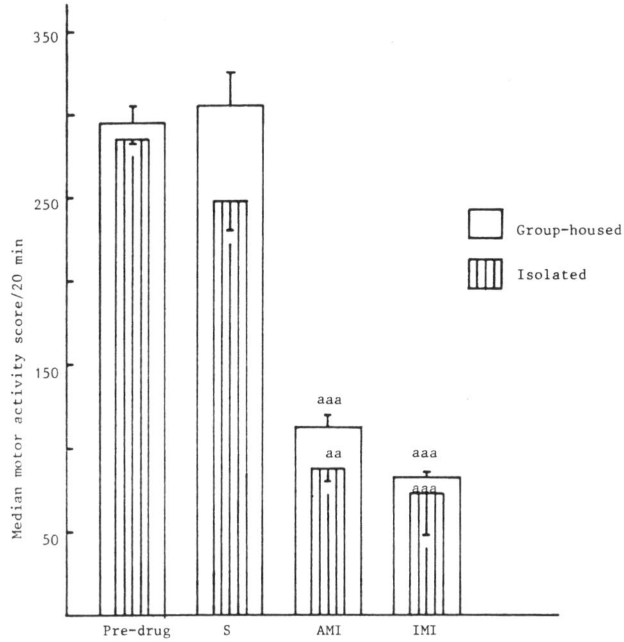

Fig. 2 Pre-drug (baseline) and post-drug median motor activity scores of rats. Pre-drug activity was determined 1 day before drug treatment. Rats were placed single 1 hour after saline S, amitriptyline HCl (15 mg/kg) [AMI], or imipramine HCl (30 mg/kg) [IMI] treatment [number of subjects = 24 (pre-drug) and 8 (post-drug)]. Difference between pre-drug and post-drug scores in each group: $^{aa}p < .02$, $^{aaa}p < .002$.

However, the antidepressant drugs blocked rather markedly the isolation induced differences in exploratory activity. Saline had no apparent effect on the motor activity of isolated and group-housed rats. On the contrary amitriptyline and imipramine significantly reduced motor activity. But this effect was not selective, since both antidepressant drugs produced a decrease in the motor activity of both isolated and group-housed rats.

DISCUSSION

The present results indicate that long term isolation of rats from weaning on induced alterations in exploratory activity. Isolated rats showed increased levels of investment in the environment. In agreement with our findings, rats, reared in both shorter (Sahakian et al., 1975; Speiser & Weinstock, 1973; Syme, 1973) and longer (Garzon et al., 1979; Garzon & Del Rio, 1981) isolation, were more active when introduced into a novel environment than rats reared in groups. The significant pre-drug exploratory activity differences between isolated and group-housed rats disappeared after treatments of imipramine or amitriptyline, at a dose of producing most pronounced action in an animal depression model (Porsolt et al., 1978). The antagonizing effect of these classical antidepressant drugs on exploratory activity in isolated animals provides evidence for the proposed animal model of depression. However motor activity, the other behavioural measure for depression in this experiment, was not shown to be affected by long term isolation. So it can be claimed that exploratory activity changes in isolated animals may be regarded as an irritability or a state of anxiety since the proposed animal model of depression was suggested to be suitable to anxious/hyperactive subtype of human depression. However, anxiolytics were found ineffective in isolation induced behavioural syndrome (Garzon & Del Rio, 1981). Thus the exploratory activity changes in isolated animals, which might be due to "individually housing" stress, may reflect a behavioural disorder different from depression.

CONCLUSION

Our results partly failed to confirm Garzon and co-workers findings, in that neither a long-term isolation induced motor hyperactivity, nor a clear selective antidepressant activity was found. Further studies, however, will help to clarify the validity of long term isolation as an animal model of depression.

REFERENCES

Garzon, J., Fuentas, J.A., Del Rio, J. (1979) Eur. J. Pharmacol. 59, 293-296.
Garzon, J., Del Rio, J. (1981) Eur. J. Pharmacol. 74, 287-294.
Katz, R.J. (1981) Neurosci. Biobehav. Rev. 5, 231-246.

Porsolt, R.D., Anton, G., Blavet, N., Jalfre, M. (1978) Eur. J. Pharmacol. 47, 379-391.
Sahakion, B.J., Robbins, T.W., Morgan, M.J., Iversen, S.D. (1975) Brain Res. 84, 195-205.
Speiser, Z., Weinstock, M. (1973) Br. J. Pharmacol. 48, 348-349.
Syme, L.A. (1973) Anim. Learn. Behav. 1, 161-163.

Animal Models in Psychopharmacology
Advances in Pharmacological Sciences

LONG-LASTING BEHAVIOURAL CHANGES AFTER A SINGLE FOOTSHOCK STRESS SESSION. A MODEL OF DEPRESSION?

H.H. van Dijken, J.A.M. van der Heyden*, J.Mos*, F.J.H.Tilders

Dept. Pharmacology, Medical Faculty, Free University, Van der Boechorststraat 7, 1081 BT Amsterdam, The Netherlands. * Dept. Pharmacology, Duphar B.V., P.O.Box 900, 1380 DA Weesp, The Netherlands.

SUMMARY: Exposure of male rats to a single brief period of footshocks induced behavioural changes as measured in the open field test that lasted for at least 3 weeks. Moreover, such rats showed a long-lasting increased responsiveness to a sudden change in environmental stimuli. These effects of footshock exposure are probably useful as an animal model of depression.

Stress seems to play an important role in the development of depression (Anisman & Zacharko,1982). Several animal models of depression are based on stress-induced behavioural deficits, e.g. learned helplessness and chronic unpredictable stress (Willner, 1984). In these models, chronic antidepressant (AD) treatment effectively restores the behavioural deficits, whereas single administration does not. However, the treatment regimes used are subject to criticism, as AD's are given prior to or concurrent with stress exposure. This raises the question whether the drugs prevent rather than reverse the behavioural changes induced by the stress procedures (Jesberger & Richardson, 1985). AD treatment starting after the development of the stress-induced behavioural deficits would resemble more closely the pharmacotherapy of human depression. In order to employ such a treatment regime in an animal model, it is a prerequisite that the induced behavioural deficits are chronic in nature.

Accordingly, the objective of the present study was to develop an animal model in which long-lasting behavioural changes are induced that can be reversed by chronic AD treatment.

MATERIALS AND METHODS

Male Wistar (Harlan, Zeist, The Netherlands) rats weighing 300-350 g. were used. These rats were housed in groups of 3 rats per macrolon cage under a 12 hours light and 12 hours dark (7.00 hr. lights on) cycle. All rats in a cage experienced the same experimental procedure. The rats (S) were stressed by inescapable and unpredictable footshocks. The stress session lasted 15 minutes, in which the S rats received 10 scrambled shocks (1 mA) of 6 seconds each according to a V.I. schedule. Control (NS) rats were placed in the same shock apparatus for 15 minutes without receiving shocks. At various intervals after footshock exposure, the behavioural response of different NS and S groups were measured in a circular (diameter=1.0 m.) open field (5 minutes test-time). Behavioural testing was performed between 8.00 and 15.00 (lights on).

In a second series of experiments rats housed individually under a reversed day/night cycle (7.00 hr. lights on and 19.00 hr. lights off) were exposed to a similar footshock session. The behavioural response of these rats to a sudden reduction in background noise was investigated in a noise test as described by Steenbergen et al. (1989) with slight modifications. Briefly, the rats were placed in an observation cage (80x60x50 cm.) lit by red light. A constant 70 dB 'white' noise was produced via a speaker above the cage. The noise was switched off after 3 minutes. Behaviour was recorded during the 3 min. preceeding and during the 3 min. following switching off the noise. The noise test was compared with a test without switching off the noise in order to subtract the specific response to the sudden reduction in background noise. Both tests were carried out 21 days after footshock procedure. testing took place between 8.30 and 13.00 hr. (dark phase). Different groups of rats were used for both tests.

The results are expressed as means and the standard errors. Statistical analysis was carried out using the Mann-Whitney U test for comparison of two independent groups of observations.

RESULTS

The behavioural response of NS and S rats was studied in the open field at different post stress intervals. It should be noted that independent groups of NS and S rats were used for each post-stress interval in order to prevent possible habituation effects. Marked inhibition of locomotion was observed in S rats as compared to NS animals at all post-stress intervals studied (Fig. 1). In addition, S rats showed significantly more defecation in the open field than NS rats (e.g. S: 6.4±0.5, NS: 2.4±0.7 at post-stress interval of 7 days; $p<0.001$). This experiment revealed that one relatively mild footshock stress session impaired open field behaviour for at least 3 weeks.

Moreover, the results of this and other experiments suggest that this effect needs time to develop and increases with increasing time interval between footshocks and behavioural test. In the experiment presented this was confirmed by Kruskal-Wallis one-way analysis of variance for the S groups ($H(3,44)=9.17$, $p=0.027$). Such an effect of the factor time between stress and test was not found for NS rats ($H(3,44)=4.95$, $p=0.176$).

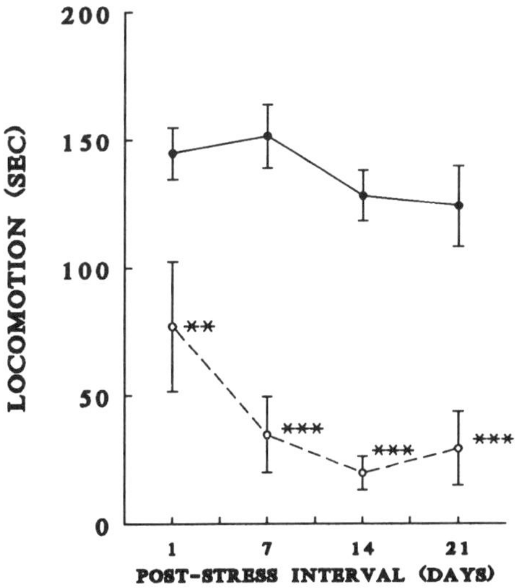

Fig. 1 Locomotion (duration in sec.) in the open field of rats previously exposed to inescapable footshocks. Different groups of NS and S rats were used for each post-stress interval. NS=filled circles (n=12); S=open circles (n=12). ** $p<0.01$, *** $p<0.001$ vs. NS; Mann-Whitney U test

The effect of footshock stress on the behavioural response to a sudden change in environmental stimuli was also investigated. The noise test was performed 3 weeks after footshock exposure. In this test, S rats were slightly more immobile than NS rats during the first 3 minutes of the test (noise on) and showed a marked increase in immobility after noise had been turned off, whereas a small increase in immobility was observed in NS rats (Fig. 2, left panel).

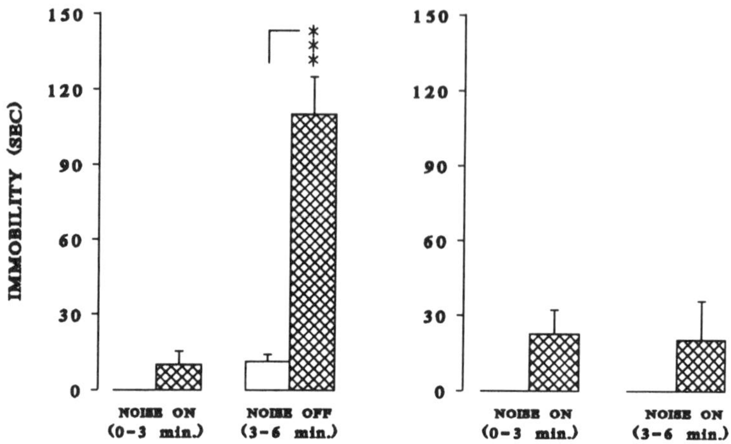

Fig. 2 Immobility (duration in sec.) performed by NS (open bars) and S (closed bars) rats in the noise test (left figure) and in the test with 6 min. constant noise (right figure) days after the footshock session. Independent groups of NS (N=10) and S (N=10) rats were used for both test situations. *** $p<0.001$ vs. NS, Mann-Whitney U test

This freezing-like behaviour of S rats was a specific behavioural response to turning off the noise since immobility did not increase in the second period of the test when the noise was continuously present (Fig. 2, right panel). S rats also showed a higher rate of defecation in the noise test compared to NS rats (S: 6.7 ± 1.1, NS: 1.9 ± 0.8, $p=0.002$). Thus, a single footshock session increased the behavioural responsiveness of male Wistar rats to a sudden change in environmental stimuli on a long-term basis (3 weeks). Similar experiments were performed with Tryon Maze Dull S3 male rats (Harlan, Zeist, The Netherlands). The results of these experiments were essentially the same as those found in Wistar rats.

We investigated whether repeated testing would allow us to monitor the development of these behavioural changes in another series of experiments. NS and S rats were placed in the noise test at a post stress interval of 14 days. The immobility response to switching off the noise was greater in S rats than in NS rats (S: 162.0 ± 11.0, NS: 60.0 ± 18.1; $p<0.001$). Re-testing of these animals in the noise test one week later resulted in the disappearance of the difference in immobility response (S: 127.9 ± 15.3, NS: 91.2 ± 18.6; $p=0.16$). However, when these animals were placed in the open field test 28 days after footshock exposure the typical behavioural deficit was found as observed in previous experiments (locomotion: NS: 111.7 ± 17.9, S: 21.8 ± 11.2; $p<0.001$).

Reversion of the order of the test situations (repeated open field test followed by a noise test) led to essentially the same results.

DISCUSSION

The present observations demonstrated that one footshock stress session induced long-lasting behavioural deficits in the open field test. Interestingly, the effects of footshock exposure progressively develops with time after the stress session. A period of rest after the stress of 7 days seemed to suffice for the full development of the behavioural deficits.

Antelman et al. (1988) reported that a single stressful event blocked the actions of diazepam for up to at least 4 weeks. In this study it was also noted that the effects of the stressful event increased with increasing time between stress and test.

The data of the noise test experiments showed that a single footshock session increased the behavioural responsiveness to a sudden reduction in background noise. The increased immobility response probably reflects increased stress sensitivity in those animals (Steenbergen et al., 1989). The experiments performed with Tryon Maze Dull S3 male rats suggest that these long-term behavioural effects of a single brief footshock session is a general phenomenon, which is not limited to a particular strain of rats.

Repeated testing in the same situation resulted in habituation, while changing the test environment revealed that the effects of footshock were still present. These experiments showed that the footshock induced behavioural changes persisted for at least 4 weeks.

In conclusion, the behavioural deficits induced by a single brief footshock session fulfill an important criterion for an animal model of depression as the deficits are long-lasting. However, a thorough pharmacological validation has to be carried out before we can decide wether the described behavioural effects of footshock exposure can be used as an animal model of depression.

REFERENCES

Anisman, H., Zacharko, R.M. (1982) Behav. Brain Sci. 5, 89-137.
Antelman, S.M., Knopf, S., Kocan, D., Edwards, D.J., Ritchie, J.C., Nemeroff, C.B. (1988) Brain Res. 445, 380-385.

Jesberger, J.A., Richardson, J.S. (1985) Biol. Psychiatry 20, 764-784.
Steenbergen, J.M., Koolhaas, J.M., Strubbe, J.H., Bohus, B. (1989) Physiol. Behav. 45, 729-733.
Willner, P. (1984) Psychopharmacology 83, 1-16.

Animal Models in Psychopharmacology
Advances in Pharmacological Sciences

ANTIDEPRESSANT SUBCLASSIFICATION BASED ON THE QUANTITATIVELY ANALYZED ELECTROCORTICOGRAM OF THE RAT

F. Krijzer, R. van der Molen, B. Olivier and F. Vollmer

Department of Pharmacology, Duphar B.V., P.O.Box 900, 1380 DA Weesp, The Netherlands

SUMMARY: Behavioural clinical data show different types of antidepressant drugs, e.g. sedating, stabilizing or activating ones. Another differentiation may be based on the presence or absence of anxiolytic activity. These different profiles can be recognized by the changes of the electrical brain activity induced by these drugs in humans and animals. Here we present a subclassification of antidepressant drugs based on the quantitative analysis of the ECoG of the rat. Subclass 1 is characterized by decrease of the power over the whole frequency range from 0 to 100 Hz indicating sedative activity. Subclass 2 is characterized by power changes characteristic for anxiolytic activity. Subclass 3 is characterized by power changes characteristic for stimulatory activity. The classification of drugs based on EcoG-profiles may be a relevant animal model to detect new antidepressants.

INTRODUCTION

Psychotropic drugs affect the electrocortical activity (ECoG) of the brain in a drug characteristic way. However, drugs belonging to the same therapeutic class, e.g. antidepressants, neuroleptics or anxiolytics share certain common ECoG effects. These common effects which characterize a certain drug class are used as markers to classify and eventually predict a possible clinical applicability of new developed drugs (Itil, 1982; Krijzer, et al., 1988).

However it is not only possible to predict roughly clinical applicability. In the human ECoG at least one can distinguish several different subclasses of antidepressant drugs. It is known from the clinic that antidepressants can be subdivided in psychomotor activating, psychomotor stabilizing or psychomotor sedating (Poldinger, 1986). Saletu et al. (1985) differentiate between two types of pharmaco-EEG profiles from antidepressants in humans: a thymeretic or a thymoleptic one, characterized by respectively an augmentation of alpha activity, increase in amplitude and eventually an attenuation of slow and fast activities, indicating activating properties (nomifensine), or by a concomitant increase of slow and fast activities and a decrease of alpha activity, suggesting sedative qualities (amitriptyline, imipramine). Trazodone, on the other hand, is described as an antidepressant with anxiolytic properties (Brogden et al., 1981).

These (sub-)classifications may be of clinical importance with respect to the prescription of drugs for certain patient populations. For elderly geriatric patients the drug of choice might be a non-sedative drug like fluvoxamine. For depressed patients with an anxiety complex antidepressants with an anxiolytic component will be of particular interest. In this contribution we will show that it is also possible to differentiate between different subclasses within a single drug class based on the analysis of the ECoG of the rat.

METHODS

In our laboratory a much improved drug analysis system based on ECoG recordings in vigilance controlled, freely moving rats has been developed. The effects of a number of antidepressant, neuroleptic, anxiolytic and stimulant drugs have been analyzed and almost all drugs could be classified in its appropriate class (Krijzer et al., 1988). Of all drugs 3 or 4 dosages have been tested, each dose in 5 rats.

The ECoG is differentially recorded between the frontal and parietal cortex in vigilance controlled rats during the 6 minutes immediately before and starting at 20 and 45 minutes after i.p. drug treatment. For each recording period a mean power spectrum is calculated. Power spectral changes after drug or saline are expressed as percentage of the pretreatment power spectrum and used as input for an analysis of variance resulting in a t-profile for each postdrug period per drug dose. The weighted mean of the t-values of all t-profiles of antidepressants belonging to a certain subclass has been calculated resulting in a so called (sub-)class profile.

RESULTS/DISCUSSION

Fig. 1 summarizes the data of the various antidepressants tested.

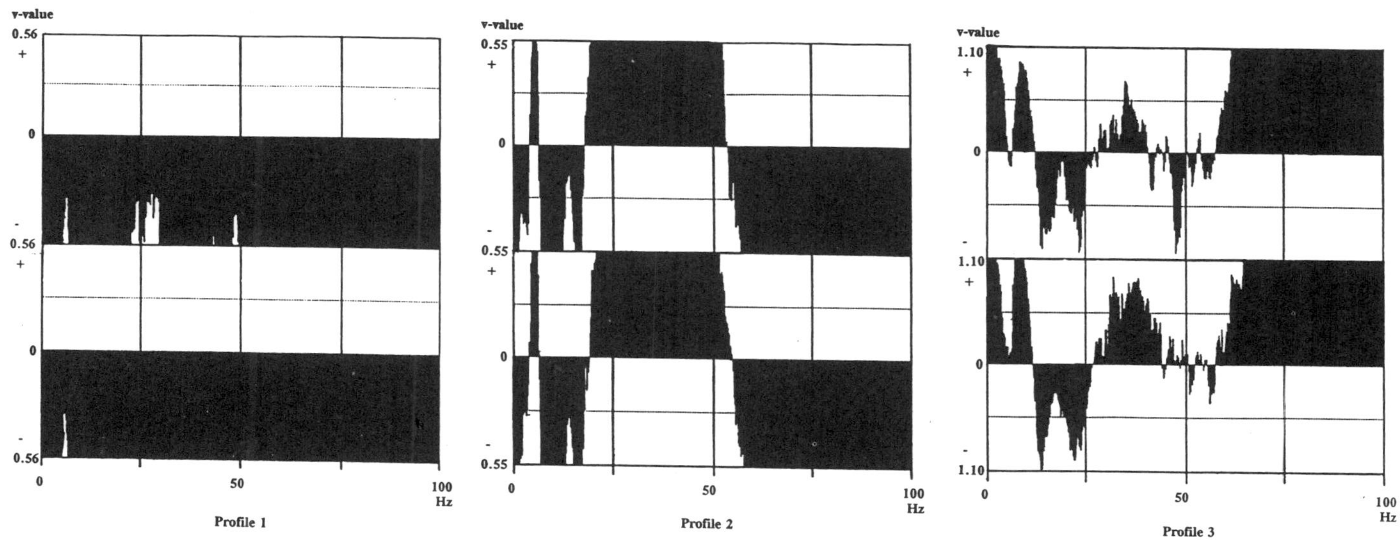

Fig. 1 Subclassification of antidepressant drugs in three groups from which the drug induced changes are presented: profile 1: sedative antidepressants, profile 2: anxiolytic antidepressants and profile 3: stimulating antidepressants.

Along the x-axis frequencies in Herz, ranging from 0 to 100, along the y-axis v-values resulting from the averaging of a number of t-profiles.

The drug profiles of the antidepressant drugs chlorimipramine, desmethylimipramine, nortryptyline, viloxazine, imipramine, clovoxamine, maprotyline, pargyline, femoxetine and nisoxetine have been averaged and are presented in profile 1. All these antidepressants are characterized by a more or less stronger decrease of the power over the whole frequency range from 0 to 100 Hz in the rat ECoG and are classified as general depressant antidepressants in our system. In this group are included a number of thymoleptics that are intrinsic mood-elevating but sedative (Kielholz, 1968). In non-hospitalized agitated or suicidal patients this may be an important advantage but in e.g. geriatric patients this will be a disadvantage.

The drug profiles of the antidepressants fluvoxamine, mianserin, zimeldine, rolipram, clorgyline, iprindole and tranylcypromine have been averaged and are presented in profile 2. This profile is characterized by a power decrease in the frequency band from 7 to 18 Hz, a power increase from 19 to 53 Hz and a power decrease from 55 to 100 Hz. In our system the power increase from 19 to 53 Hz is characteristic for anxiolytics as it is typically shown after benzodiazepines and 5-HT_{1A} agonists. This power increase is also seen in humans in the so-called beta band (all frequencies higher than 13 Hz) after acute treatment with e.g. mianserin (Fink et al., 1977) and fluvoxamine (Itil et al., 1977). Indeed a number of antidepressant drugs are claimed to be better anxiolytics in certain patient populations than the benzodiazepines. Typical examples are fluvoxamine and mianserin.

The drug profiles of nomifensine and bupropion have been averaged and are presented in profile 3. This profile is characterized by a power increase from 0 to 14 Hz, a power decrease from 14 to 25 Hz, a power increase from 25 to 50 Hz and a power increase from 60 to 100 Hz. The power increase from 60 to 100 Hz is thought to be characteristic for psychostimulants as this increase is typically shown after drugs like d-amphetamine, caffein, pipradrol and phencyclidine. Indeed from both nomifensine and bupropion there have been reports of mild central stimulating effects.

Thus three subgroups of antidepressants can be distinguished based on changes in the rat ECoG. This preclinical subclassification seems to fit well with the clinical subclassification although there remain exceptions. It seems to be possible to use this drug profiling technique based on changes in the ECoG of the rat to detect potential new psychoactive drugs and to predict their clinical applicability.

REFERENCES

Brogden, R.N., Heel, R.C., Speight, T.M. (1981) Drugs 21, 401-429.

Fink, M., Irwin, P., Gastpar, M., De Ridder, J.J. (1977) Psychopharmacology 54, 249-254.
Itil, T., Bhattachyaryya, A., Polvan, N., Huque, M., Menon, G. (1977) Prog. Neuro-Psychopharmacol. 1, 309-322.
Itil, T., Itil, T.(1982) In: EEG in drug research. (W.Herrmann, Ed), Gustav Fischer, New York, pp 131-158.
Kielholz, P. (1968) Medicinae scientia donat, 1, 5, 3 - 9, (MSD-Ned.)
Krijzer, F. (1988) In: Depression, anxiety and aggression (B.Olivier, J.Mos, Eds), Medidact B.V., Houten, Holland, pp 43-49.
Poldinger, W. (1986) Adv. Pharmacother. 2, 25-33.
Saletu, B., Grunberger, J. (1985) Meth. Find. Exp. Clin. Pharmacol. 7, 137-151.

Animal Models in Psychopharmacology
Advances in Pharmacological Sciences

ANIMAL MODELS AND DRUG SCREENS FOR DEPRESSION: PRAGMATISM AND THE VALIDITY REQUIREMENT

Trevor Archer

Dept. of Psychology, Univ. of Goteborg, P.O.Box 14158, S-400 20 Göteborg, Sweden

The search for animal models of depressive illness presents a most difficult task even though it may periodically appear that suitably reliable compounds for the treatment of depressive illness are within grasping distance. Research strategies for understanding the etiology, pathogenesis and treatment of depressive disorders remain among the most varied, confused and strenuous pursuits in the neuropharmacology of drug development. Any attempt to produce an animal model must be considered very much a pragmatic effort but this consideration in no way decreases the necessity of continued search: pragmatism has generally provided great benefits for the suffering. Among the many reasons for the failures to obtain completely reliable and valid animal models of depression, one may consider the enormous complexity of the disorder, as for example illustrated by the distinction between bipolar and unipolar illness in affective disorders. Thus, the bipolar disorder is characterized by episodes of both major depression and mania whereas the unipolar disorder is characterized by episodes of major depression. Other types of affective disorders related to depressive illness (as outlined and reviewed in DSM III) include cyclothymic disorder, involving varying periods of depression and hypomania, dysthymic disorder, a chronic mood disorder less severe than major depression, and, seasonal affective disorders. The existing test models and bioassays simply cannot cope with complexity of this order: one must therefore weight the term 'animal model' most stringently before applying it and in most cases make a preferred reference to 'drug screens'. In spite of these misgivings, the papers by Panksepp, Porsolt and Willner provide both thoughtful and broad reviews of the status of animal models in depression. Above all it is to the credit of the editors that these three chapters complement each other exceedingly well.

Panksepp describes a separation procedure, the social-isolation induced despair in chicks, making some effort not only to relate to other procedures but also to evaluate certain biochemical concomitants (cf. Panksepp et al., 1980). The chronic social isolation of young domestic chicks causes home-separation-induced distress vocalisations (DVs) as well as those attributable to social-separation. Each chick was isolated at 5 days of age and elevated levels of distress calls were immediately apparent; these calls would continue even if the young birds were replaced in a social context. However, they appeared to grow normally otherwise. Low doses of reserpine (0.25 - 0.5 mg/kg) decreased the DVs of chicks that had undergone 5 days of isolation to a greater extent than their socially-housed counterparts. These data offer some confirmation for the effects of imipramine in primate separation models (see Suomi et al. 1978; Porsolt et al. 1984), but a rather wide range of effective and ineffective compounds have been assessed by Lehr (1989). Acute treatment with imipramine (5-10 mg/kg) effectively increased DVs that had been reduced by reserpine, and even sustained the DV-reduction occurring in socially-housed chicks after prolonged separation (2-5 hours). The effects of separation appear to be rather more severe in day-old chicks: one-third of them lost weight rapidly and died within 2-4 days whereas the remaining two-thirds continued healthy although the deterioration of DV in response to isolation was more drastic than in the case of the five-day chicks. Imipramine, here too, reliably reversed the index of 'despair'. Panksepp and co-workers (e.g. Panksepp et al., 1986; Panksepp, 1990) have also found that a single day's isolation in Swiss-Webster mice at 17-22 days of age produced 'despair-like' effects. The duration of immobility in the Porsolt swim test was notably enhanced; however, if isolation was imposed at 26 days of age no effects in the swim test were obtained. Thus, the experiments both with chicks and mice underline the important aspect of 'critical periods' in development and their relationship to the affective states.

Porsolt's chapter offers an explicit definition of the requirements of a pharmacological model of depression, i.e. those procedures which have been conceived specifically with the aim of detecting the action of antidepressants in experimental animals. Utilizing the validation criteria outlined by McKinney (1977), he then continues to outline certain requirements of these 'depression' models: (1) a maximum of sensitivity and selectivity (as indexed by the absence of false negatives and false positives, respectively), (2) the possible similarity/analogy to particular clinical syndromes of depressive illness is considered an advantage in avoiding the "me-too" label, (3) a rapid and simple test model is advisable in the screening of large numbers of potential antidepressant compounds but the final necessity of a chronic rather than acute action ought to considered. Thus, the type of 'cook book recipe' provided here will be an invaluable guide for researchers interested in quickly applying one or another drug screen.

Essentially, Porsolt favors the pragmatic approach without totally disregarding the wearisome realities characterized by DSM III. In doing so, he has provided an exhaustive review of the efficacy and practicability of a list of animal models/screens that are currently used. Porsolt's review complements that of Willner most fortuitously since the latter, in listing the most prominent features of some 'animal models' promotes those features his own research has indicated to be of utility.

The chapter by Willner is more explicitly concerned with clinical implications than the chapters discussed above and to lend substance to this endeavor two 'psychological' procedures are described, stress induction and social isolation. For example, inescapable footshock (commonly) or immobilization (less commonly) stress will generally cause more or less robust alteration of a wide range of behaviours, such as spontaneous motor activity and exploration, and learning performance in different tasks. Both acute stress and chronic exposure to mild, unpredictable stressors alter the behavioural concomitants of hedonia, which may be indexed by the decreased consumption of preferred substances and lowered preference for contextual cues associated with drug reward. The social isolation procedure seems to have been applied, to a large extent, to tests of social learning performance in tests of social cooperation to obtain reward. Thus, Willner hypothesizes an interesting dissociation whereby the anhedonia effects of stress induction reflect dysfunctions of dopaminergic neurotransmission in the nucleus accumbens whereas the performance deficits induced by social isolation may reflect alterations of serotonergic neurotransmission. Thus, for example, the poor performance of singly housed rats in the social cooperation task may be restored by chronic antidepressant treatment (Willner et al., 1989). Two other major features of this chapter are important and should be attended to: Table 2 listing the major symptoms of certain chosen models and the section on patterns of depressive symptomatology.

Their three chapters offer a refreshingly varied and timely approach to the issue of animal models in depression and one may be considered fortunate in being instructed to discuss them. Panksepp's chick and mouse isolation procedures are potentially of great interest but much important experimentation must remain. What are the precise implications of the waning in distress vocalisations? Have tests of learning performance been applied to these preparations? Have other antidepressant compounds apart from imipramine been assessed? Such evidence would allow some preliminary consideration of the neurochemical substrates. Certainly, the changes measured in brain CGRP and substance-P 24 hours after social isolation are worth noting, especially in view of reliable alteration of substance-P levels following chronic antidepressant treatment (e.g. after repeated injections of imipramine, alaproclate or zimeldine, cf. Brodin et al., 1984; Bartfai et al., 1986). Porsolt circumvents the essential issue of animal model validity by using the distinction between pharmacological and psychological models. However, by definition the pharmacological models may, perhaps, be more expediently referred to as drug screens. His assessment of a number of drug screens in terms of the requirements set is both informative and instructive although one may show concern over the overly psychological labels applied to so-called pharmacological models, "behavioural despair" and "learned helplessness". Willner's application of a serotonin-dopamine dissociation to account for the performance deficits as opposed to anhedonia remains appealing although one ought not to be overly sanguine. Understandingly, the problem of mood is sidestepped: might this be a noradrenergic-serotonergic modulation? (see Commissaris and Fontana, 1990; Plaznik et al., 1989).

The issue of individual differences is becoming increasingly attended to and must be eventually incorporated into both drug screens and putative animal models; certainly it is evident from clinical evidence that many individual variation seen clinically may stem from neuroendocrine changes (as for example is observed in post-menstrual mood changes and decreased cognitive performances). This particular endocrine feature of affective disorders, that seems not to be discussed in any of above papers, refers for example to basal levels of circulating corticosteroid hormones, an important aspect in the olfactory bulbectomy procedure which is quite consistently applied, especially for the testing of new selective 5-HT uptake inhibitors. Brain glucocorticoid and mineralocorticoid receptors are differentially involved in the homeostatic mechanisms involved in adaptive response to stress at limbic-hypothalamic sites (De Kloet, 1990). The individual differences occurring in limbic mineralocorticoid and glucocorticoid receptors are subject to alteration by developmental experiences resulting in an increased complexity in the functional correlates applied as the dependent variables, i.e. the test procedures. Consideration for neurotransmitter-hormone and neurotransmitter-neurotransmitter interactions in models of depression are necessary in view of the coexistence phenomenon (e.g. Lundberg and Hökfelt, 1983). Repeated administration of amitriptyline over two weeks was found to both elevate spinal cord tissue levels and ex vivo release of substance-P (SP) and thyrotropin releasing hormone (TRH) (Bartjai et al., 1986). Chronic treatment with imipramine was chosen to increase SP and TRH immunoreactivity in ventral horn terminals (Kitayama et al., 1987) and also glucocorticoid receptor immunoreactivity in the locus coeruleus and cell groups of the rostral ventromedial medulla without altering basal levels of corticosterone and aldosterone (Kitayama et al., 1988). Finally, repeated treatment over 14 days with either zimelidine, alaproclate and imipramine altered both tissue levels of SP and the immunoreactivity of some tachykinins, e.g. neurokinin A and B (Brodin et al., 1985, 1987). Taken together, these threads indicate a much greater level of complexity is required if sufficient status is to be obtained by animal models of depression and the clinical realities adequately reflected, as for example the reduced number of glucocorticoid receptors in the lymphocytes of depressed patients (Whalley et al., 1986).

Bearing in mind the general observation that antidepressant compounds do not act immediately to alleviate depressive illness, it is necessary to consider, as suggested by Porsolt (this volume), the chronic vs acute effectiveness of drug screens in assessing validity. He suggests an exit from the dilemma by adapting (or adopting) an experimental model that indicates both the acute and chronic effectiveness of antidepressant compounds but that shows greater effectiveness after the chronic regime. Again, this avenue provides a pragmatic response to one of the strongest constraints on the usefulness of depression models and may be the most reasonable course open. However, one may pose the question: how closely do the behavioural, biochemical and neuroendocrine correlates analyzed in existing models/screens (if indeed they are done so reliably) compare with their clinical counterparts in persons with depressive disorders? Perhaps the test models applied in the laboratory never achieve the degree of severity, in respect to the broad spectrum of symptoms, that is registered in the clinic. The insistence on chronic rather than acute efficacy may therefore be misleading and even counterproductive. Thus, it may be legitimate to uphold the basic assumption that in the application of certain drug screens the acute effects may correspond to the delayed clinical effect.

Seiden and co-workers (e.g. O'Donnell and Seiden, 1982, 1983; Seiden et al., 1985; Marek and Seiden, 1988) have provided over several years the definitive work on the reliability and usefulness of the differential-reinforcement-of-low-rates-of-responding 72-seconds (DRL72s) schedule of reinforcement as a drug screen of antidepressant and potential antidepressant drugs. Despite the drawback of an extended period of training in the operant conditioning schedule in order to obtain a stable baseline (between 10 and 14 weeks), the DRL72s procedure has much evidence to commend it on the basis of sensitivity and selectivity of drug action. Generally, three dependent variables are

Kitayama, I., Janson, A.M., Fuxe, K., Agnati, L.F., Cintra, A., Ögren, S.O., Härfstrand, A., Eneroth, P. Tsutsumi, T., Jansson, G., Steinbusch, H.W.M., Visser, T.J. (1987) J. Neural Transm. 70: 251-285.

Kitayama, I., Janson, A.M., Cintra, A., Fuxe, K., Agnati, L.F., Ögren, S.O., Härfstrand, A., Eneroth, P., Gustafsson, J.Å. (1988) J. Neural Transm. 73: 191-203.

Lehr, E. (1989) Psychopharmacology 97, 145-146.

Lundberg, J.M., Hökfelt, T. (1983) Trends in Neurosci. 6: 325-333.

Marek, G.L., Seiden, L.S. (1988) J.Pharmacol. Exp. Ther. 244, 650-658.

Mc Kinney, W.T. (1977) In: Animal Models of Psychiatry (E.Usdin and I.Hanin, Eds), Pergamon, Oxford, pp 117-126.

O'Donnell, J.M. Seiden, L.S. (1982) Psychopharmacology 78, 214-218.

O'Donnell, J.M., Seiden, L.S. (1983) J. Pharmacol. Exp. Ther. 241, 68-75.

Panksepp, J. (1990) In: Psychobiology of Stress (S.Puglisi-Allegra and A.Oliverio, Eds), Dordrecht, The Netherlands. (In press).

Panksepp, J., Meeker, R., Bean, N.J. (1980) Pharmacol. Biochem. Behav. 12, 437-443.

Panksepp, J., Normansell, J.L., Herman, B., Bishop, P., Crepeau, L. (1988) In: The Physiological Central of Mammalian Vocalizations (J.D.Newman, Ed), Plenum Press, New York, pp 263-299.

Plaznik, A., Kostowski, W., Archer, T. (1989) Prog. Neuro-Psychopharmacol. & Biol. Psychiat. 13, 623-633.

Porsolt, R.D., Roux, S., Jalfre, M. (1984) Pharmacol. Biochem. Behav. 20, 979-981.

Seiden, L.S. Dahms, J.L., Shaughenessy, R.A. (1985) Psychopharmacology, 86, 55-60.

Suomi, S.J., Seaman, S.F., Lewis, J.K., De Lizio, R.D., McKinney, W.I. (1978) Arch. Gen. Psychiatry, 35, 321-325.

Whalley, I.J., Borthwick, N., Copolow, D., Dick, H., Christic, J.E., Fink, G. (1986) Br. Med. J. 292: 859-862.

Willner, P., Sampson, D., Phillips, G., Fichera, R., Foxlow, P., Muscat, R. (1989) Behav. Pharmacol. 1, 85-90.

part 3:

PSYCHOSIS

Animal Models in Psychopharmacology
Advances in Pharmacological Sciences

PHARMACOLOGICAL MODELS IN THE DEVELOPMENT OF ANTIPSYCHOTIC DRUGS - NEW STRATEGIES

B. Costall, A.M. Domeney, M.E. Kelly and R.J. Naylor

Postgraduate Studies in Pharmacology, The School of Pharmacy, University of Bradford, Bradford, BD7 1DP, United Kingdom

SUMMARY: In order to advance new therapeutic approaches to the treatment of psychosis it has been necessary to develop new animal models which diverge considerably from the traditional tests for neuroleptic activity. The latter generally reflected drug action on all cerebral dopamine receptors, but were not discrete or persistent in nature. In developing new tests it has been considered that the disturbances of schizophrenia are discrete to the mesolimbic system, and are persistent. The discrete disturbances have been mimicked by intracerebral locations using stereotaxic surgery and persistency of dopamine disturbance by intracerebral infusions from osmotic minipumps. Examples are given of the behavioural consequences of dopamine disturbance in mesolimbic terminal areas, the nucleus accumbens and amygdala, and of disturbance in the ventral tegmental area to initiate the release of endogenous dopamine. Examples of the ways in which the new test strategies have been used have focused around the new class of potential antypsychotic agents, the 5-HT_3 receptor antagonists, and have thus allowed the investigation of serotonin involvement in limbic dopamine dysfunction. In the presence of a raised limbic dopamine activity caused by the injection or infusion of dopamine agonists into the nucleus accumbens or amygdala, or direct stimulation of dopamine cells in the ventral tegmental area, rats and marmosets demonstrate an increased locomotor activity. 5-HT_3 receptor antagonists such as ondansetron, zacopride or ICS205-930 administered locally into the limbic nuclei or peripherally antagonised the behavioural hyperactivity. Neuroleptic agents caused a similar inhibition but, unlike the 5-HT_3 receptor antagonists, depressed activity to below normal values. The behavioural models reveal that 5-HT_3 receptor antagonists can inhibit a raised mesolimbic dopamine function in the total absence of an effect on normal levels of behavioural responding. The profile of action of the 5-HT_3 receptor antagonists is indicative of a novel antipsychotic potential. These compounds will therefore serve to test the validity of the new test procedures, and data from their trials in the clinic is eagerly awaited.

Major initiatives in the development of novel antipsychotic agents have been hindered by a lack of suitable animal models, and a lack of novel 'tools' to challenge those models. Thus, for many years the behavioural backing to the development of neuroleptic agents was designed about the activities of these agents in their own right - their ability to induce catalepsy, to inhibit stereotypies, to induce circling behaviour, to inhibit climbing, and so forth (see review by Costall and Naylor, 1980). The key element of all of these tests is now evident - they require a striatal dopamine blockade, possibly combined with an inhibition of psychomotor initiative in the limbic system. There now appears a limited rationale for such tests, although they did serve a most valuable role in the detection of new chemical classes of neuroleptic drugs. The basis for development was dopamine blockade with a lack of selectivity to any particular brain region. Indeed, the classic tests measured ability to modulate striatal dopamine function - but the importance of this action relative to the dopamine disturbance of schizophrenia, which is believed to influence the limbic and cortical circuitry, remained uncertain. Indeed, that neuroleptic agents antagonise at striatal dopamine receptors reflects their undoubted ability to induce extrapyramidal side effects - dystonias, parkinsonism, tardive dyskinesias (see review by Marsden et al., 1986). Our approaches to the development of neuroleptic agents was necessarily pragmatic; we searched for dopamine antagonist action in the brain per se, that would in turn help to reveal the prime therapeutic target sites.

The dopamine hypothesis of schizophrenia: The dopamine hypothesis of schizophrenia has remained a keystone in the search for novel antipsychotic therapy (see review by Costall and Naylor, 1986). The differences in thinking are now in terms of how this hypothesis can be refined. Firstly, as commented on above, it is now considered essential to locate the disturbance of dopamine in schizophrenia to a discrete brain circuitry. To administer a dopamine agonist peripherally to influence all cerebral dopamine receptors is an unlikely reflection of the disease state - it is believed that any 'model' of psychosis should concentrate on local disturbance of the mesolimbic/cortical circuitry. To this end, a battery of tests has been developed via which a target site in this circuitry is located by stereotaxic techniques, whether this be an area of cell bodies supplying the so-called mesolimbic/ cortical areas (the ventral tegmental area) or a terminal region such as the nucleus accumbens or amygdala.

A second major consideration, in addition to the discrete enhancement of mesolimbic dopamine function, is the persistency of the dopamine disturbance. Whilst the treatment of schizophrenia tends to be directed to an acute or chronic presentation, in no situation is it an event of a few minutes or hours. Thus, to disturb a discrete mesolimbic nucleus persistently to cause a dopamine excess in the area would appear to be the most ideal 'model' for assessment of antipsychotic drug activity. This has been achieved experimentally by use of Alzet osmotic minipumps, located subcutaneously in the scapula region of an animal, and connected to chronically indwelling injection units in a mesolimbic region, nucleus accumbens or amygdala (Costall et al., 1982; Bradbury et al., 1985). This technique allows one to effect a slow and persistent release of dopamine into the selected terminal site for many days. The consequent measures can be simple or complex, as will be described later in this chapter.

A criticism which may be levelled at any technique using endogenously administered dopamine or dopamine agonist is that this may provide a different behavioural profiling to that which may be achieved should the dopamine excess be endogenous. To approach this problem a series of substance P analogues, and initially substance P itself, have been injected by intracerebral route discretely into the ventral tegmental area - the biochemical consequence is an increase in dopamine release in terminal regions, including the amygdala and nucleus accumbens (Hagan et al., 1987; 1990). The behavioural consequence is enhanced psychomotor drive. Both can be measured to allow an appreciation of drug activity not only on the changed behaviour but also on its biochemical correlate.

New test procedures: Taking these approaches together, a battery of useful tests has been assimilated. Each test in its own right may have limitations, but together the tests overcome many of the criticisms of earlier methods. All approach the mesolimbic system as a discrete structure, via the cell body area, or the terminal structures which have an important influence on psychomotor drive, mood and emotion (nucleus accumbens and amygdala respectively), and take a combination of acute or, more ideally, persistent disturbances in dopamine function. Also important, the tests have been extended from rodent to primate. Hence, the full battery of tests now used to profile both existing or novel potential antipsychotic compounds is as follows:

1. Acute administration of amphetamine into the mesolimbic nucleus accumbens of rat
2. Continuous infusion of dopamine into the nucleus accumbens of rat
3. Continuous infusion of dopamine into the amygdala of rat
4. Continuous infusion of dopamine into the nucleus accumbens/ ventral striatum of marmoset
5. Acute administration of the neurokinin agonist DiMe-C7 into rat ventral tegmental area

Overall, with many novel compounds, the aim has been to develop approaches to modulate the mesolimbic dopamine excess without direct influence on dopamine receptors, particularly those outside the mesolimbic circuitry. Generally, therefore, the search has been for compounds which lack side effect potential and, therefore,

- do not cause catalepsy
- do not inhibit normal locomotor activity
- do not antagonise stereotypies
- do not displace [^{3}H]spiperone from its binding sites
- do not elevate prolactin
- do not cause circling behaviour in the classic Ungerstedt model, or even when placed directly into one striatum
- do not influence spontaneous or dopamine agonist-induced climbing behaviour

A caution may be added in taking such a strict approach since many investigators believe that a mix of a neuromodulating compound with some dopamine blocking action may be more ideal. Higher doses of such compounds may better control the initial overt behavioural problems of schizophrenia, and the neuromodulatory component may subsequently dominate as the dose is lowered, to provide a chronic control without the side effect potential.

The ability of novel potential antipsychotic agents to inhibit a raised limbic dopamine function: An example of how compounds may modify dopamine function independent of a dopamine receptor blockade is best achieved by reference to a group of novel psychopharmacological agents now classified widely as the 5-HT$_3$ receptor antagonists. These compounds have been established as having ability to correct disturbed neurotransmitter function, including an ability to inhibit a raised mesolimbic dopamine activity (Costall et al., 1987). The 5-HT$_3$ binding sites in the brain are not dense, but are very discretely located. "Hot spots" in the amygdala, hippocampus and limbic cortex are particularly prominent (Kilpatrick et al., 1987) and may provide the necessary sites of drug action.

The 5-HT$_3$ receptor antagonists, including ondansetron, zacopride, ICS205-930 and granisetron, are all able to inhibit the behavioural consequences of a raised mesolimbic dopamine function (see review by Costall et al., 1990). The actions of ondansetron and zacopride have been most widely explored, and their profiles have been established using a number of neurochemical approaches in different laboratories. Zacopride has proved to be one of the best radioligands of a series which have been produced (Barnes et al., 1988) and has been used to define the topography of receptor distribution in the brain, where the structures identified, e.g. amygdala, cortex, can be linked well with predicted requirements for locus of activity in psychosis.

In the behavioural tests ondansetron, zacopride and other 5-HT_3 receptor antagonists can effect changes reflective of those achieved using known antipsychotic agents, chlorpromazine, fluphenazine, haloperidol, clozapine, sulpiride. Thus, in the model where amphetamine (20μg bilateral) is injected into the rat nucleus accumbens a hyperactivity develops after a delay of some 10 minutes and then persists for a maximum of 1 hour. This is dose-dependently antagonised by the compounds mentioned and by ondansetron (Costall et al., 1987). It is worthy of note that whilst in all other tests it will be evident that the 5-HT_3 receptor antagonists are more potent that the classical or atypical neuroleptics, against amphetamine hyperactivity they are required, like the traditional compounds, in low mg per kg doses. The intra-accumbens amphetamine model using rat is particularly valuable to demonstrate the influence of the 5-HT_3 receptor system. Thus, whilst 5-HT_3 receptor antagonists can inhibit the amphetamine response, the 5-HT_3 agonist 2-methyl-5HT can enhance the response, and this enhancement may be antagonised by an agent such as ondansetron, in ng per kg doses (Costall et al., 1987).

The amphetamine model gives valuable first clues as to the potential inhibitory effects of any new group of compounds, but the intensity of response induced often renders consistent inhibition difficult, and so a second approach was adopted in our laboratories to determine the behavioural consequences of continuously infusing dopamine (25μg/24h, 0.48μl/h) into the nucleus accumbens of rat. The behavioural consequences are robust and peaks of hyperactivity develop in a phasic manner over a 13 day period of dopamine infusion (Costall et al., 1982). A continuation of infusion time can extend the occurrence of peak responding, and all can be inhibited by classical neuroleptics, atypical compounds such as clozapine, and by 5-HT_3 receptor antagonists such as ondansetron (Costall et al., 1984; 1987; 1990). The major differences lie in the doses required to effect a control of the behavioural problems of the dopamine infusion: these are in the mg per kg range for agents such as fluphenazine and in the μg per kg range for an agent such as ondansetron. Further differences are the effects of high dose intervention - the classical neuroleptics cause undue motor depression and sedation, whilst at high doses of 5-HT_3 receptor antagonists efficacy is lost. There is presently no explanation for this loss of efficacy of the 5-HT_3 antagonist compounds at high dosage, but the so-called bell-shaped response curve has become a characteristic of the first compounds developed in the series. For some compounds, such as granisetron, the bell-shaped curve is tight, and the effective dose range is narrow. However, the effective dose ranges are generally extremely wide, and it is interesting to note that whilst the safety margin for the classical neuroleptics (before sedation or motor depression occurs) is in the range 2.5-10, for the 5-HT_3 receptor antagonists such as ondansetron and zacopride (before loss of efficacy) the safety margin is in the order of 1000 (see Costall et al., 1990).

The intra-accumbens dopamine infusion model has been valuable in detecting the actions of other non- or atypical-neuroleptic agents having antipsychotic potential. Thus two compounds desenkephalin-γ-endorphin (dEγE) and Org 5222, a tetracyclic compound structurally related to mianserin and having mixed dopamine and 5-HT_2 antagonist action, are effective in the dopamine infusion model (Costall et al., 1989; 1990). Preliminary clinical evidence indicates that Org 5222 has antipsychotic potential (Tonnaer et al., 1990) and studies investigating the antipsychotic potential of dEγE are under way.

The chronicity of the dopamine infusion model lends it several advantages. One can follow the disturbances in cognitive and mood performance during the infusion period, and these changes, and drug influence on them, will be the subject of further reviews. Also important is the fact that the model can attempt to answer questions about the influence of prior neuroleptic treatment on the activity of a new compound. This is an important opportunity since any new compound being placed in clinical trial is most likely to be given to patients receiving or withdrawn from classical neuroleptic therapy. If the behavioural consequences of dopamine infusion are inhibited throughout a 13 day period by haloperidol, which is then abruptly withdrawn, an intense and prolonged rebound hyperactivity develops (Costall et al., 1984). This can be inhibited by a reinstatement of the classical agent, but at even higher doses than those which were used to inhibit the dopamine infusion effects. In contrast, a compound such as ondansetron can inhibit the rebound hyperactivity caused by withdrawing neuroleptic cover of a mesolimbic dopamine excess at μg per kg doses. Further, a single challenge with ondansetron effects an inhibition of the rebound response for several days. Indeed, when ondansetron was given at 0.1 mg/kg the inhibition of the rebound response was significant for 10 days (Costall et al., 1987). An inevitable, and tempting, conclusion is that the more severe the mesolimbic disturbance and the behavioural consequences, the more effective an agent such as ondansetron would appear to be. Whatever, the animal data would indicate that prior exposure to a neuroleptic agent does not appear to compromise the antipsychotic potential of a 5-HT_3 receptor antagonist.

The same procedures as described in the rat can be replicated in the marmoset, which extends the findings from rodent to a primate species (Costall et al., 1987). The phasic increases in psychomotor drive initiated by dopamine infused into marmoset ventral striatum, the primate area equivalent to rodent nucleus accumbens, is generally less phasic, but pronounced. This increased psychomotor drive is particularly sensitive to blockade by the neuroleptic agents, indeed, it is very difficult to effect a blockade of the dopamine hyperactivity response in marmoset without reducing locomotion to levels below control values. As in the rat, 5-HT_3 receptor antagonists can inhibit the behavioural consequences of the mesolimbic dopamine excess induced in the marmoset, again in the μg per kg dose range. Ondansetron and zacopride have been shown to be particularly effective, but also other novel compounds such as dEγE and Org 5222, which are detected in the above rodent tests, have demonstrated activity in the marmoset model (Costall et al., 1989; 1990).

The dopamine infusion technique may be extended from the nucleus accumbens to the amygdala. However, the function of the amygdala appears to be considerably more complex in that this is lateralised (see Bradbury et al., 1985). Thus, whilst dopamine infusions can be made bilaterally into the nucleus accumbens to increase psychomotor drive, this behavioural change can only be effected from the left amygdala when this is located in the subdominant hemisphere (the dominancy being determined in a simple turn preference test). When dopamine is infused slowly and persistently into the left subdominant amygdala the ensuing psychomotor drive, which is phasic in nature, although with fluctuations, can be inhibited by typical neuroleptics and by 5-HT_3 receptor antagonists, the former in mg per kg doses, the latter in μg per kg doses. It is interesting that in this particular test procedure in rat the loss of efficacy of the 5-HT_3 receptor antagonists is not so evident (Costall et al., 1987).

It is interesting when commenting on the apparently unusual laterality of the amygdala that there is considerable evidence in the literature to support this concept. Possibly the most compelling biochemical evidence was generated by Reynolds (1983) who studied the catecholamine neurotransmitter concentrations in bilaterally dissected post-mortem brain of schizophrenics. He reported marked differences in the dopamine content of the right and left amygdala, the levels being elevated approximately two-fold in the left hemisphere. Usually one is able to level criticism at post-mortem findings in that the biochemical changes may reflect the consequences of a long period of neuroleptic treatment, often in combination with other psychotherapeutic agents. However, this could not be expected to be the situation in the studies reported by

Reynolds since both hemispheres would have responded to the consequence of neuroleptic/psychotropic drug challenge.

The complexity of the inter-amygdala 'communication' system is intriguing, and it is relevant to mention that both the traditional neuroleptics and the 5-HT_3 receptor antagonists are able to inhibit the behavioural consequences of intra-amygdala dopamine infusion when given peripherally, intracerebrally into the amygdala of dopamine infusion or into the opposite, right amygdala (Costall et al., 1987). There are several points to note here, most obvious is that the right and left amygdalae must be differentiated, particularly for neurochemical analyses. Also important, the data confirms that one hemisphere of the brain can alter its function in response to disturbances in the other. It is hoped that findings of this kind will emphasise the caution required when, behaviourally or biochemically, manipulations are made on one side of the brain and the other side is used for 'control' analyses.

To complement the tests which have relied on exogenously applied dopamine to discrete mesolimbic nuclei, procedures were developed whereby a stable substance P analogue [DiMe-C7, [$pGlu^5$, $MePhe^8$, Sar^9]SP-(5-11)] is injected into the cell body area of the rat mesolimbic dopamine system, the ventral tegmental area (Hagan et al., 1987; 1990). The consequence is a release of endogenous dopamine in terminal areas, including the nucleus accumbens and amygdala, which leads to a behavioural response of increased psychomotor drive, measured as an increase in locomotor activity. This behaviour develops rapidly after the intracerebral injection, peaks after 10 to 20 minutes, and persists for some 60 minutes. It is inhibited by a classical neuroleptic such as haloperidol, and by novel agents such as ondansetron or zacopride, the former being effective in mg per kg doses, the latter in μg per kg doses. It is important that the 5-HT_3 receptor antagonists can be shown to at least partially correct the biochemical disturbance caused by the DiMe-C7 injection: thus, a DOPAC/DA ratio which increases by some 70-80% above control values is corrected back to an order of 25% by ondansetron (50μg/kg).

Questions raised by the new behavioural strategies - The nature of the 5-hydroxytryptamine-dopamine interaction and future studies: The ability of 5-HT_3 receptor antagonists to inhibit an excessive dopamine function is indicative of a permissive role of 5-HT on the dopamine system. The precise nature of the 5-HT-dopamine interaction remains to be defined but may involve a regulation of the release of dopamine in the mesolimbic/cortical systems. Thus, measurement of dopamine release in the rat nucleus accumbens, using trans-cerebral dialysis techniques, indicates that the peripheral administration of ICS205-930 prevents the stimulation of dopamine release induced by haloperidol, nicotine, morphine and other drugs that are known to stimulate the firing activity of dopamine neurones (Imperato and Angelucci, 1989; Carboni et al., 1989). It is possible that 5-HT might exert an excitatory effect on cells located in the ventral tegmental area to facilitate dopamine cell firing. A 5-HT_3 receptor antagonist effect at such sites would provide a midbrain site of 5-HT-dopamine interaction. Lack of action of the 5-HT_3 receptor antagonists on normal behaviour also correlates with a lack of ICS205-930 to modify basal dopamine release in the nucleus accumbens (Carboni et al., 1989).

In additional and important studies using the dialysis technique, Imperato et al. (1990) have reported that ICS205-930 prevents stress induced activation of limbic and cortical dopamine release with again, no effect on basal release. The failure of diazepam to prevent such changes indicates that the effects of ICS205-930 are unlikely to be achieved via an anxiolytic action. Given the importance of stress in activating the dopamine system and precipitating deterioration in schizophrenic illness, this particular action of ICS205-930 is worthy of much more detailed investigation using other 5-HT_3 receptor antagonists. It may involve a midbrain site of action but there are alternative possibilities.

Perhaps the most important alternative is an action of the 5-HT_3 receptor antagonists in the forebrain structures to directly modify dopamine release. Blandina et al. (1988) were the first to report that 5-HT and 2-methyl-5-HT induced a release of dopamine from rat striatal slices which was blocked by ICS205-930. However, such data cannot be immediately related to a functional role of the 5-HT_3 receptors since the 5-HT_3 receptor antagonists have no known ability to modify striatal function. Also, although the more selective 5-HT_3 agonist phenylbiguanide causes a concentration-dependent release of dopamine from striatal slices, ICS205-930 and another 5-HT_3 receptor antagonist, MDL73,147, failed to antagonise the dopamine induced release (Schmidt and Black, 1989). It was concluded that the release observed in these experiments was Ca^{++} independent and mediated via a carrier process. It would be of interest to extend these 'in vitro' studies to mesolimbic tissue, and to include 'in vivo' microdialysis experiments using local drug additions and sampling, to more clearly define the effect of 5-HT_3 agonist/antagonist action on dopamine release in the limbic forebrain.

Evidence for a forebrain site of 5-HT_3 agonist/antagonist interaction relevant to antipsychotic action has also been obtained from microiontophoretic studies. Ashby et al. (1989a, b) have reported that the microiontophoretic application of 2-methyl-5-HT to medial prefrontal cortical cells suppresses cell firing: this effect is blocked by ICS205-930 and granisetron, and by the atypical neuroleptic agent clozapine. These observations are particularly interesting and, whilst difficult to reconcile with the well known actions of 5-HT_3 receptors to mediate depolarising responses (see Elliott and Wallis, 1990), are clearly worthy of detailed investigation in different limbic-cortical structures.

Further developments in the behavioural studies are also desirable. The potential of 5-HT_3 receptor antagonists to modify a behavioural perturbation in the dopamine system was measured as changes in locomotor activity. However, there is also a loss of sensorimotor gating following the peripheral administration of dopamine agonists in rodents which probably involves the mesolimbic dopamine system. Thus, the injection of dopamine into the rat nucleus accumbens caused a dose-dependent decrease in prepulse inhibition of an acoustic startle response (Swerdlow et al., 1990), and such observations may be relevant to the deficits in prepulse inhibition shown in schizophrenic patients (Bratt et al., 1978). Abnormal dopamine activation in the limbic system may modify the initiation and organisation of appetitive behaviour by allowing increased amounts of sensory information to activate lower mechanisms. A loss of limbic gating may result in information overload and the possibility of cognitive fragmentation. In future behavioural studies investigating a 5-HT_3 receptor role in moderating limbic dopamine function it may be highly pertinent to consider drug action on gating mechanisms.

CONCLUSIONS

The new strategies which have been described for a search for novel antischizophrenic drugs have departed from simple behavioural approaches. It is accepted that the new tests require considerable experimental skills and are time consuming. However, they have diverted away from the classical me-too tests, tests relying heavily on side-effects of drugs, and do provide an alternative way forward. They have been used to reveal novel mechanisms for the control of limbic dopamine function and provide a basis for continued research into our understanding of the mechanisms underlying the disturbances of psychosis.

REFERENCES

Barnes, N.M., Costall, B., Naylor, R.J. (1988) J. Pharm. Pharmac. 40, 548-551.

Blandina, P., Goldbarb, J., Green, J.P. (1988) Eur. J. Pharmacol. 155, 349-351.

Bradbury, A.J., Costall, B., Domeney, A.M., Naylor, R.J. (1985) Neuropharmacology 24, 1163-1170.

Bratt, D., Stone, C., Callaway, E., Geyer, M., Glick, I., Bali, L. (1978) Psychophysiology 15, 339-343.

Carboni, E., Acquas, E., Frau, R., DiChiara, G. (1989) Eur. J. Pharmacol. 164, 515-519.

Costall, B., Domeney, A.M., Kelly, M.E., Naylor, R.J., Tomkins, D.M. (1990) Pharmacol. Biochem. Behav. 35, 607-615.

Costall, B., Domeney, A.M., Naylor, R.J. (1982) Neuropharmacol. 21, 327-335.

Costall, B., Domeney, A.M., Naylor, R.J. (1984) Neuropharmacol. 23, 287-294.

Costall, B., Domeney, A.M., Naylor, R.J. (1986) In: The neurobiology of dopamine systems (W.Winlow, R.Markstein, Eds), A. Wheaton and Co. Ltd., Exeter, pp 253-265.

Costall, B., Domeney, A.M., Naylor, R.J. (1989) Neuropharmacology 28, 1223-1229.

Costall, B., Domeney, A.M., Naylor, R.J., Tyers, M.B. (1987) Br. J. Pharmacol. 92, 881-894.

Costall, B., Naylor, R.J. (1980) Rev. Pure Appl. Pharmac. Sci. 1, 3-83.

Costall, B., Naylor, R.J. (1986) In: The psychopharmacology and treatment of schizophrenia (P.B.Bradley and S.R.Hirsch, Eds), Oxford Medical Publications, Oxford, pp 132-165.

Costall, B., Naylor, R.J., Tyers, M.B. (1990) Pharmac. Ther. 49, 181-202.

Elliott, P., Wallis, D.I. (1990) Analysis of the actions of 5-hydroxytryptamine on the rabbit isolated vagus nerve. Naunyn-Schmiedeberg's Arch. Pharmacol. 341, 494-502.

Hagan, R.M., Bulter, A., Hill, J.M., Jordan, C.C., Ireland, S.J., Tyers, M.B. (1987) Eur. J. Pharmacol. 138, 303-305.

Hagan, R.M., Jones, B.J., Jordan, C.C., Tyers, M.B. (1990) Br. J. Pharmacol. 99, 227-232.

Imperato, A., Angelucci, L. (1989) Neurosci. Lett. 101, 214.

Imperato, A., Puglisi-Allegra, Z., Zocchi, A., Scrocco, M.G., Casolini, P., Angelucci, L. (1990) Eur. J. Pharmacol. 175, 211-214.

Kilpatrick, G.J., Jones, B.J., Tyers, M.B. (1987) Nature 330, 746-748.

Marsden, C.D., Mindham, R.H.S., Mackay, A.V.P. (1986) In: The psychopharmacology and treatment of schizophrenia (P.B.Bradley and S.R.Hirsch, Eds), Oxford Medical Publications, Oxford, pp 340-402.

Schmidt, C.J., C.K. Black (1989) Eur. J. Pharmacol. 167, 309-310.

Swerdlow, N.R., Bratt, D.L., Masten, V.L., Geyer, M.A. (1990) Psychopharmacology 101, 414-420.

Tonnaer, J.A.D.M., Room, P., Van Delft, A.M.L., Farde, L., Vrijmoed-de Vries and Sitsen, J.M.A. (1990) 17th Congress of Collegium Internationale Neuro-Psychopharmacologicum, Kyoto, Sept. 10th-14th, Abst. 013-3-7.

Animal Models in Psychopharmacology
Advances in Pharmacological Sciences

THE ETHOLOGICAL ANALYSIS OF MONKEYS IN A SOCIAL SETTING AS AN ANIMAL MODEL FOR SCHIZOPHRENIA

B.A. Ellenbroek

Psychoneuropharmacological Research Unit, Dept. of Pharmacology, University of Nijmegen, P.O.Box 9101, 6500 HB Nijmegen, The Netherlands

INTRODUCTION

Research into the etiology, pathology and therapy of schizophrenia has sofar not led to many breakthroughs. Despite the discovery, in the early fifties of the efficacy of neuroleptic drugs, few major advances have since occurred. This is of course in part due to the heterogeneity in the schizophrenic symptomatology but certainly also to the lack of adequate animal models.

The development of animal models has been hampered by the nature of the schizophrenic symptoms. Since most of the symptoms are of a cognitive (non-motor) nature it is hard (if not impossible) to mimic them in an animal model. For this reason, the animal models with face validity (i.e. those with symptom similarity) are based almost exclusively on the few motor symptoms present, i.e. catatonia and stereotypy. However, since none of the symptoms of schizophrenia are pathognomic (i.e. specific for the disease), it is dangerous to develop animal models based on only one symptom.

Most of the animal models for schizophrenia are, therefore, based on the principle of pharmacological isomorphism. This implies that the models are based on the effects of drugs known to be therapeutically effective. In the case of schizophrenia this implies that these models are based on the effects of neuroleptic drugs. Among these models are intracerebral self stimulation (Wauquier, 1979), conditioned avoidance behaviour (Arnt, 1982), the catalepsy test (Janssen et al., 1965) and the paw test (Ellenbroek et al., 1987; Ellenbroek & Cools, 1988). These animal models have the disadvantage that new drugs, discovered in these tests will probably work through a mechanism comparable to the reference compound used. Thus major new advances in the therapy are unlikely to occur using these models. Moreover, since these animal models are, in general, unrelated to schizophrenia they will not increase our insight into the etiology and pathology of the disease.

The most promising animal models are the ones that aspire construct validity. These try to model the psychopathological construct underlying the disease and are closely related to the models with cross species psychological processes (Matthysse, 1986). Most of these models are based on the construct that schizophrenic patients are unable to distinguish relevant from irrelevant stimuli. Several of these models, like latent inhibition and the startle response have been discussed in a recent review (Ellenbroek & Cools, 1990). A disadvantage of these models is that they are all based on a construct which may be valid for the acute phase of the disease but are probably not valid for the more chronic phase. The disturbance of latent inhibition, for instance, occurs in the acute but not in the chronic phase of schizophrenia (Baruch et al., 1988).

Animal models for the chronic phase of schizophrenia are extremely rare. This is important because patients in the acute and chronic phase differ in their symptomatology. Thus whereas positive symptoms predominate in the acute phase of the disease, negative symptoms are more prominent in the chronic phase. Positive symptoms are features which occur as a result of the disease. The most important positive symptoms in schizophrenia are hallucinations and delusions. Negative symptoms, on the other hand are features normally present but which are, as a result of the disease, strongly reduced or even absent. The most important negative symptoms in schizophrenia are flat affect, anhedonia and social withdrawal. Developing animal models for the more chronic phases of schizophrenia are of major importance because the common neuroleptic drugs are generally ineffective in this phase of the disease (see below).

THE BEHAVIOUR OF ANIMALS IN A SOCIAL SETTING

Although humans, like most other primate species are social animals, most neuro- and psychopharmacological research is done in isolated animals. Animal models for schizophrenia are also mainly based on the behaviour of single animals, despite the fact that changes in the social behaviour are a well known characteristic of schizophrenic patients. It might therefore be worthwhile to study the behaviour of animals in a social setting. This is all the more important because it has already been shown that the sensitivity of monkeys for drugs is greatly influenced by the animal's environment. Thus, Vrijmoed-de Vries & Cools (1985) have shown that the threshold for inducing motor deficits in Java monkeys (Macaca fascicularis) after intracaudate injections of carbachol is much lower when the animals are in their normal social setting.

Table 1 The ethogram of nonsocial behavioural items.

Code	Behavioural Item	Description/remarks
BAS		**Cluster of basic postures & locomotion**
SUP	Sitting upright	Normal sitting posture
STA	Standing	Standing on all four legs
STU	Standing upright	Standing on hindlegs only
HAN	Hanging	Hanging upside down from the ceiling
LYI	Lying	
LOC	Locomotion	All forms of locomotion
SELF		**Cluster of self-oriented behav. items**
STRE	Stretching	
RUB	Rubbing	With the back of its hand
SHA	Shaking	
YAW	Yawning	
SCRA	Scratching	Usually fast movements without looking
GOS	Self-grooming	Usually slower movements with looking
ANI	Anogenital insp.	Inspecting the anogenital region
OBJ		**Cluster of object-oriented behav. items**
INO	Inspecting object	Includes gnawing on the cage
SCA	Scanning	Looking around without fixating
SWIN	Swinging	Swinging while hanging on ceiling
DRIN	Drinking	
EAT	Eating	

Although the social behaviour of different species has been studied, including mice, rats and guinea pigs, the most consistent results have been obtained in the monkey. This is most likely due to the clearly identifiable social structure and the very rich repertoire in social behaviour (at least to the human observer). In order to study the behaviour of monkeys (or any other animal) in a systematic way, it is important to have a proper ethogram in which the various social and nonsocial behaviours are unequivocally defined. Many different ethograms have been compiled, ranging from

Table 2 The ethogram of the social behavioural items.

Code	Behavioural Item	Description/remarks
LOO		Subcluster looking
LOO	Looking	Clearly looking at one or more monkeys
TOC	Touching	
GROO		Subcluster grooming
GROO	Grooming	Other animals
HUDL	Huddling	Sitting closely together, usually with one arm over the shoulder of the other
WRES	Wrestling	always in a friendly way
SEX		Subcluster sexual behavioural items
INSP	Inspecting	Inspecting anogenital region
PRES	Presenting	Showing the anogenital region by lifting the tail. Directed to another
MON	Mounting	
COPU	Copulating	
GRAB	Grabbing	The female grabs the hindleg of the male during the copulation
AGGRES		Subcluster aggressive behavioural items
FIX	Fixating	Strongly focussing another animal
POI	Pointing	Movement of the head to other animal
BOB	Bobbing	Movement of torso towards other animal
LUN	Lunging	Movement of torso and forelegs towards other animals
FROW	Frowning	
OMO	Open mouth threat	
BOU	Bouncing	Jumping with four legs on the ground
PUL	Pulling	Usually at the fur of other monkeys
PUSH	Pushing	
MIS	Missing	Making a hitting movement, but missing on purpose
HIT	Hitting	
BITE	Biting	
SUBMIS		Subcluster submissive behavioural items
FAW	Facing away	Looking away from a threating monkey
GRI	Grinning	Retraction of the upper and lower lips
OGI	Open mouth grin	Grinning, but with mouth open
SMA	Smacking	rapidly opening and closing the mouth, usually with protruded lips

very broad categories to a very detailed description which can lead to more than one hundred different behavioural items (see van den Bercken, 1979). We have used a medium sized ethogram (see tables 1 and 2) in which individual behavioural items are clustered into five behavioural clusters. These include 1) Basic Postures and Locomotion, which are always present; 2) Self Oriented Behaviours, for which the animals needs neither his environment nor other monkeys; 3) Object Oriented Behaviours, for which a monkeys needs his environment but not other monkeys; 4) Active Social Behaviour, which are directed towards other monkeys, and 5) Passive Social Behaviours, which are received from other monkeys. The social behavioural clusters are further subdivided in certain subclusters (see table 2) to investigate the effects of drugs on specific social behaviours, such as affiliative, sexual or agressive behaviours. With the help of this ethogram it has been possible to study the effects of drugs on the behaviour of Java monkeys in detail.

THE EFFECTS OF AMPHETAMINE ON SOCIALLY HOUSED MONKEYS

Many different drugs influence the social behaviour of monkeys (see Smith & Byrd, 1983). However, for our discussion on animal models for schizophrenia, only the effects of psychostimulants and psychotomimetics seem relevant. Most studies have evaluated the effects of amphetamine on the behaviour of monkeys. This is not surprising given the fact that amphetamine primarily releases dopamine (Robinson & Becker, 1986) and that the dopamine hypothesis of schizophrenia is still the prevailing hypothesis (see McKenna, 1987; Carlsson, 1988).

This dopamine hypothesis is based not only on the finding that neuroleptic drugs, which primarily block the dopamine D_2 receptors are therapeutically effective in schizophrenia, but also on the finding that amphetamine can induce a syndrome closely resembling schizophrenia (Connell, 1958). Moreover, in schizophrenic patients amphetamine worsens the existing symptomatology without apparently inducing any additional symptoms (Janowsky & Davis, 1974). For that reason, the effects of amphetamine on the social and nonsocial behaviour of monkeys may be of value for schizophrenia.

Amphetamine has profound and complex effects on the nonsocial behaviour of various species of monkeys. In the Java monkey (Macaca fascicularis) amphetamine (0.5 mg/kg i.m) induces a decrease in the duration of self oriented behaviour and an increase

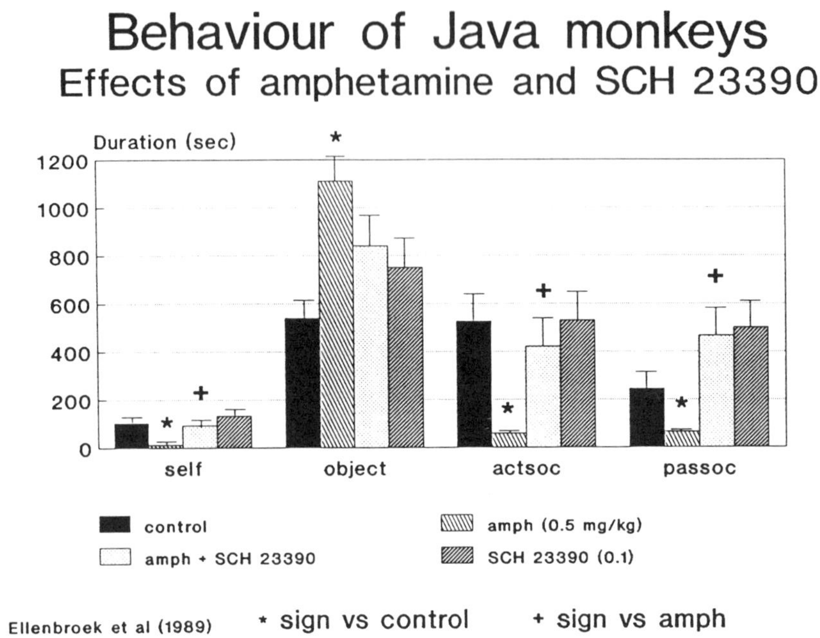

Fig. 1 Effects of amphetamine and SCH 23390 on the duration of behaviours in Java monkeys.

in object oriented behaviour (fig. 1, Ellenbroek et al., 1989). A closer inspection of these data shows that the increase in object oriented behaviour was entirely due to an increase in scanning the environment. The other items in this cluster (like inspecting object) were decreased. This increase in scanning should not be confused with the well known checking phenomenon seen in monkeys treated with amphetamine or apomorphine (Scraggs & Ridley, 1978; Schlemmer et al., 1980; Miczek & Yoshimura, 1982). Checking is defined as rapid eye and head movements. Schlemmer et al. (1980) have suggested that these movements result from an increase in vigilance. Ridley & Baker (1983) have, however, refuted this opinion. These authors placed marmosets (Callithrix jacchus) in a drum with windows that gave a view at the other monkeys. If checking is due to hypervigilance, the authors reasoned, then amphetamine should increase the time spent at the window. However, at 4 mg/kg i.p. a decrease rather than an increase in time spent near the windows was seen. Since this dose of amphetamine induces checking, the authors concluded that checking was unrelated to hypervigilance. However, since this dose of amphetamine also induces social isolation, the decrease in time spent at the window may be more related to social isolation than to checking. On the other hand, Schlemmer & Davis (1983) provided data to suggest that checking is also not a form of stereotypy, since, unlike stereotyped movements tolerance develops rapidly to checking (see also Ridley et al., 1979).

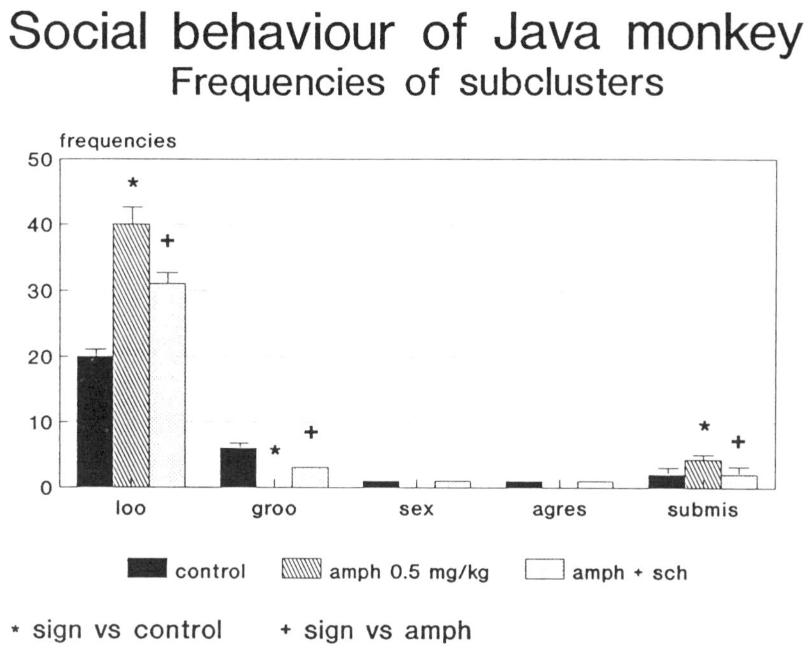

Fig. 2 Effects of amphetamine and SCH 23390 on the frequencies of behaviour in Java monkeys.

In rodents, amphetamine is well known for its locomotor stimulating effect. In monkeys, amphetamine also influences locomotor activity, but in a more complex manner. In some studies amphetamine was found to decrease locomotor activity (Ridley et al., 1979; Nielsen & Lyon, 1982; Ellenbroek et al., unpublished data), whereas others have found increases in locomotion after amphetamine (Miczek & Yoshimura, 1982) or methamphetamine (Crowley et al., 1974). Miczek & Gold (1983) have offered an explanation for these seemingly divergent data. These authors showed that amphetamine increased locomotion in subdominant and subordinate squirrel monkeys (Saimiri sciureus), while at the same time decreasing locomotor activity in dominant monkeys. Interestingly, the dominant monkeys are the animals with the highest level of spontaneous motor activity, whereas the subordinate monkeys have the lowest basal level of locomotor activity. Miller (1976) also reported an increase in locomotor activity after amphetamine treatment in the two lowest ranking rhesus monkeys (Macaca mulatta).

The most prominent change in nonsocial behaviour seen in monkeys after amphetamine treatment is stereotypy. Although amphetamine induces stereotypy in a large variety of animal species (Randrup & Munkvad, 1967), studying this phenomenon in monkeys in a social setting has some inherent advantages. Using this approach, Miczek & Gold (1983) showed that, in contrast to the effects on locomotor activity, amphetamine induces stereotypy in all monkeys, regardless of sex, age or social status. Interestingly, in contrast to many rodent species, a large number of different behaviours can become stereotyped. Among these are normal behaviours, such as self grooming (Schiorring, 1977; Schlemmer & Davis, 1983) and vocalisations (Bellarosa et al., 1980; Ellenbroek et al., 1989), but also abnormal behaviour such as rocking, defined as rhythmical movements of the torso (Ellenbroek et al., 1989). However, not only nonsocial but also social behaviours can become stereotyped, such as social grooming (Schiorring, 1977) and presenting behaviour (Bellarosa et al., 1980). Interestingly, Schlemmer & Davis (1983) have suggested that higher ranking stumptailed macaques (Macaca arctoides) develop more social stereotypies than lower ranking monkeys.

Table 3 Monkey species in which amphetamine or methamphetamine(*) induces social isolation after acute or chronic administration.

Species	Acute/Chronic	Reference
Macaca nemestrina*	Acute	Crowley et al. (1974)
Macaca arctoides	Chronic	Schlemmer et al. (1976)
Macaca mulatta	Chronic	Haber et al. (1977)
Cercopithecus aethiops	Acute	Schiorring (1977)
Callithrix jacchus	Acute	Scraggs & Ridley (1979)
Saimiri sciureus	Acute	Miczek & Yoshimura (1982)
Cebus apella	Acute	Lifshitz et al. (1987)
Macaca fascicularis	Acute	Ellenbroek et al. (1989)

Summarising the data on nonsocial behaviour, amphetamine seems to (1) increase scanning of the environment; (2) induce checking behaviour; (3) have a differential effect on locomotor activity; and (4) induce stereotyped behaviour, which can take the form of normal and abnormal behaviour and which can be of social or nonsocial form.

Like the effects on nonsocial behaviour, amphetamine has clearcut effects on the social behaviour. Indeed one of the most consistent findings in the behavioural pharmacology of amphetamine is probably its strong depressant effect on social behaviour in general. This has been observed in many primate species after amphetamine, methamphetamine or apomorphine (see table 3). This social isolation is most clearly observed as a decrease in the duration of social behaviour (see fig. 1) and an increase in spatial distance, the so called proximity (see fig. 3). However, a closer examination of the effects of amphetamine on social behaviour shows that amphetamine also has more subtle effects.

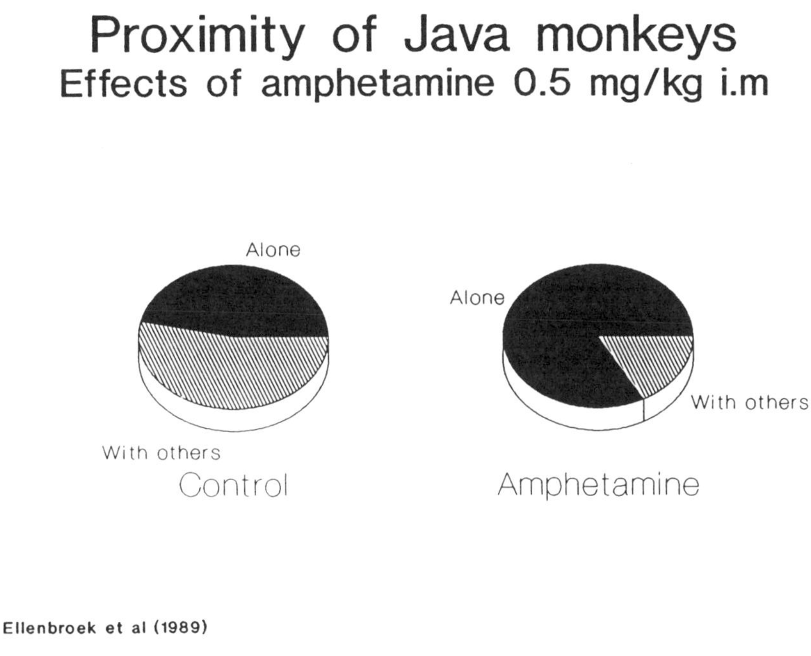

Fig. 3 Effects of amphetamine on the duration of proximity in Java monkeys.

This is already indicated in fig. 2. From this figure it can be seen that although amphetamine produced a strong decrease in the duration of both active and passive social behaviour, it did not reduce the frequency of these behaviours. In fact, amphetamine tended to increase the frequency of the active social behaviour. Since duration measures are dominated by long lasting behavioural items, whereas frequency measures are mainly determined by short lasting behavioural items, these data suggest that amphetamine predominantly reduces a specific subset of long lasting social behavioural items.

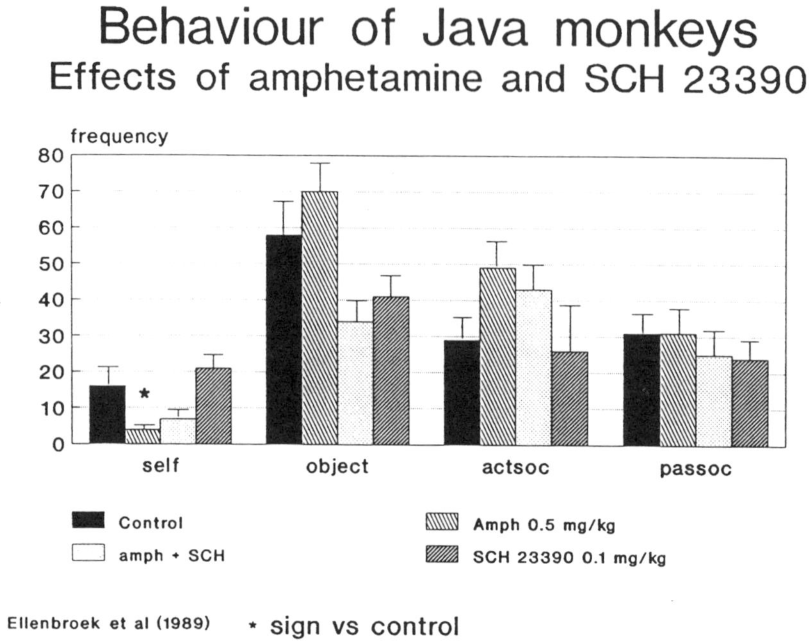

Fig. 4 Effects of amphetamine and SCH 23390 on the frequencies of the subclusters of social behaviour in Java monkeys.

Fig. 4 shows that this is indeed the case. In this figure we have analysed the effects of amphetamine on the frequency of the specific subclusters of active social behaviour. This figure shows that amphetamine decreases the frequency of affiliative behaviours (i.e. grooming), whereas a strong increase is seen in looking at other monkeys. Only a minor increase was seen in submissive behaviour whereas few sexual and aggressive encounters occurred in our animal group. These data are partly in agreement with the literature whereas in other respect they deviate from the results obtained by others. As discussed above, animals treated with amphetamine seem to respond more to changes in the environment (for instance the increase in scanning). This might explain the increase in looking at other monkeys, also reported by Haber et al., (1981). A strong increase in submissive gestures has been reported to occur after treatment with methamphetamine (Crowley et al., 1974) or amphetamine (Haber et al., 1977; Schlemmer & Davis, 1983). The, only minor, increase in submissive behaviour seen in our experiment (fig.

3) may be due to the fact that we treated mainly females in our experiment. Females generally show much less submissive behaviour than males.

The effects of amphetamine on aggressive behaviour are complex and equivocal. Thus Haber et al. (1977), Smith & Byrd (1983) and Lifshitz et al. (1987) found increases in aggressive behaviour, whereas Miczek & Gold (1983) found decreases in aggressive behaviour. Smith & Byrd (1985) studied the effects of amphetamine on aggressive behaviour in more detail and found that aggressive behaviour towards nonadults increased whereas it decreased towards adults. Moreover, they showed that aggression towards kin-related monkeys increased, whereas it decreased towards nonkin-related animals. Thus these data suggest that the effects of amphetamine on agonistic behaviour depend upon characteristics of the treated monkey as well as on the other group members. Moreover, the monkey species seems to be of importance. Rhesus monkeys, for instance are known to be particularly aggressive (see Haber et al., 1977), whereas in Java monkeys aggressive behaviour is much less frequently observed.

In conclusion, amphetamine induces a strong social isolation, observed as a decrease in affiliative behaviour and an increase in the time spent alone. Moreover, amphetamine increases the frequency of looking at other monkeys and of submissive behaviours. The effects of amphetamine on sexual and agonistic behaviours, on the other hand are less clearcut.

THE EFFECTS OF AMPHETAMINE AS AN ANIMAL MODEL FOR SCHIZOPHRENIA

After having described the effects of amphetamine on social and nonsocial behaviours of monkeys, we will next turn to the usefullness of these effects as an animal model for schizophrenia. In order to evaluate the validity of animal models it is necessary to develop specific criteria. As discussed by Willner (1984) and by Ellenbroek & Cools (1990) it is useful to subdivide these criteria into three subsets: criteria for predictive, face and construct validity (see also introduction).

We will start our discussion with face validity. These criteria analyse the symptom similarity of the model. Although, as already mentioned in the introduction, the nature of the schizophrenic symptoms make modelling difficult, some similarities are nevertheless apparent. Most obvious is, of course the occurrence of stereotypy in monkeys and in schizophrenic patients. The fact that both in schizophrenic patients and in amphetamine treated monkeys these stereotypies can occur in many different forms suggests that the stereotypy in monkeys is a better model than that in rats (which only show a very small repertoire of stereotyped behaviour). The second obvious parallel is the strong social isolation which occurs both after amphetamine treatment in monkeys and in schizophrenic patients. This similarity is enhanced by the finding that in monkeys the social isolation is an active social isolation (Annett et al., 1989). Thus amphetamine treated monkeys not only initiate less social interactions, but they also break social relations more often. Such an active social isolation is also characteristic for schizophrenic patients who are known to actively withdraw from their social environment. Thus two characteristic features of schizophrenia (stereotypy and social isolation) are modelled in one animal model. Moreover, these features represent typical examples of positive (stereotypy) and negative (social isolation) symptoms. Thus, the effects of amphetamine on monkeys allow us to study both the positive and negative symptoms of schizophrenia in one animal model.

Before turning to the criteria for predictive validity, it is necessary to recall that whereas the positive symptoms are sensitive to neuroleptic treatment, negative symptoms seem to be much more resistant (Johnstone et al., 1978; see table 4). In agreement with this it has been shown that the amphetamine induced stereotypy, but not the social isolation can be reversed by neuroleptic drugs, such as haloperidol and chlorpromazine (Miczek & Yoshimura, 1982; Schiorring, 1977). Although it is still unclear whether checking is a stereotyped behaviour (see previous section), haloperidol is also known to reduce the amphetamine induced increase in checking (Scraggs & Ridley, 1979). Several other criteria have also been developed for testing the predictive validity of an animal model for schizophrenia (Ellenbroek & Cools, 1988; 1990). Thus, the effects of neuroleptic drugs should not be reduced by anticholinergic drugs, and chronic treatment should also not reduce the effects of neuroleptics (see also Matthysse, 1981). Unfortunately, few studies have specifically looked at these criteria.

In contrast to the positive symptoms of schizophrenia, the negative symptoms are relatively insensitive to neuroleptic treatment (Johnstone et al., 1978). Therefore, the criteria for predictive validity as discussed by Ellenbroek & Cools (1990) are not valid in this case, since all these criteria are based on the effectiveness of neuroleptic drugs. Since no effective treatment for the negative symptoms has sofar been found, the analysis of the predictive validity rests solely upon negative results. Thus, it has been reported that several drugs are ineffective in treating the negative symptoms of schizophrenia. Among these drugs are adrenergic antagonists, opiate antagonists, benzodiazepines and neuroleptic drugs. As indicated in table 4, these drugs are also ineffective in antagonising the amphetamine induced social isolation in monkeys. The only drug which has sofar been found to prevent the amphetamine induced social isolation is the selective dopamine D_1 receptor antagonist SCH 23390 (Ellenbroek et al., 1989). Moreover, this drug has been found to not only reverse the effects of amphetamine, but to more or less completely normalise social behaviour. Unfortunately SCH 23390 has a very short duration of action and can therefore not be used in clinical practice.

Table 4 The comparable ineffectiveness of drugs to interfere with the negative symptoms of schizophrenia and with the amphetamine induced so-cial isolation in monkeys. *: SSRG: Scottish Schizophrenia Research Group.

Drug	Negative symptoms	Social isolation
Neuroleptics	Johnstone et al. (1978)	Schiorring (1977)
	Angrist et al. (198O)	Scraggs & Ridley (1979)
	Kucharski et al. (1984)	Miczek & Yoshimura (1982)
	SSRG* (1987)	
α-Adrenergic antagonists	Albus et al. (1986)	Scraggs & Ridley (1979)
β-Adrenergic antagonists	Myers et al. (1981)	Scraggs & Ridley (1979)
Benzodiazepines	Csernansky et al. (1988)	Scraggs & Ridley (1979)
Opiate antagonists	Githin et al. (1981)	Winlow & Miczek (1988)

The data sofar suggest that the effects of amphetamine on the behaviour of monkeys in a social environment represent an animal model with at least some face and predictive validity. In order to evaluate the construct validity of this model, it is essential to know the psychopathological construct underlying schizophrenia. Although this is, obviously, still largely unknown, there are many indications that schizophrenic patients have a lowered threshold for external stimuli. Moreover, these patients seem to be less able to differentiate between relevant and irrelevant stimuli (for a more complete discussion see Ellenbroek & Cools, 1990). This psychopathological disturbance is though to underly at least the positive symptoms. The disturbance underlying the negative symptoms is even less clear, although it has been suggested that these are due to a compensatory mechanism aimed at reducing this "sensory overload" (Hemsley, 1988).

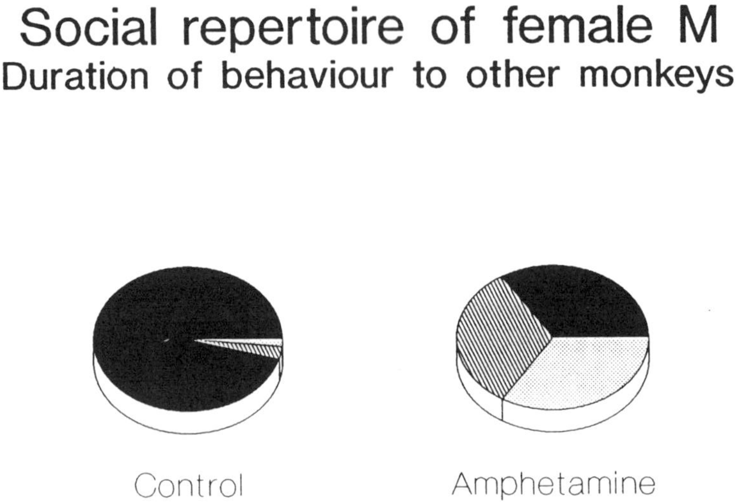

Fig. 5 Effects of amphetamine on the relative duration of social behaviours of monkey M directed to the other members of the group.

There are several indications that the amphetamine model also has some construct validity. Thus, the increase in scanning as well as looking at other monkeys suggests that the amphetamine treated animals have a heightened perception of the environment. Moreover, our results also show that the animals seem to be less able to differentiate between different stimuli. This is illustrated in Fig. 5, which represents the duration of active social behaviour of a female monkey directed to the other three monkeys of a group of 4 Java monkeys. In the left part of Fig. 5 it can be seen that, under control situations, this female has a clear preference for one monkey (in this case the α-male). However after amphetamine treatment (0.5 mg/kg i.m) this preference disappears and the monkeys shows an equal amount of social behaviour towards all monkeys. This pattern has been observed in all monkeys treated with amphetamine sofar. These data suggest that amphetamine both quantitatively and qualitatively influences the social behaviour of monkeys. This idea is further strenghtened by the finding of increases in submissive behaviour (see Fig. 4 and Haber et al., 1977 and Schlemmer & Davis, 1983). Since this increase in submissive gestures is not accompanied by an increase in aggressive behaviour directed to the treated monkey, it appears that the animals have difficulty in correctly interpreting communicative signals. In other words, the amphetamine treated animals seem to interpret a normal situation as threatening. If this is true, this increase in submissive gestures may represent an animal model for paranoid ideation (Haber et al., 1977). Interestingly, in contrast to the social isolation, the increase in submissive gestures seems to be prevented by neuroleptic drugs, such as haloperidol and pimozide (Schlemmer & Davis, 1981). Another indication for this incorrect interpretation of communicative signals is the strong increase in presenting behaviour seen in monkeys treated with amphetamine (Bellarosa et al., 1980). Again this was not accompanied by an increase in sexual behaviour by the other monkeys. Moreover, presenting behaviour (especially of males) has been suggested to be another form of submissive behaviour (see Bellarosa et al., 1980), and this might therefore be a variant of the increase in submissive behaviour reported by Haber and her collegues (1977).

With respect to the construct validity of the social isolation, the situation appears less clear. Bellarosa et al. (1980) showed that low doses of amphetamine increase social behaviour, whereas higher doses decrease it. This may suggest that the social isolation may indeed occur as a result of overstimulation. However, more research is needed to confirm this hypothesis.

INVESTIGATIONS INTO THE NEURONAL MECHANISM

After delineating the model, and discussing its validity the most important question is: "What is the neuronal mechanism underlying the effects of amphetamine, and does that give us insight into the pathology and etiology of schizophrenia?" Unfortunately, very few data are available on the neuronal structures involved in the effects of amphetamine on social and nonsocial behaviour of monkeys. It may not be surprising (given the complexity of social behaviour), that lesions of many different structures disrupt social behaviour. Among these structures are lesions of the frontal cortex (Butter & Snyder, 1972), temporal cortex (Franzen & Myers, 1973; Raleigh & Steklis, 1981), hippocampus (Kolb & Nonneman, 1974), amygdala (Dicks et al., 1969; Kolb & Nonneman, 1974), and septum (Jonason & Enloe, 1971; Kolb & Nonneman, 1974). In the review of Kling & Steklis (1976) the differential effects of a number of lesions on the social behaviour have been compared. However, the usefullness of these data for our present discussion are questionable. First of all, the use of lesions already poses a serious problem. Most of the lesions are made electrolytically, which means that they not only destroy all cell bodies in the area, but also all fibers passing through this area. Moreover, even with specific chemical lesions (for instance by using 6-hydroxydopamine to selectively destroy catecholamine neurons) the problem of post lesion compensatory mechanisms is still a serious drawback. Finally, a common outcome does not necessarily imply a common mode of action. In other words even if lesions of for instance the prefrontal cortex were to lead to the same disturbances in social behaviour as seen after amphetamine treatment, this not necessarily means that amphetamine works through an interaction with the prefrontal cortex.

A better approach is therefore to use the intracerebral injection technique. Using this approach, it may be possible to induce some (or all) of the symptoms seen after systemic injections of amphetamine by applying this drug locally into specific brain regions. Using this approach Dubach & Bowden (1983) found that injections of dopamine into the nucleus accumbens disrupted social grooming. Comparable injections of dopamine into the caudate nucleus did not lead to any overt change in behaviour. Likewise, Annett et al., (1983) found social isolation after injections of amphetamine into the nucleus accumbens, but not after injections into the caudate nucleus of marmosets. In a nonsocial situation, Dill et al. (1979) found a strong increase in submissive postures after application of dopamine into the nucleus accumbens of squirrel monkeys. Although these data point to a role of the nucleus accumbens in the effects of amphetamine, a recent study by Annett et al. (1989) seem to be at variance with this view. These authors directly compared the effects of systemic amphetamine and locally applied amphetamine. They found that, although amphetamine indeed induces social isolation when injected into the nucleus accumbens, it was different from that seen after systemic amphetamine. After systemic amphetamine there was a decrease in the number of initiated social contact and an increase in the number of breaking social contact. With locally applied amphetamine both the number of contacts initiated and contacts broken was increased. These authors therefore suggested that the increased social isolation seen after intra-accumbens amphetamine might be due to an overall increase in locomotion, and may therefore be unrelated to the social isolation seen after systemic injections of amphetamine.

CONCLUSIONS AND FUTURE PERSPECTIVES

Studying the social behaviour of animals is a complex and laborious process. Nevertheless, such research can lead to a great insight into our understanding of the functioning of the nervous system. In the present review we have tried to indicate that the effects of amphetamine on the behaviour of monkeys in a social setting may be an interesting animal model for schizophrenia. As indicated in table 5, many features characteristic for schizophrenia seem to have an analogue in monkeys treated with amphetamine. Moreover, both positive and negative symptoms seem to be modelled in the same animal model. Therefore, this animal model seem to be particularly suited for studying the etiology, pathology and psychophysiology of schizophrenia.

Table 5 Comparison of the effects seen after amphetamine treatment in monkeys and the features of schizophrenia.

Schizophrenic feature	Possible animal analogue
Decreased threshold for external stimuli	Increased scanning Increased looking at other monkeys
Inability to differentiate between relevant and irrelevant stimuli	Decreased selectivity for one monkey
Paranoid ideation	Increased submissive behaviour
Social withdrawal	Decreased affilitative behaviour Decreased proximity
Stereotypy	Stereotypy

With respect to the validity of the model it seems that apart from the face validity, some predictive and construct validity seems to exist. However, additional research is necessary at this point. It seems worthwhile to test the predictive validity of the positive symptoms on the basis of the criteria as discussed by Ellenbroek & Cools (1988, 1990), since these criteria are developed on the basis of clinical experience. With respect to the predictive validity of the analogues of negative symptoms the situation is less clear. No effective treatment has sofar been found, which makes testing of the model difficult. More research seems also necessary with respect to the construct validity of the model. As suggested above, monkeys treated with amphetamine seem to have difficulty in correctly interpreting social signals. This was suggested to be an animal analogue of paranoid ideation. It would be of great importance if this could be confirmed in other experiments. One of the ways in which to test this hypothesis might be to use the so called cooperative conditioning apparatus as designed by Robert Miller (1967). In this paradigm two monkeys are conditioned to press a lever when a light comes on, in order to avoid shock. When both monkeys have reached a stable level of responding, one monkeys is left in one room with the light (the stimulus monkey) whereas the other monkey (the responder monkey) is left in another room with the lever. Both monkeys can see each other through a mirror. In order to avoid shocks, it is necessary for the stimulus monkey to transmit the information that the light has come on, and for the responder monkey to correctly interpret this information in order to press a lever. If amphetamine indeed alters the interpretation of signals emitted by other monkeys, treating the responder monkey should alter the pressing on the lever, either increasing the pressing between sessions, or decreasing the pressing during the session. Although Miller and his colleagues (1973) have tested the effects of amphetamine in this paradigm, they used a low dose, which hardly affected social behaviour. This may explain why they did not find clear effect of amphetamine in this cooperative conditioning paradigm.

A more complex problem is how to analyse the construct validity of the negative symptoms. Since no clear construct seems to be apparent in the literature, no easy way of testing the construct of social isolation in monkeys is available. As pointed out, the only suggested construct is the compensatory mechanism theory, which suggests that negative symptoms occur as a result of a compensatory mechanism activated to protect the organism from the sensory overload. If this construct is true then one would expect that preventing the occurrence of sensory overload should prevent negative symptoms (and thus social isolation) from becoming manifest. This might be one way of evaluating this construct.

When the animal model as described in this review is indeed a good animal model then research into the neuronal mechanisms of the amphetamine induced effects in monkeys might lead to important clues to the structures involved in schizophrenic symptoms. There are probably two different approaches to tackle this problem. The first one is to directly apply amphetamine into different brain areas and to compare the observed effects with the effects of systemically applied amphetamine. The second approach would be to systemically apply amphetamine and to try to block the effects with local application of antagonists. Given the fact that we recently found that the selective dopamine D_1 receptor antagonist SCH 23390 is able to completely prevent the effects of amphetamine, we now have the tools to take this second approach.

In conclusion, we have tried to illustrate that despite the complexity of animal social behaviour, it seems a worthwhile approach to study the pharmacology of social behaviour. When studied in a systematic way, using an appropriate ethogram, it can be shown that the effect of amphetamine on the social behaviour of monkeys is a good animal model for schizophrenia, in which a large number of characteristic features of schizophrenia are modelled.

REFERENCES

Albus, M., Botschev, C., Müller-Spahn, F., Naber, D., Münch, U., Ackenheil, M. (1986) Pharmacopsychiatry 19, 101-105.
Angrist, B., Rotrosen, J., Gershon, S. (1980) Psychopharmacology 72, 17-19.
Annett, L., Ridley, R., Gamble, S., Baker, H. (1983) Psychopharmacology 81, 18-23.
Annett, L., Ridley, R., Gamble, S., Baker, H. (1989) Psychopharmacology 99, 222-229.
Arnt, J. (1982) Acta Pharmacol. Toxicol. 51, 321-329.
Baruch, I., Hemsley, D., Gray, J. (1988) J. Nerv. Ment. Dis. 176, 598-606.
Bellarosa, A., Bedford, J., Wilson, M. (1980) Pharmacol. Biochem. Behav. 13, 221-228.
Butter, C., Snyder, D. (1972) Acta Neurobiol. Exp. 32, 525-565.
Carlsson, A. (1988) Neuropsychopharmacol. 1, 179-186.
Connell, P. (1958) Amphetamine psychosis, Maudsley Monograph nr 5, Oxford Univ. Press, New York.
Crowley, T., Stynes, A., Hydinger, M., Kaufman, C. (1974) Arch. Gen. Psychiatry 31, 829-838.
Csernansky, J., Riney, S., Lombroso, L., Overall, J., Hollister, L., Arch. Gen. Psychiatry 45, 655-659.
Dicks, D., Myers, R., Kling, A. (1969) Science 165, 69.
Dubach, M., Bowden, D. (1983) In: Ethopharmacology: Primate models of neuropsychiatric disorders (K.Miczek, Ed), Alan Liss Inc. New York, pp 157-184.
Ellenbroek, B., Cools, A. (1988) Life Sci. 42, 1205-1213.
Ellenbroek, B., Cools, A. (1990) Behav. Pharmacol. (In press).
Ellenbroek, B., Peeters, B., Honig, W., Cools, A. (1987) Psychopharmacology 93, 343-348.
Ellenbroek, B., Willemen, A., Cools, A. (1989) Neuropsychopharmacol. 2, 191-199.
Franzen, E., Myers, R. (1973) Neuropsychologia 11, 141-157.
Githin, M., Gerner, R., Rosenblatt, M. (1981) Psychopharmacology 74, 51-53.
Haber, S., Barchas, P., Barchas, J. (1977) In: Animal models in psychiatry and neurology (I.Hanin and E.Usdin, Eds), Pergamon Press. New York, pp 107-115.
Haber, S., Barchas, P., Barchas, J. (1981) Biol. Psychiatr. 16, 181-196.
Hemsley, D. (1988) In: Essentials of adult abnormal psychology (E.Miller and J.Cooper, Eds), Churchill Livingstone, London, pp 101-127.
Janowsky, D., Davis, J. (1974) In: Neuropsychology of monoamines and their regulatory enzymes (E.Usdin, Ed), Raven Press. New York, pp 317-324.
Janssen, P., Niemegeers, C., Schellekens, K. (1965) Arzneim. Forsch. 15, 104-117.
Johnstone, E., Crow, T., Frith, C., Carney, M., Price, J. (1978) Lancet, i, 848-851.
Jonason, K., Enloe, L. (1971) J. Comp. Physiol. Psychol. 75, 286-301.
Kling, A., Steklis, H. (1976) Brain Behav. Evol. 13, 216-238.
Kolb, B., Nonneman, A. (1974) Physiol. Behav. 13, 637-643.
Kucharski, L., Alexander, P., Tune, L., Coyle, J. (1984) Psychopharmacology 82, 194-198.
Lifshitz, K., O'Keeffe, R., Lee, K. (1987) Psychopharmacol. Bull. 23, 487-492.
McKenna, P. (1987) Br. J. Psychiatr. 151, 288-301.

Matthysse, S. (1981) In: The neurobiology of the nucleus accumbens (R.Chronister, J.DeFrance, Eds), Haer Institute for Electrophysiological Research, Maine, pp 351-359.
Matthysse, S. (1986) Progress in Brain Research 65, 259-270.
Miczek, K., Gold, L. (1983) Psychopharmacology 81, 183-190.
Miczek, K., Yoshimura, H. (1982) Psychopharmacology 76, 163-171.
Miller, M. (1976) Psychopharmacology 47, 71-74.
Miller, R. (1967), In: Social communication among primates, (S.Altmann, Ed), Univ. Press, Chicago.
Miller, R., Levine, J., Mirsky, I. (1973) J. Pers. Soc. Psychol. 28, 396-405.
Myers, D., Campbell, P., Cocks, N., Flowerdew, J., Muir, A. (1981) Br. J. Psychiatry 139, 118-121.
Nielsen, E., Lyon, M. (1982) Biol. Psychiatr. 17, 423-435.
Raleigh, M., Steklis, H. (1981) Exp. Neurol. 73, 378-289.
Randrup, A., Munkvad, I. (1967) Psychopharmacology 11, 300-310.
Ridley, R., Baer, H., Scraggs, P. (1979) Biol. Psychiatr. 14, 753-765.
Ridley, R., Baker, H. (1983) In: Ethopharmacology: Primate models of neuropsychiatric disorders (K.Miczek, Ed), Alan Liss Inc, New York, pp 101-135.
Robinson, T., Becker, J. (1986) Brain Res. Rev. 11, 157-198.
Schiorring, E. (1977) In: Cocaine and other stimulants (E.Ellinwood and M.Kilbey, Eds), Plenum Press. New York, pp 481-522.
Schlemmer, R., Casper, R., Siemsen, F., Garver, D., Davis, J. (1976) Psychopharmacology, Comm. 2, 49-59.
Schlemmer, F., Davis, J. (1981) Pharmacol. Biochem. Behav. 14 (Suppl 1), 95-102.
Schlemmer, F., Davis, J. (1983) In: Ethopharmacology: Primate models of neuropsychiatric disorders, (K.Miczek, Ed), Alan Liss Inc. New York, pp 33-78.
Schlemmer, F., Narasimhachari, N., Davis, J. (1976) J. Pharm. Pharmacol. 32, 285-289.
Scottish Schizophrenia Research group, (1987) Br. J. Psychiatr. 150, 331-344.
Scraggs, P., Ridley, R. (1978), Psychopharmacology 59, 243-245.
Scraggs, P., Ridley, R. (1979), Psychopharmacology 62, 41-45.
Smith, E., Byrd, L. (1983) In: Ethopharmacology: Primate models of neuropsychiatric disorders (K.Miczek, Ed), Alan Liss Inc. New York, pp 1-31.
Smith, E., Byrd, L. (1985) Pharmacol. Biochem. Behav. 22, 135-139.
VandenBercken, J. (1979) Information statistics analysis of social interaction in Java monkeys applied in the neuro-ethology of the caudate nucleus, PhD thesis, Nijmegen.
Vrijmoed-de Vries, M., Cools, A. (1985) Exp. Neurol. 87, 58-75.
Wauquier,A. (1979) Int. Rev. Neurobiol. 21, 335-403.
Willner, P. (1984) Psychopharmacology 83, 1-16.
Winlow, J., Miczek, K. (1988) Psychopharmacology 96, 493-499.

Animal Models in Psychopharmacology
Advances in Pharmacological Sciences

THE EFFECT OF HALOPERIDOL AND CLOZAPINE ON THE BEHAVIOURAL CONSEQUENCES OF STIMULATING MESOLIMBIC AND NIGRO-STRIATAL DOPAMINERGIC PATHWAYS IN THE RAT

N.R. Oakley and M.J. Sheehan

Neuropharmacology Department, Glaxo Group Research Ltd, Ware, Herts, SG12 0DP, United Kingdom

SUMMARY: Mesolimbic and nigro-striatal dopamine systems were activated by injections of muscimol into the VTA and SNR respectively. The behavioural patterns produced (LMA, sniffing, rearing and grooming) were the same in each case. Clozapine, but not haloperidol, selectively antagonised the effects of VTA injected muscimol.

INTRODUCTION

Antipsychotic drugs have been divided into typical (e.g. haloperidol) and atypical compounds (e.g. clozapine) according to whether or not they cause extrapyramidal side effects (EPS) in man. The induction of EPS is thought to be due to a blockade of nigro-striatal dopamine systems, whereas the positive antipsychotic effect is regarded as a manifestation of the blockade of mesolimbic dopamine systems (Meltzer et al., 1989). In an effort to predict the potential antipsychotic and EPS-inducing ability of drugs we have developed behavioural models, that rely on selective stimulation of mesolimbic and nigro-striatal dopamine pathways. This has been achieved by direct central injections of the GABA-A agonist, muscimol, into either the ventral tegmental area (VTA) or the substantia nigra zona reticulata (SNR) of rats.

Muscimol injections into these regions have previously been reported to induce behavioural stimulation (Arnt & Scheel- Kruger, 1979; Scheel-Kruger et al., 1977). Due to the universal inhibitory nature of GABA, the stimulation induced by muscimol is likely to be via an indirect mechanism, which is not inconsistent with the known neuro-anatomy of the two regions (Fallon & Loughlin, 1985).

METHODS

Male Wistar-derived rats (AHA, 180-220g) were implanted with bilateral guide cannulae terminating 3 mm above the VTA (coordinates; AP -4.8, -5.3, Lat 1, D 8.5) or SNR (AP -4.8, -5.8, Lat 2.5, D 8.5, Paxinos & Watson, 1982).

After 7 days recovery, rats were injected bilaterally via a 30 gauge injection cannula advanced into the VTA or SNR with either vehicle (artificial CSF) or muscimol (10-200ng) in a volume of 0.5 ml over a period of 1 min. In all studies rats were placed in clear perspex solid bottomed cages, to which they had been allowed 30 min habituation, and observed for 30-40 min after muscimol administration. The following behaviours were scored for 2 min in every 10 min time period:- Locomotor activity (LMA) - line crossings/min Rearing - number/min Sniffing - Intensity (0-absent, 1-slight, 2-moderate, 3-intense) and bout duration (s) Grooming - intensity and bout duration as for sniffing.

For antagonist studies, rats were pretreated with haloperidol (0.01-0.25mg/kg i.p.) or clozapine (0.1-12.5mg/kg i.p.) 30 min before muscimol (100ng) administration. ED_{50} values were calculated using the method of Litchfield & Wilcoxon (1949). Rats were used no more than 3 times with at least 7 days being left between each testing. The locations of injection cannulae were verified histologically at the conclusion of each series of experiments.

RESULTS

Injections of muscimol into both the VTA and SNR produced dose-related increases in LMA (Fig. 1). Similar increases were seen in rearing behaviour (data not shown) and sniffing intensity (Fig. 1).

Bout duration of sniffing was also increased in a dose related fashion after VTA administration of muscimol (e.g. CSF 0.5 $\pm$ 0.4s, Musc. (30ng) 5.4 $\pm$ 1.0s, Musc. (200ng) 19.2 $\pm$ 3.2s), but not to the same degree after SNR administration (e.g. CSF 1.4 $\pm$ 0.7s, Musc. (30ng) 5.1 $\pm$ 0.7s, Musc. (200ng) 7.4 $\pm$ 2.3s). Grooming behaviour was seen but was inconsistent and, therefore, not fully analysed. These behavioural effects were back to control levels (CSF injections) 30-40 minutes after administration.

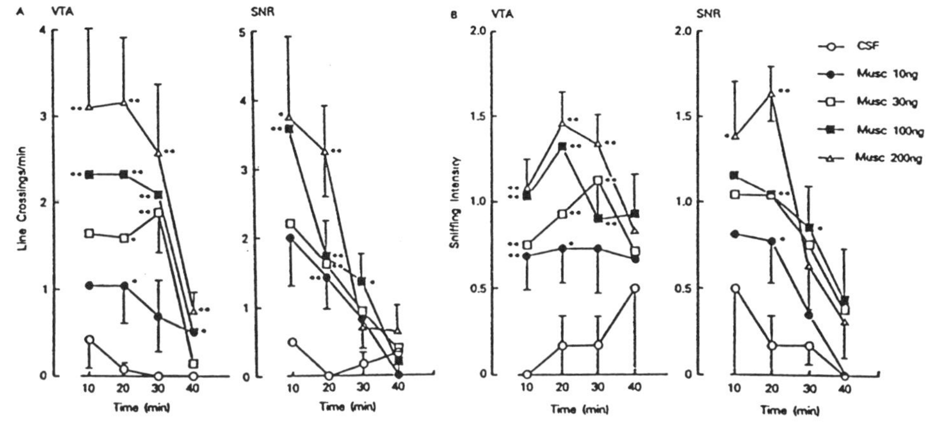

Fig. 1 The effect of muscimol injected into the VTA or SNR on locomotor activity (A) as measured by line crossings on a 4-square grid or sniffing intensity (B). Results are means ± SEM, n=6-14. *p<0.05, **p<0.01 vs CSF control.

Table 1 The effect of haloperidol and clozapine on the behavioural consequences of bilateral injection of muscimol into the VTA or SNR. n = 6-10 per dose level used, *p<0.05 vs VTA

	ED50 (95% confidence limits) mg/kg i.p.			
	Haloperidol		Clozapine	
	VTA	SNR	VTA	SNR
LMA	0.03 (0.01-0.06)	0.03 (0.01-0.8)	0.24 (0.06-0.96)	9.70* (3.00-31.40)
Rearing	0.02 (0.01-0.05)	0.03 (0.01-0.13)	0.94 (0.24-3.71)	0.80 (0.21-3.03)
Sniffing (Intensity)	0.03 (0.01-0.08)	0.03 (0.01-0.10)	0.58 (0.19-1.77)	6.10* (2.53-14.70)
Sniffing (Bout Duration)	0.01 (0.01-0.04)	0.02 (0.01-0.10)	0.14 (0.03-0.66)	7.60* (1.67-34.66)

In pharmacological studies, whereas all the behaviours induced by muscimol injected into either the VTA or SNR were equally sensitive to inhibition by haloperidol, clozapine blocked the behavioural consequences of VTA-administered muscimol (with the exception of rearing) much more readily than it did those induced by SNR-administered muscimol (Table 1).

DISCUSSION

In these experiments, we have used discrete injections of muscimol in an attempt to produce selective activation of mesolimbic and nigro-striatal dopamine pathways in the rat. The rapid onset and relatively short duration of these effects indicate that it is probable that they are due to an action of muscimol at the injection site and not due to spread.

Interestingly, the range of behaviours produced by injections of muscimol into either VTA or SNR (namely LMA, rearing, sniffing and grooming) were exactly the same. These was also little or no quantitative difference between the two sets of behaviours.

The behaviours induced by muscimol were very sensitive to the effects of haloperidol. However, clozapine was 10-50 times less potent at blocking the effects of SNR rather than VTA administered muscimol (with the exception of rearing). The exact reason for this differential effect of clozapine is unclear, but may be due to an antagonistic effect at dopamine D_1 and 5-HT_2 as well as dopamine D_2 receptors (Meltzer et al., 1989).

It may also reflect the lack of propensity of clozapine to induce EPS in man and thus suggests that these behavioural models may be useful in determining the relative antipsychotic activity and EPS-inducing ability of neuroleptic drugs.

REFERENCES

Arnt, J., Scheel-Kruger, J. (1979) Life Sci 25, 1351-1360.

Fallon, J.H., Loughlin, S.E. (1985) In: The Rat Nervous System (G.Paxinos, Ed), Academic Press, Sydney, pp 353-373.

Litchfield, J.T., Wilcoxon, F. (1949) J. Pharmacol. Exp. Ther. 96, 99-113.

Meltzer, M.J., Matsubara, S., Lee, J-C. (1989) J. Pharmacol. Exp. Ther. 251, 238-246.

Paxinos, G., Watson, C. (1982) The rat brain in stereotaxic coordinates. Academic Press, New York.

Scheel-Kruger, J., Arnt, J., Magelund, G. (1977) Neurosci. Lett. 4, 351-356.

Animal Models in Psychopharmacology
Advances in Pharmacological Sciences

THE GLUTAMATERGIC SYSTEM AS A TARGET OF ANTIPSYCHOTIC DRUG ACTION

W.J. Schmidt

University of Tübingen, Dept. Neuropharmacology, Mohlstrasse 54/1, D-7400 Tübingen, Germany

SUMMARY: Inhibition of glutamatergic neurotransmission induces psychomotor stimulation in humans suggesting that glutamate may be critically involved in schizophrenia. In rats, antagonists of the NMDA-subtype of glutamate receptors also produce psychomotor stimulation, e.g. stereotyped running and sniffing. Several classical (e.g. haloperidol) and atypical (e.g. clozapine) antipsychotics specifically antagonize these stereotypies.

INTRODUCTION

The glutamate hypothesis of schizophrenia suggests that a deficient glutamatergic transmission underlies this disease (Kim et al., 1980). Progress in the understanding of the role of glutamate receptors and their interaction with dopamine has added a great deal of supporting evidence to this hypothesis.

Behavioural studies: Phencyclidine (PCP) and related drugs (e.g. ketamine, dizocilpine (MK-801), SKF 10 047) are antagonists of glutamatergic transmission at the NMDA receptor subtype; they block the ion channel gated by the NMDA receptor. In humans, PCP evokes psychotic symptoms that are regarded as the best pharmacological model for schizophrenia (Allen and Young, 1978). An existing psychosis even exacerbates under PCP (Luby et al., 1959) and its effects last beyond the drug intoxication. The PCP-evoked psychosis can be counteracted by neuroleptic drugs. Recent evidence suggests that the psychotomimetic effects of several sigma ligands are also mediated by the PCP binding site of the NMDA receptor complex (Balster, 1990).

Also in rats, drugs such as PCP and dizocilpine exert psychomotor stimulant effects: stereotyped running, sniffing and an attentional deficit; grooming is reduced (Tiedtke et al., 1990, Mansbach and Geyer, 1989). As pointed out by Carlsson and Carlsson (1990) this stimulant effect is produced in mice, even in the almost complete absence of brain dopamine.

The cortico-striatal glutamatergic neurons seem to be critically involved since stereotypy is also evoked by kynurenic acid, a broad spectrum NMDA-antagonist, or by the selective competitive antagonist AP-5 when injected directly into the antero-dorsal striatum (Schmidt, 1986). The stereotypies induced by either systemic or intrastriatal administration of NMDA-antagonists are antagonized by neuroleptics (Schmidt 1986, Tiedtke et al., 1990).

Thus, dopamine and glutamate within the anterodorsal striatum, seem to exert opposite effects upon behaviour. If this assumption is true, then the converse pharmacological treatment should produce the opposite effects. Recently, this has been shown: Akinesia and rigidity (catalepsy), induced by neuroleptics, is reversed by NMDA-antagonists (Schmidt and Bubser, 1989, Schmidt et al., 1990). However, it is important to note here, that dopamine and glutamate do not exert opposite effects on learning and memory tasks in 8-arm and T-mazes: The amnesic effects of dizocilpine are even augmented by the neuroleptics haloperidol and clozapine (Schmidt and Bischoff, 1988).

Lesion studies indicate that in the striatum, dopamine and glutamate interact at medium sized spiny neurons on which dopamine and NMDA receptors are located. Neurotoxic axon-sparing lesions performed with NMDA-agonists, for example quinolinic acid, destroy medium sized spiny neurons that use GABA and substance P as their transmitter (Di Figlia, 1990). Such a lesion produces stereotyped running and sniffing in rats, reminiscent of the effects of a transient blockade of NMDA receptors with AP-5 or dizocilpine. Further, such a lesion counteracts neuroleptic induced catalepsy (Schmidt and Bischoff, 1988).

Electrophysiological and biochemical studies: Dopamine and glutamate exert opposite effects on brain field potentials in the freely moving rat: The NMDA-antagonists dizocilpine, ketamine, and PCP affect brain field potentials in the same way as activation of dopaminergic transmission does (Dimpfel and Spüler, 1990). Further, the glutamatergic input and the dopaminergic input have opposite effects on the firing rate of striatal neurons; this effect is reflected by antagonistic effect of these neurotransmitters on the state of phosphorylation of DARPP-32 within these neurons (for ref. see Halpain et al., 1990).

Receptor binding studies: Markers for glutamatergic transmission are especially high in regions thought to be involved in the pathogenesis of schizophrenia, e.g. the frontal cortex, the entorhinal region and the hippocampus. Dizocilpine binding sites are increased in several regions in postmortem schizophrenie brains reaching significance in the putamen (Kornhuber et al., 1989). Also increases in [^{3}H]-kainate binding sites in cortical regions have been reported (Toru et al., 1988). These findings may tentatively be interpreted as an upregulation of glutamate receptors as a consequence of presynapic glutamatergic hypofunction in schizophrenia.

Histological studies: It has recently been shown that treatment with PCP, dizocilpine, ketamine and tiletamine leads to vacuolization in pyramidal shaped neurons in the posterior cingulate and retrosplenial cortices. This evidence of neuronal loss due to suppression of NMDA receptor functioning led Olney et al. (1990) to speculate that such a process might be operative in schizophrenia: Chronic suppression of NMDA receptor function may produce psychotic symptoms in humans and a loss of a select subpopulation of cingulate - retrosplenial neurons.

METHODS AND RESULTS

Stereotyped sniffing was induced by dizocilpine (0.16 mg/kg i.p.). The classical antipsychotic drug haloperidol (0.025 mg/kg i.p.) did not change spontaneous sniffing but specifically antagonized dizocilpine-induced sniffing. Higher doses (0.05 and 0.1 mg/kg i.p.) reduced spontaneous, but more potently dizocilpine-induced sniffing. The atypical neuroleptic drugs clozapine (2.5 and 5 mg/kg i.p.) sulpiride (30 mg/kg i.p.) and umespirone (2.5, 5 and 10 mg/kg p.o.) did not reduce spontaneous sniffing, but specifically and potently antagonized dizocilpine-induced sniffing. Promethazine (0.3 mg/kg i.p.) that lacks antipsychotic activity at low doses as well as buspirone (0.8 and 4.2 mg/kg i.p.) reduced spontaneous and dizocilpine-induced sniffing to about the same extent. Midazolam (0.3 mg/kg i.p.) a sedative and tranquilizing benzodiazepine, reduced spontaneous but not dizocilpine-induced sniffing.

DISCUSSION:

Antagonism of dizocilpine-induced sniffing discriminates antipsychotics from the non-antipsychotics tested. While non-antipsychotics reduce spontaneous sniffing, antipsychotics selectively antagonize dizocilpine-induced abnormal sniffing. If these findings are further corroborated, this test may be a useful tool in the search of antipsychotics. It may be superior to dopamine-induced stereotypies since the latter give false negative results for clozapine.

Classical neuroleptics may counteract dizocilpine-induced sniffing by a blockade of inhibitory dopamine receptors located on medium spiny neurons. The mechanism of action of atypical neuroleptics is unknown; besides a dopamine antagonism, 5-HT_2 antagonism has been discussed. Perhaps clozapine has also glutamate-mimetic actions by direct interaction with the NMDA receptor. Recently it has been reported that clozapine inhibits [^{3}H]-dizocilpine binding to the NMDA receptor gated ion channel.

Conclusion: The data presented are in accordance with the glutamate hypothesis of schizophrenia which states that a deficient glutamate activity underlies this disease. Thus, treatments that enhance glutamatergic activity may exert antipsychotic effects. Although it appears to be very difficult to develop drugs that enhance glutamate-activity in the brain without producing unwanted neurotoxic effects, the regional heterogenity of glutamate receptors and the multiple sites for pharmacological intervention at the NMDA receptor complex and at the non-NMDA receptors may provide the basis for the development of a new class of antipsychotic drugs.

Acknowledgements: With financial support by the Deutsche Forschunsgemeinschaft SFB 307.

REFERENCES

Allen, R.M., Young, S.J. (1978) Am. J. Psychiatry 135, 1081-1084.
Balster, R.L. (1990) Psychopharmacology, Suppl. 101 S.
Carlsson, M., Carlsson, A. (1990) Trends Neurosci. 13, 272-276.
Di Figlia, L. (1990) Trends Neurosci. 13, 286-289.
Dimpfel, W., Spüler, M. (1990) Psychopharmacology 101, 317-323.
Halpain, S., Girault, J.-A., Greengard, P. (1990) Nature 343, 369-372.
Janowsky, A., Berger, S.P. (1989) Schizophrenia Res. 2, 189.
Kim, J.S., Kornhuber, H.H., Schmid-Burgk, W., Holzmüller, B. (1980) Neurosci. Lett. 20, 379-382.
Kornhuber, J., Mack-Burkhardt, F., Riederer, P., Hebenstreit, G.F., Reynolds, G.P., Andrews, H.B., Beckmann, H. (1989) J. Neural Transm. 77, 231-236.
Luby, E.D., Cohen, B.D., Rosenbaum, G., Gottlieb, J.S., Kelley, R. (1959) Arch. Neurol. Psychiat. 81, 363-369.
Mansbach, R.S., Geyer, M.A. (1989) Neuropsychopharmacology 2, 299-308.
Olney, J.W., Labruyere, J., Price, M.T. (1990) Science 247, 221.
Schmidt, W.J. (1986) Psychopharmacology 90, 123-130.
Schmidt, W.J., Bischoff, C. (1988) Psychopharmacology 96, S51.
Schmidt, W.J., Bubser, M. (1989) Pharmacol. Biochem. Behav. 32, 621-623.

Schmidt, W.J., Bubser, M., Hauber, W. (1990) Trends Neurosci. 13, 46.
Tiedtke, P., Schmidt, W.J. (1990) J. Neural. Transm. 81, 173-182.
Toru, M., Watanabe, S., Shibuya, H., Nishikawa, T., Noda, K., Mitsushio, H., Ichikawa, H., Kurumaji, A., Takashima, M., Mataga, N., Ogawa, A. (1988) Acta Psychiatr. Scand. 78, 121-137.

part 4:

AGGRESSION AND SEXUAL BEHAVIOUR

Animal Models in Psychopharmacology
Advances in Pharmacological Sciences

CONCEPTS IN ANIMAL MODELS FOR PATHOLOGICAL AGGRESSIVE BEHAVIOUR IN HUMANS

J. Mos and B. Olivier

Duphar B.V., CNS Pharmacology, P.O.Box 900, 1380 DA Weesp, The Netherlands

SUMMARY: Animal models for psychiatric disorders should ideally fullfil the following criteria: the behavioural abnormalities should be measurable in an appropriate context and the neurobiological causes for this deviant behaviour should be known. In case these criteria are met, these models might be called homologous, in a similar way as for instance used in functional morphology. It is clear that these conditions are not met by the current models for aggressive behaviour in animals.

The most obvious reasons are that there is neither a precise description of what the real aggressive abnormalities are, nor a good understanding of the causes of clinically maladaptive aggressive behaviour. Since maladaptive aggression does not occur as a separate diagnostic entity, but is found as one of the symptoms of many disorders, it is likely that many different causes may exist for pathological aggression. With regard to the behavioural manifestation, it will be argued that pathological aggression might best be described by the frequency of occurrence, the intensity (verbal versus physical) and the context in which aggression occurs. Although continuous progress has been made in these areas, the state of the art does not permit the development of homologous animal models.

Instead, the concept is presented to study functional models of aggressive behaviour in animals with ethologically valid procedures. Functional models do require a good insight into the mechanisms and conditions by which the behaviour is induced as well as the results of the aggressive interactions. Territoriality is a concept which is often used in relation to animal models of aggression and which has been argued to play an important role in humans as well. Resident-intruder aggression and maternal aggression in rats serve as examples for the concepts, although this is necessarily a limitation of the current knowledge.

Apart from these more naturalistic assessments of aggression several models will be discussed in which aggression is induced by various treatments or by electrical brain stimulation. Moreover, models reflecting defensive aspects of agonistic behaviour will be discussed.

Using a wide array of models with functional relevance to aggression, insight is gained into the causes and consequences of aggression. It is expected that these models will be useful in finding new pharmacological treatments of pathological aggression in which at least the same neural circuits are involved as in the adaptive forms of aggression.

Data will be presented on the effects of the serotonergic compound eltoprazine, neuroleptics and benzodiazepines. The factors and behavioural processes mediating these effects have been analysed by ethopharmacological methods and form a good example of the feasibility of the approach outlined above.

INTRODUCTION

It is obvious that before animal studies can be performed for any disorder, it should be known what the essential features are that should be modelled. What are these essential features for pathological aggression in man? Despite the many studies on aggression, both preclinical and clinical, this is exactly the area which is desperately in need for an overall theory which can be tested and verified. Pathological aggression is not a DSM-III disorder for which criteria are set as to what is normal or abnormal. Neither is a full or even partial knowledge of underlying biological factors available which might form a satisfactory basis for an entry to rational research. Thus a discussion about the nature and characteristics of pathological aggression is needed before a succesful attempt can be made to develop animal models. It should follow that the limitations of the animal models are clear and well described.

Despite the drawbacks adherent to aggression research noted above, there is an increasing knowledge of the effects of psychoactive compounds on aggressive behaviour, not only in animals, but also in clinical settings. Thus there are two roads which seem to develop rather independently, one concerning the fundamental causes of aggression and dysfunctions, the other comprising the modification of behaviour by pharmacological treatment. In this paper these two roads cannot be brought together since neither of the approaches is sophisticated enough to allow generalizations to the other scope of research. However, as the ideas and the conclusions from both approaches should affect each other, as many intersections as possible will be crossed.

CAN PATHOLOGICAL AGGRESSION BE DEFINED?

The multitude of definitions of aggression (Moyer, 1968) underlines the complexity of this set of behaviours. Aggression serves many purposes, which not necessarily have a negative connotation. The variety of situations in which aggression may occur and the differences between species do not render biology as the prime candidate to come with definitions concerning pathological or normal aggression. However, the study of the biology of aggression has revealed that some basic characteristics are shared in many situations. Aggression in animals often occurs in situations of competition. Many subjects can be the cause of the competition. In human situations it is also important to have a clear vision of the area of conflict. Conflict in itself may be unavoidable, but the ways to handle conflict may vary. In this sense the context in which aggression may occur and the intensity of aggression may deviate from longterm functional behaviour. With regard to the context and the intensity table 1 summarizes some variables which are important in human conflicts.

Table 1: Variables which affect the functionality of aggressive behaviour

Context of aggression:	
	- perception of threat
	- perception of area of conflict
	- perception of ways to resolve the conflict
Intensity of aggression:	
	- frequency of aggressive acts
	- character of aggression

In animal studies (see for example Huntingford and Turner, 1987) the immediate biological needs for food, shelter, nesting place, mate etc are usually so obvious that the area of conflict is generally well-defined. Althoough in humans this may be less evident in some cases, there are often clearly demarcated areas over which conflict may arise. A wrong perception of the area of conflict may result in non-functional aggression. If there is no genuine area of conflict, there is no need for aggression. In many psychiatric patients the perception of reality may be disturbed, which may lead to a misjudgement of areas of conflict. Similarly the perception of threat may be unrealistic and lead to aggression in situations which would not escalate in persons which did not experience inappropriate threat. The ways to resolve potential areas of conflict may also play a role in defining whether aggression is appropriate or not. If non-aggressive forms of conflict resolution are effective, then aggressive options are not functional, if only for the risks of serious damage that both parties run. The context of the conflict thus to a large extent determines the functionality of aggressive behaviour. However, since the perception of situations is often difficult to measure, let alone be evaluated, common sense judgements and cultural settings of moral standards are usually the ultimate definitions of normal or pathological aggression.

The intensity of aggression can at least be measured more objectively, albeit that this is often troublesome in everyday situations since overt aggression occurs at

aggressive behaviours. In this sense, at least the specificity of anti-aggressive effects can be evaluated by comparing the drug effects on different behaviours. One of the most naturalistic models to test aggression is one employing the concept of territoriality. Malmberg (1980) is one of the few authours who has studied and convincingly shown that terrioriality is of prime importance in humans as well. Apart from emphasis on these models, we have also used models in which aggression can be studied in a context which would not directly evoke aggression. The approach we have chosen to develop specific anti-aggressive compounds will exemplify these choices in more detail.

STRATEGY FOR THE DEVELOPMENT OF ANTI-AGGRESSIVE DRUGS (SERENICS)

The strategy to develop compounds that specifically reduce aggression strongly depends on the available methodologies to study agonistic behaviour. Traditionally, isolation-induced aggression in mice has received attention for pharmacological studies, but was not primarily used for the study of aggression (Janssen et al., 1960; Valzelli et al., 1967; Yen et al., 1959). This led to views of isolation-induced aggressive behaviour (IIA) as 'abnormal' or 'compulsive behaviour' (Valzelli, 1973). These interpretations have contributed to the use of this test as a model for human aggression. However, where Garattini and Valzelli (1981) consider IIA as a model for affective disorders in humans, others have proposed IIA as a screening test for antipsychotics (Janssen et al., 1960), anxiolytics (Ross and Ogren, 1976) or for antidepressants (Malick, 1979).

In contrast, the ethological approach views isolation aggression in mice as normal, ie resembling the behavioural patterns of a territorial mouse living behaviourally isolated from other males (Blanchard et al., 1979; Brain, 1975). After a certain isolation period male mice may show a range of aggressive, defensive or flight behaviour (Krsiak, 1974,1975). Thus IIA can be viewed as a naturally occurring behaviour, which can be described in terms of all acts and postures earlier described in the full behavioural repertoire of agonistic interactions (Grant and Mackintosh, 1963; Olivier and van Dalen, 1982; Poshivalov, 1981).

This example stresses the importance of how behaviour is viewed and studied, i.e. what concepts and approaches are used to study, analyze and interpret the observed behavioural changes after drugs. In many pharmacological studies only single measures are used such as latency to attack, duration of attack, percentage of animals fighting, aggressiveness scores etc. By contrast, ethopharmacology 'applies the methods and concepts of ethology to the study of drug-induced changes in behaviour', to quote one of the pioneers of ethopharmacology, Keith Dixon. Ethology states that the organism is well equipped with the necessary characteristics to survive and reproduce succesfully in a specific environment. The environments are usually not static and diverse compartments are needed for e.g. resting, feeding, pup care, etc. The adaptative capacities of animals also comprise changes over time like diurnal, circannual and climate rhythms in order to survive. All these challenges are met with different behavioural strategies, among which aggression is one set of behaviours.

Aggression is not the only behaviour that occurs in situations of conflict and thus the whole stream or flow of continuous activities should be monitored. Especially after drug treatment this continuous recording of behaviour is useful to analyse the full profile of behavioural changes, whether these are specific or interfere with the expected behaviour in this context.

Ethological studies, for example that of agonistic behaviour of male rats, always start with describing the behavioural elements, the ethogram. Behavioural elements must be defined in such a way that both the begin and the end are clearly recognized, such that quantitative measures can be obtained about the frequency, the duration and the sequential pattern of behavioural elements. Ethograms are based on very extensive studies of different behavioural situations in order to describe the full behavioural repertoire (e.g. Lehman and Adams, 1977; Olivier, 1981). Since highly artificial circumstances may lead to limited or distorted ethograms, it is important to create testing environments and conditions that are as natural as possible. This is a vital element in ethopharmacological studies, which to a large extent determine the usefulness and the interpretation of the obtained behavioural responses, as well as the nature of the drug effects.

The foregoing implies that there is structure in behaviour, a number of behavioural elements makes a behavioural system, e.g. agonistic behaviour. The association of an element with other agonistic elements and the typical sequential organization over time often determine the classification into behavioural categories. Usually such behavioural categories can be divided into appetitive, the more variable introductory elements and the more stereotyped consummatory elements. These theoretical approaches and the methods to quantify and structure behavioural observations have greatly contributed to the study of aggression and the influence of drugs on animals behaviour. The subsequent progress in the search for specific anti-aggressive agents will be outlined below.

ANIMAL MODELS FOR AGONISTIC BEHAVIOUR

Often a distinction is made between offensive and defensive agonistic behaviour (Adams, 1979). While offensive agonistic behaviour is characterized by the initiative of the aggressor and damage to the opponent (Blanchard et al., 1977a,b, 1978; Mos et al., 1984; Olivier et al., 1983), defensive behaviour, in contrast, lacks active approach (initiative) and no wounds (or incidental ones only) are inflicted by the defensive animal. Several models focus (although not exclusively) upon the 'offensive' components of agonistic interactions. Other models reflect the more defensive aspects of agonistic behaviour. We have used both type of models and will separately discuss the main characteristics. As examples of the effects of drugs, the behavioural profile of the serenic compound eltoprazine will be given. Reference drugs are neuroleptics (haloperidol, chlorpromazine) and benzodiazepines (chlordiazepoxide, diazepam, oxazepam).

MODELS OF OFFENSIVE BEHAVIOURS

Isolation-induced offensive behaviour: A manipulation often used to induce aggression is isolation of male animals (typically mice) for several weeks. A significant proportion of such isolated animals upon encountering another will reliably exhibit attack behaviour (Valzelli, 1969). This isolation-induced aggression model is one of the most frequently used aggression models in behavioural pharmacology (Malick, 1979). Table 2 shows the results of various compounds in this model. Clearly eltoprazine and haloperidol inhibit aggression at low doses, whereas benzodiazepines only affect attacks at higher doses.

Because these isolated male mice show a full repertoire of agonistic behaviours (Miczek and Krsiak, 1979), ethologic techniques can be used to allow detecting of very specific drug effects (Olivier and van Dalen, 1982). After a relatively short isolation period to increase the propensity of attacks, a short test in a neutral arena reveals a diverse pattern of activities, because the situation delivers a mix of offensive-defensive behaviour. The model is very interesting because it shows properties of compounds which are revealed only partially, or not at all, by common pharmacological test models (Miczek and Barry, 1976; Miczek and Krsiak, 1979; Olivier, 1981). An example of the results obtained in this model is shown in figure 1.

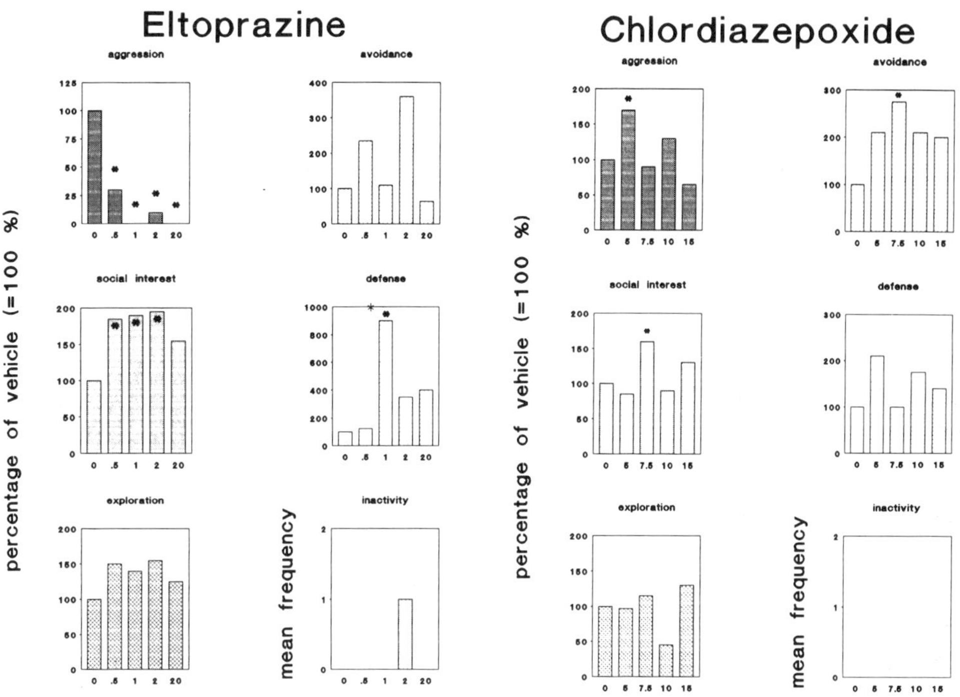

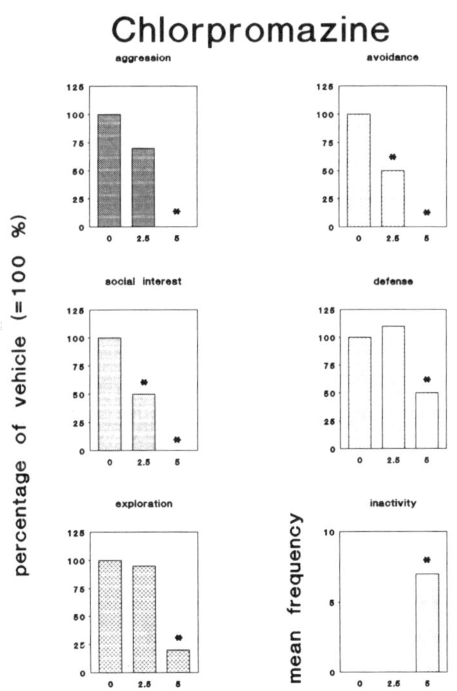

Fig. 1 Effects of eltoprazine (0.5 - 20 mg/kg po), chlordiazepoxide (5 - 15 mg/kg po) and chlorpromazine (2.5 - 5 mg/kg po) on the frequency of occurrence of 6 behavioural categories in intermale aggression in mice. For inactivity the mean frequency has been depicted, for the other categories the frequency is shown as percentage of vehicle (0 mg/kg).

Table 2 Effect of various psychoactive drugs on isolation-induced aggression in male mice

Drug	Mean ED_{50} value in mg/kg (po)
Eltoprazine	0.4
Haloperidol	0.8
Chlorpromazine	4.7
Diazepam	12
Chlordiazepoxide	73

The results show that eltoprazine decreases aggression without sedatory effects. as exploration is not reduced and social interest even increased. Chlordiazepoxide did not affect aggression and only incidental effects were observed in other behavioural cateegories. Chlorpromazine inhibited attacks in this model, but these effects were non-specific as many active behaviours decrease and inactivity increased.

Resident-intruder offensive behaviour: This model, used with increasing frequency in psychopharmacology, uses the resident animal's typical response to a conspecific intruder (Adams, 1976). In this model, a male rat is housed with a female. a situation resembling the natural situation in which animals establish and defend territories (Barnett, 1975; Lore and Flannelly, 1977). When such terrritorial males meet a strange intruder in their territory, heavy fighting may occur. This apparently offensive behaviour can be considered quite natural (Blanchard and Blanchard. 1977; Miczek, 1979). This paradigm differs both from isolation-induced aggression in mice and intermale aggression in rats, because there is no isolation, which may lead to behavioural abnormalities (Valzelli, 1973). Moreover, resident-intruder paradigms have a very wide species generality (van Hooff, 1977; Wilson, 1975), probably including man (Malmberg, 1980). Isolation-induced aggression is restricted to certain species (Miczek and Krsiak, 1979). This model discriminates effectively the quality and behavioural mechanisms of action of several drugs with some anti-aggressive properties (Olivier, 1981; Olivier et al., 1984a,b).

Examples of drug-induced changes of anti-aggressive compounds in this model are given in figure 2. Eltoprazine exerted a dose-dependent decrease in offensive aggression. This coincided with a slight increase in social interest and an increase in exploration. Avoidance was somewhat enhanced at 1.25 and 2.5 mg/kg, but this had returned to baseline at 5 mg/kg, whereas inactivity was somewhat increased at the highest dose.

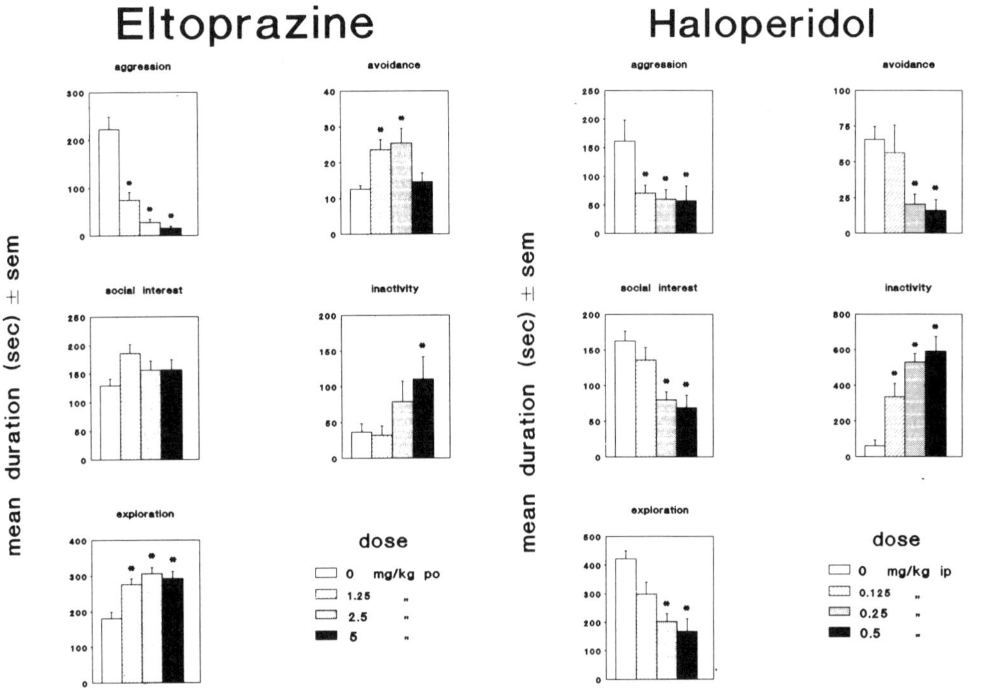

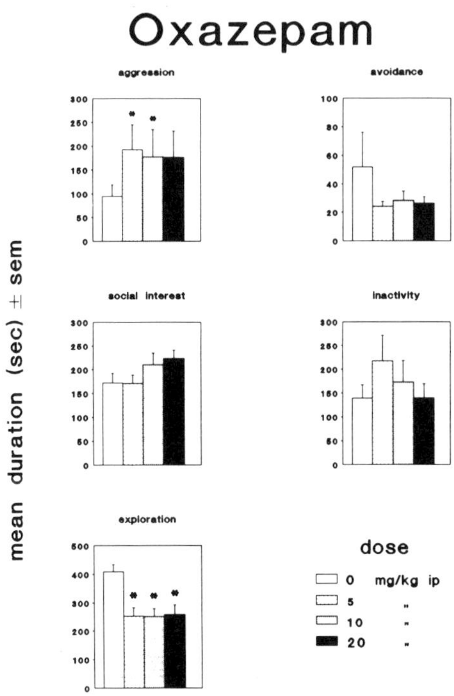

Fig. 2 The effects of eltoprazine (1.25-5 mg/kg po), haloperidol (0.125-0.5 mg/kg ip) and oxazepam (5-20 mg/kg ip) are shown on the mean duration of 5 representative behavioural categories in resident-intruder (territorial) aggression in male rats. Dose were given ip 30 min before testing and orally 60 min before testing. Each drug was tested in 12 rats.

This profile of anti-aggressive actions is specific for the serenics. Haloperidol for instance non-specifically decreased aggression: inactivity increased at all doses and social interest and exploration decreased. Oxazepam even increased aggression at the doses tested, but did neither affect social interest nor inactivity. The doses tested are either below the active dose in animal models for psychosis (haloperidol) or well within the anxiolytic range (oxazepam).

This resident-intruder model is a critical step in the strategy to develop specific anti-aggressive compounds. First of all, this model, combined with ethopharmacological approaches allows the detection of specific anti-aggressive effects, i.e. a dose-dependent decrease in aggression which is not accompanied by sedation, stereotypy, muscle relaxation or other interfering behavioural responses. Second, the concept of territoriality, as a major factor in aggression, either adaptive or pathological, shows a great species generality. That this concept extends to humans was clearly demonstrated by Malmberg (1980).

Offensive behaviour following electrical stimulation of the brain: Behaviour similar to that of offensive territorial males can be induced by electrical stimulation in the hypothalamus of male and female rats (Kruk et al., 1979; Kruk and van der Poel, 1980). The current threshold required to elicit attack behaviour is used as the index of anti-aggressive potency of a test compound. Because stimulation of the same electrode can be used to assess the effects of drugs not only upon potentially aversive qualities of the current, but also upon sedative or stimulating effects (via locomotion), the model allows estimation of the specificity of the anti-aggressive effects of test drugs.

Figure 3 shows examples of drug-induced changes in this model. The changes in thresholds for aggression, teeth-chattering and locomotion after eltoprazine, chlordiazepoxide and haloperidol are given in percentage. Eltoprazine clearly enhanced the threshold for aggression and only slightly for teeth-chattering, whereas the threshold for locomotion was even decreased, indicating the specificity of action on aggression. Haloperidol enhanced aggression, teeth-chattering (slightly) and locomotion thresholds (which could not even be measured) at the same time. This stresses the non-specific effect of haloperidol. Chlordiazepoxide had no effect on aggression and teeth-chattering thresholds at lower doses, but enhanced the thresholds for locomotion and aggression moderately at the highest dose, presumably indicating the presence of muscle relaxation.

One of the more attractive features of this model is that the rat performs offensive aggression in a context that does not normally evoke aggression. The test arena is not the home cage of the experimental animal and before or after the electrical stimulation social investigation and exploration occur. Thus the induction of aggression by electrical stimulation occurs outside the normal context, but the topography of the attacks strongly resembles very intense aggressive behaviour. If a compound reduces this form of aggression without deleterious side effects, it forms a strong argument for potential use in human pathological aggression, in which we have argued the functionality of aggression depends upon the context. From an ethopharmacological viewpoint this model is more artificial than for instance resident-intruder aggression, but it finds its place within the other models especially for the reasons mentioned.

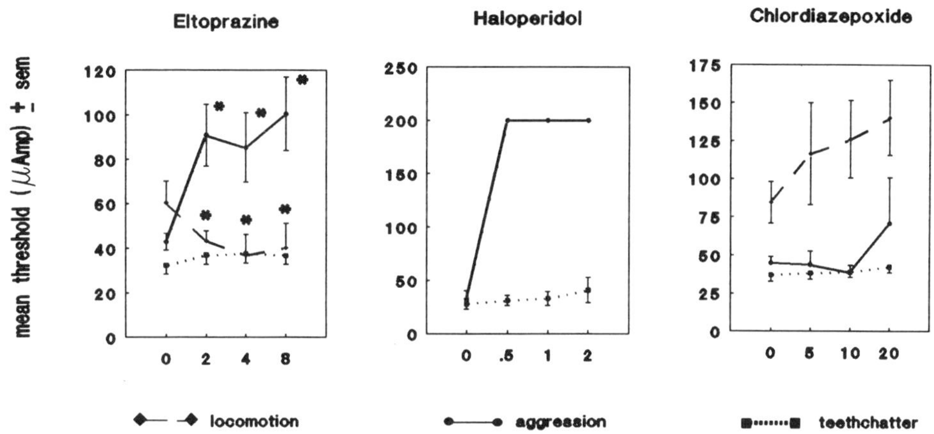

Fig. 3 The mean thresholds (± SEM) in μA are given for the three behaviours evoked by electrical stimulation from the hypothalamus of male rats: aggression, teeth-chattering and locomotion. The effects of eltoprazine, haloperidol and chlordiazepoxide are given on these thresholds. Significant difference (*: $p<0.05$) from 0 mg/kg is depicted. For more details see Olivier et al. (1990).

Maternal offensive behaviour: Although males of a species are generally more apt than females to exhibit attack behaviours in a variety of situations (Moyer, 1976), females will also exhibit offensive behaviours. The maternal aggression model, for example, depends upon the fact that a lactating female rat with young will exhibit offensive behaviours toward a wide variety of intruders. This behaviour is most pronounced during the first part of the lactating period (Erskine et al., 1978; Olivier and Mos, 1986a). Because the critical stimulus is clearly the proximity of some threatening object to the female's young, some authors (Moyer, 1976) have suggested the behaviour is essentially defensive. However, these behaviours can be considered as offensive (van der Poel et al., 1984). We concur with this view because female behaviour in this model is clearly self-initiated, pro-active and not necessarily in reaction to any threat initiated by the intruder.

We developed this model further for use in psychopharmacology (Mos et al., 1990) and have tested many different compounds from various therapeutic and chemical classes. To exemplify the results of these studies figure 4 shows how eltoprazine, haloperidol and chlordiazepoxide affect the female's behaviour. Eltoprazine reduced aggression but did not significantly alter other behaviours. Haloperidol was tested at low doses only and did not significantly decrease aggression, although inactivity already increased at the highest dose tested. Chlordiazepoxide, however, increased aggression at low doses. The bell shaped curve after chlordiazepoxide suggests that at higher doses counterforces like muscle relaxation prevent the increased aggression. It is clear from these examples that different profiles of action can be obtained from different classes of drugs in this model of naturalistic and adaptive aggression. Haloperidol is sedatory, but no doubt anti-aggressive at higher doses and chlordiazepoxide even has pro-aggressive actions, which also applies to other benzodiazepines (Mos and Olivier, 1989). A more specific profile of action is observed after serotonergic compounds, like eltoprazine, which are acting at the $5\text{-}HT_{1A/1B}$ receptor. One of the special advantages of this model is the fact that such a clearly different neurochemical and environmental background subserves this behaviour. Yet, this intense form of aggression reacts quite similar to drug treatment as observed in for instance resident-intruder aggression. It thus adds significantly to the available toolbox of animal models to study aggression.

MODELS OF DEFENSIVE BEHAVIOURS

Pain-induced defensive agonistic behaviour: A widely used 'aggression model' in pharmacology is foot-shock or pain-induced aggression (Ulrich and Azrin, 1962), in which agonistic behaviour is elicited typically by delivering an electrical shock to the hind paws of a pair of rats or mice. While either or both animals may attack, the behaviour is now conceived as defensive (Blanchard et al., 1977a,b; 1978), in part because the animals mutually exhibit typical, upright defensive postures and squealing. Although these reactions are well integrated, no complete sequences of fighting and no offensive threat displays are present.

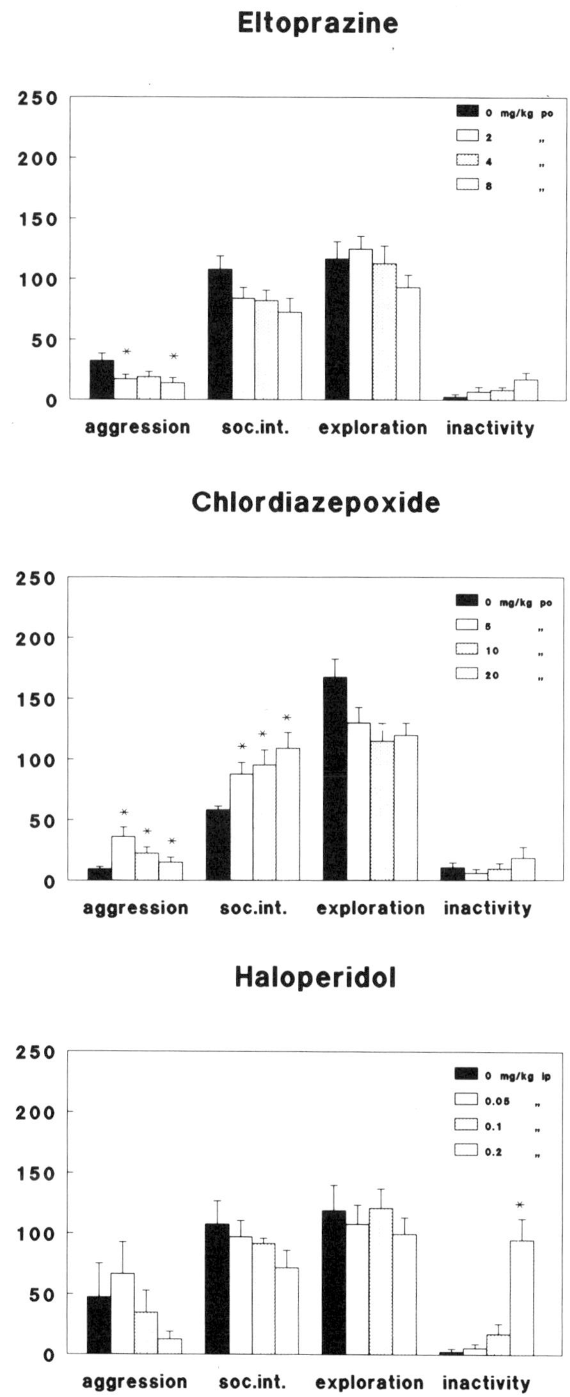

Fig. 4 Effects of eltoprazine, haloperidol and chlordiazepoxide on four behavioural categories recorded in maternal aggression by rats. Dose on the X-axis (mg/kg) is plotted against the time in seconds spent on the various behavioural categories. * depicts significant difference from vehicle.

Although this model has been extensively used to assess anti-aggressive activity of drugs, a confounding factor in the model is that the behaviour-releasing stimulus (pain) can be masked by analgesic properties of psychoactive drugs. This fact, together with the very restricted behavioural repertoire in this paradigm, limits the utility of this defensive model in characterising the anti-aggressive properties of drugs.

However, a useful feature of the model is that the induced defensive behaviour may be used to discriminate drug effects in this highly artificial model from naturally occurring behaviours. Table 3 summarizes the results of our experiments on prototypical compounds in a modification of the method described by Tedeschi et al. (1959). Eltoprazine did not suppress the defensive fighting and had no effects on paralysis measured as the ability of mice, hanging by their forelimbs from a thin bar, to bring their hindlimbs on to the bar. Diazepam and chlordiazepoxide did reduced defensive fighting at doses roughly equal to those leading to muscle relaxation. A similar effect was observed after chlorpromazine.

It is noteworthy that eltoprazine differs so much from the other compounds in the sense that neither defense nor muscle relaxation are affected. This stresses the unique profile of eltoprazine in suppressing offensive aggression leaving defense intact. This profile of action is also clear from the following model of defense in a more natural context.

Table 2 Effect of drugs on foot-shock induced defence (D) and paralysis (P) in male mice. The ED_{50}-values for anti-defensive activity or paralysis are given as well as the ratio. A higher ratio indicates a more specific effect. n.d. = not determinable. For details see Olivier et al. (1990).

	Oral ED_{50}-value (mg/kg)		
Compound	defense	paralysis	ratio P/D
Eltoprazine	>46.4	>46.4	n.d.
Diazepam	6.8	2.5	0.4
Chlordiazepoxide	15.4	11.7	0.76
Chlorpromazine	8.2	7.3	0.91

The intruder model of defensive behaviour: A more natural model of defensive behaviour is that which considers the behaviour of the intruder in the resident-intruder or maternal aggression. Study of this defensive behaviour reveals that defending rats have special tactics to protect the more vulnerable parts of their bodies. In unconstrained situations, animals on the defense usually flee from the territory of the residential male or lactating female, but when this is impossible, as is often the case in laboratory settings, they defend themselves by flight, crouching, upright defensive postures, emission of ultrasounds and submissive postures. Generally these behaviours aim at protecting the back, the area where most wounds are inflicted by attacking rats (Mos et al., 1984).

Although this model involves at least two rats, an offensive and a defensive one, it gives the opportunity to measure the capacities of the defending animal when it is treated with a drug. When the behaviour of the defender is assessed by ethological methods, this model gives a very powerful and subtle way to describe the effects of drugs. Despite its obvious advantages, this defensive model has only found limited use in psychopharmacological research in aggression (Olivier and Mos, 1986b).

Figure 5 shows the effects of acute treatment of the intruders with eltoprazine, haloperidol and chlordiazepoxide on their behaviour when confronted with a lactating female. Treatment of the intruder did not modify the number of attacks by the lactating female. Thus while treatment of the female modifies the attacks, the females do not change their strategy if the opponent's behaviour is modified. In the right column of the figure, the behaviour of the intruder is depicted in three categories. Eltoprazine did not significantly change the behaviour of the intruder, i.e. the animals were capable of adequate defense. Immobility tended to increase, but this could also be part of freezing as part of defensive behaviour. Chlordiazepoxide did not modify the behaviour of intruders, whereas it does change the lactating female if she is treated. Haloperidol, when administered to the intruders changed their behaviour. There was a strong, dose-dependent increase in immobility and a trend to decreased defense although this did not reach significance.

The data suggest that eltoprazine, while inhibiting attacks in offensive aggression models, does not interfere with adaptive defensive responses. The intruders are not attacked more often than vehicle treated counterparts, nor is their defensive repertoire consisting of various behavioural acts like upright posture, flight, crouching etc affected. The effectiveness of eltoprazine to suppress offense is not necessarily linked to defensive capacities, like observed for haloperidol, which sedates animals and consequently interferes with almost any behaviour that is likely to occur in the given situation. This more naturalistic assessment of defense clearly shows the differentiation of actions by serenics and other compounds that affect aggression. Summarizing, it can be concluded that serenics do not suppress defensive aggression, whereas other compounds may disrupt defensive behaviours at non-sedating doses, but clearly do not improve defensive capacities.

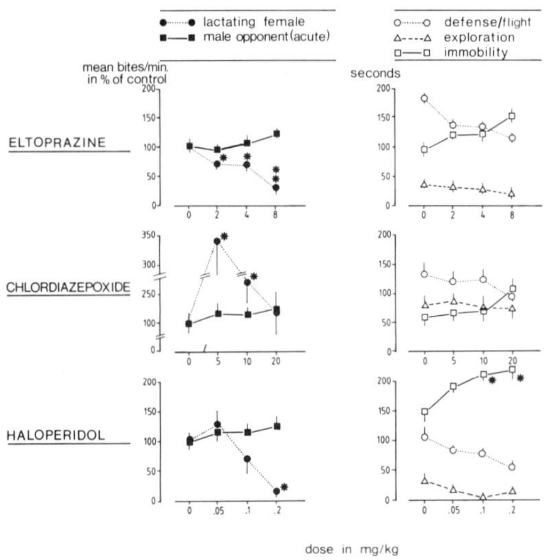

Fig. 5 The effect of 3 psychotropic drugs on offensive behaviour of lactating females (left-hand panel) or defensive behaviour of the male intruder (right-hand panel). In the left-hand panel either the lactating female is treated (direct drug effects) or the male opponent (indirect drug effect), but in both cases the behaviour of the female is recorded. In the left-hand panel the mean bites/min (± SEM) as percentage of control (=100%) are used, in the right-hand panel mean duration (± SEM) in sec.
*: ($p<0.05$) denotes significant difference from vehicle (0 mg/kg).

GENERAL DISCUSSION AND CONCLUSIONS

In general, the animal models most readily accepted as a basis for predicting responses in man are those which are homologous; i.e., in which both the condition being observed and its aetiology are demonstrably similar to that in man. Examples might include suppression of bacterial infections by antibiotics. Needless to say that few, if any animal models of psychiatric dysfunction can be considered homologous if only because the aetiology of these conditions in man is unknown. Lacking homologous models, isomorphic models - in which the observed condition is apparently similar even if the aetiology is not - may be fairly readily accepted. However, even this is not the case for pathological aggression. Finally, there are many models in which neither the condition nor the aetiology can be clearly linked with the disease being modelled, but in which there is emperical value either for the disorder or some aspect of its therapy. In psychopharmacology, the evidence is usually the discovery that agents with some therapeutic activity in man consistently co-vary with some response in an animal model, for example neuroleptics. For anti-aggressive compounds, indeed there are no homologous or isomorphic models, but two important facts need to be considered. First, it is clear that those compounds that are effective in 'controling' pathological human aggression, like neuroleptics and high doses of benzodiazepines are also effective in reducing aggression in models with functional aggression. Second, the use of ethopharmacological approaches has enabled the development of compounds which do have significant anti-aggressive actions, but which do not disturbe normal ongoing behaviour like exploration, social interest etc. Moreover, these compounds do not lead to sedation or muscle relaxation or other non-specific blockade of aggression, nor do they disrupt adequate defensive responses to being attacked.

Although these facts do not necessarily predict the efficacy of novel anti-aggressive compounds in human, they form a solid base to develop compounds with a profile of anti-aggressive action superior to current treatment. Finally it should be realized that the generality of the concept of territoriality in animals and humans strongly supports the idea that at least the relevant neurobiological mechanisms are involved. Whether these mechanisms are 'abnormal' in pathological aggression and consequently these animal models are predictive for human efficacy remains to be established for the serenics currently tested in clinical settings.

REFERENCES

Adams, D.B. (1976) Behaviour 56, 286-297.
Adams, D.B. (1979) Behav. Brain Sci 2, 201-241.
Barnett, S.A. (1975) The rat. A study in behaviour. Chicago and London: The University of Chicago Press.
Blanchard, R.J., Blanchard, D.C., Takahashi, L.K. (1977a) Aggr. Behav. 3, 145-155.
Blanchard, R.J., Blanchard, D.C., Takahashi, L.K. (1978) Biol. 23, 291-305.
Blanchard, R.J., Blanchard, D.C., Takahashi, T., Kelley, M.J. (1977b) Anim. Behav. 25, 622-634.
Blanchard, R.J., Blanchard, D.C. (1977) Behav. Biol. 21, 197-224.
Blanchard, R.J., O'Donnell, V., Blanchard, D.C. (1979) Aggr. Behav. 5, 341-352.
Brain, P.F. (1975) Life Sci. 16, 187-200.
Cherek, D.R. (1981) Psychopharmacol. 75, 339-345.
Erskine, M.S., Denenberg, V.H., Goldman, B.D. (1978b) Behav. Biol. 23, 52-66.
Garattini, S., Valzelli, L. (1981) Progr. Neuro-Psychopharmacol. 5, 159-165.
Grant, E.C., Mackintosh, J.H. (1963) Behaviour 21, 246-259.
Hooff, van J.A.R.A.M. (1977) In: Aggressive Behaviour - Causes and functions (P.R.Wiepkema and J.A.R.A.M.Van Hooff, Eds), Bohn, Scheltema en Holkema, Utrecht, pp 5-27.
Huntingord, F., Turner, A. (1987) Animal conflict. Chapman and Hall. London and New York.
Janssen, P.A.J., Jageneau, A.H., Niemegeers, J.E. (1960) J. Pharmacol. Exp. Therap. 129, 471-475.
Krsiak, M. (1974) Res. Comm. Chem. Pathol. Pharmacol. 7, 253-257.

Krsiak, M. (1975) Br. J. Pharmacol. 55, 141-150.
Kruk, M.R., Van der Poel, A.M., De Vos-Frerichs, T.P. (1979) Behaviour 70, 292-322.
Kruk, M.R., Van der Poel, A.M. (1980) Prog. Brain Res. 53, 385-390.
Lehman, M.N., Adams, D.B. (1977) Behaviour 61, 238-275.
Lore, R., Flannelly, K. (1977) Sci. Am. 236, 106-116.
Malick, J.B. (1979) Curr. Dev. Psychopharmacol. 5, 1-27.
Malmberg, (1980) Human territoriality: survey of behavioural territories in man with preliminary analysis and discussion of meaning. Mouton, The Hague.
Miczek, K.A. (1987) In: Handbook of Psychopharmacology vol 19. (Iversen, Iversen and Snyder, Eds), Plenum Publishing Corporation, pp 183-328.
Miczek, K.A., Barry, H. (1976) In: Behavioural Pharmacology (S.D.Glick, J.Goldfarb, Eds), St. Louis: Mosby p 176-257.
Miczek, K.A., Krsiak, M. (1979) In: Advances in Behavioural Pharmacology. (T.Thompson, P.B.Dews, Eds), New York, Academic Press 1979; Vol. 2: 87-162.
Miczek, K.A. (1979) Psychopharmacology 60, 253-259.
Mos, J., Olivier, B. (1989) Psychopharmacol. 97, 152-153.
Mos, J., Olivier, B., Van Oorschot, R. (1990) Aggres. Behav. 16, 145-163.
Mos, J., Olivier, B., Van Oorschot, R., Dijkstra, H. (1984) Physiol. Behav. 32, 453-456.
Moyer, K.E. (1968) Comm. Behav. Biol. 2, 65-87.
Moyer, K.E. (1976) The psychobiology of aggression. Harper and Row, Publishers: New York Hagerstown, San Francisco, London.
Olivier, B., Mos, J. (1986a) In: Cross-Disciplinary studies on Aggression (P.F.Brain, J.Martin Ramirez, Eds), University of Sevill Press, pp 73-111.
Olivier, B., Mos, J. (1986b) Stress Medicine 2, 197-209.
Olivier, B., Mos, J., Van der Poel, A.M., Krijzer, F.N.C., Kruk, M.R. (1984a) In: Biological Perspectives on aggression (D.C.Blanchard, K.J.Flannelly, R.J.Blanchard, Eds), Alan R. Liss, New York, pp 261-279.
Olivier, B., Olivier-Aardema, R.L., Wiepkema, P.R. (1983) Behav. Brain Res. 9, 59-81.
Olivier, B., Van Aken, H., Jaarsma, I., Van Oorschot, R., Zethof, T., Bradford, L.D. (1984b) In: Ethopharmacological aggression research (K.A.Miczek, M.R.Kruk, B.Olivier, Eds), Alan R Liss, New York, pp 137-156.
Olivier, B., Van Dalen, D. (1982) Aggr. Behav. 8, 163-168.
Olivier, B. (1981) Pharmacol. Biochem. Behav. 14, Suppl 1, 61-77.
Poshivalov, V.P. (1981) Pharmacol. Biochem. Behav. 14, Suppl 1, 53-59.
Ross, S.B., Ögren, S.O. (1976) J. Pharmacy Pharmacol. 28, S90-S92.
Taylor, S.P. (1967) J. Pers. 35: 297-310.
Tedeshi, P.E., Tedeschi, D.H., Mucha, A., Cook, L., Mattis, P.A., Fellows, E.J. (1959) J. Pharmacol. Exp. Therap. 125, 28-34.
Ulrich, R.E., Azrin, N.H. (1962) J. Exp. Anal. Behav. 5, 511-520.
Valzelli, L. (1969) Aggr. Behav. pp 70-76.
Valzelli, L. (1973) Psychopharmacologia 31, 305-320.
Valzelli, L., Giacalone, E., Garattini, S. (1967) Eur. J. Pharmacol. 2, 144-146.
Van der Poel, A.M., Mos, J., Kruk, M.R., Olivier, B. (1984) In: Ethopharmacological Aggression Research (K.A.Miczek, M.R.Kruk, B.Olivier, Eds), Alan R Liss. New York, pp 115-135.
Wilson, E.A. (1975) Sociobiology. Belknap Press of Harvard University Press, Cambridge, Massachusetts.
Yen, C.Y., Stanger, R.L., Millnam, N. (1958) Arch. Int. Pharmacodyn. Ther. 123, 179-185.
Yudofsky, S.C., Silver, J.M. (1986) Am. J. Psychiat. 143, 34-39.

Animal Models in Psychopharmacology
Advances in Pharmacological Sciences

PSYCHOPHARMACOLOGY AND SEXUAL BEHAVIOUR: USES AND LIMITATIONS OF ANIMAL MODELS

Barry J. Everitt

Department of Anatomy, University of Cambridge, Downing Street, Cambridge CB2 3DY, England

SUMMARY: The use of a variety of behavioural procedures to study discrete psychological processes and neural mechanisms underlying integrated patterns of sexual behaviour are described. Such procedures have especial power when used in conjunction with drugs having specific actions, but they are in no sense "models" of human sexual behaviour and would appear to have but limited value in the development of drugs which may be effective in a therapeutic context.

The sexual behaviour of experimental animals has not been studied as a model of human sexual behaviour to any great extent within the pharmaceutical industry. Of the reasons why this may be so, two perhaps stand out: firstly, disorders of sexual behaviour in men and women are not regarded with the same urgency and importance as are affective, psychotic or other psychiatric disorders - which is somewhat surprising given their widespread occurrence and pervasive impact on relationships, families and society. Secondly, the aetiology of such disorders is very complex and it is frequently difficult to distinguish between "organic" and "psychogenic" features of their clinical presentation. Furthermore, it is hard to imagine a generally successful drug treatment of a psychogenic sexual disorder such as premature ejaculation or impotence in men, since both may present as partner-specific; an affected individual may have no problem with one partner, but find it impossible to perform with another.

In a clinical psychopharmacological setting, disorders of sexual response are more frequently encountered as side effects of drugs being given therapeutically for another disorder, such as depression. The interest in a "model" of human sexual behaviour is, therefore, one much more concerned with the elimination of such effects, although unusual spin-offs from such observations are not unknown. For example, the tricyclic antidepressant clomipramine is well known to impair the sexual potency of men and may induce orgasmic dysfunction in women (Beaumont, 1973). However, this undesirable side effect of the drug has been made use of therapeutically to treat men complaining of premature ejaculation (Eaton, 1973). The mechanism of action of clomipramine in delaying ejaculation appears to be related to its α-adrenergic blocking activity (indeed, the α_2-antagonist, yohimbine, has also been used with some success - Morales et al., 1982; Reid et al., 1987) and not related to the inhibition of serotonin reuptake which is the better characterised action of the drug (Everitt, 1977).

The search for "models" of human sexual behaviour to be used as a screen for therapeutic agents with which to treat specific disorders of sexual behaviour would seem unlikely to meet with success if the "model" is intended to resemble the clinically presenting disorder except, perhaps, in certain quite specific circumstances such as the control and treatment of erectile failure. However, one of the most remarkable developments in the treatment of such erectile impotence in men, which involves the intracavernosal injection of smooth muscle relaxants such as papaverine and phenoxybenzamine (Brindley, 1986) - a treatment which has transformed the sex lives of a considerable number of men (see Bancroft, 1989) - was neither studied nor developed in animal species at all.

Only primate species would appear to be truly appropriate to study in an attempt to understand better the neuroendocrine determinants of sexual behaviour in humans, since only here is the relative emancipation of sexual behaviour from the tight hormonal regulation seen in non-primate species similar to that seen in men and women (see Bermant & Davidson, 1974; Everitt & Keverne, 1986; Sachs & Meisel, 1988), while the attendant greater flexibility in sexual responses presumably reflects the considerable degree of limbic forebrain and neocortical development. Further, the social determinants of, and constraints on, sexual responses are only possible to study with any degree of certainty in primate species (see Keverne, 1983). For example, the social subordination of female Talapoin monkeys results in amenorrhoea, infertility (the blockade of oestrogen-induced luteinizing hormone release) and hyperprolactinaemia - changes similar to those seen in women under social stress; bromocryptine reverses the hyperprolactinaemia and restores reproductive function - both in female monkeys and women (Bowman et al., 1978; Bancroft, 1989). However, the difficulty and expense of conducting studies of reproductive behaviour in socially living groups of monkeys are likely to combine to render such studies unlikely in the context of therapeutic drug development. Even in heterosexual pairs of monkeys, the interpretation of the effects on behaviour of a drug or hormone treatment may be both difficult and complex. This is well exemplified by examining the effects of systemically administered progesterone on the sexual behaviour of female rhesus monkeys. This hormone results in increased numbers of sexual solicitations which are displayed in order to stimulate a male to mount. This effect of the hormone is, however, not a direct consequence of its action within the central nervous system to alter behaviour, as might be thought, since it also follows topical, intravaginal application in doses not measurable in the systemic circulation. The actual mechanism of action involves a change in the female's vaginal odour which makes her less attractive to a male; her increase in solicitations, therefore, is an indirect consequence of a change in the male's behaviour as she attempts to compensate for his declining sexual interest in her (Baum et al., 1977a,b).

As a model behavioural system with which to investigate the mechanism of action of novel psychoactive drugs, animal sexual behaviour would appear not to be particularly useful because of the complex and serial nature of the sexual response repertoire being studied, together with the fact that at least two, interacting animals must be used. This may severely complicate the interpretation of any behavioural effects of a drug. The measurement of highly reproducible and easily quantifiable sensory and/or motor reflexes - such as limb extensor and flexor reflexes, head twitches and other simple behavioural responses familiar as drug screens in the pharmaceutical industry - seems a much more useful means of examining any commonality in the mechanisms of action of novel drugs. But there are instances where sexual reflexes have been employed in this way - for example, $5HT_{1A}$ receptor agonists appear to have the common property of facilitating elements of masculine sexual behaviour (reducing the number of intromissions prior to ejaculation and ejaculation latencies - see Ahlenius & Larsson, 1987, 1988). However, other treatments also may have this effect (e.g. intra-hypothalamic naloxone, see below and Everitt, 1990a; Hughes et al., 1987) and it is far from clear whether categorization of the specificity of drugs for the $5HT_{1A}$ receptor is better using this complicated, as opposed to other simpler, behavioural test procedures involving single animals.

The "models" of sexual behaviour that I will discuss in the remainder of this paper are not really "models" of human sexual behaviour at all. I have made it clear that I am not convinced of the utility of such models in any case. Instead, they are behavioural paradigms which facilitate the study of the underlying psychological processes and neural substrates of the integrated pattern of sexual behaviour - and there is reason to believe that these may be similar across a range of non-primate and primate species, including man. It is here that psychoactive drugs with more or less discrete mechanisms of action have played an invaluable role, especially when placed in discrete neural loci, since they have made it possible to begin to unravel the chemically defined neural mechanisms within areas of the brain that may be involved with the control of sexual behaviour.

The discussion will be restricted to a consideration of masculine sexual behaviour. It is generally true that, in the majority of studies hitherto, copulatory responses and the latencies to make them represent the most frequently quantified measures of masculine sexual behaviour. From such measures, inferences have been made about underlying psychological processes, usually within a conceptual framework laid down by Beach (1956) and elaborated by Sachs and Barfield (1976) and Sachs (1978). This framework identifies a "sexual arousal mechanism" and an "intromission-ejaculatory mechanism", the former reflected primarily in the latencies to mount and intromit, the latter in copulatory measures such as numbers of mounts and intromissions, the rates at which they occur, as well as the latency to ejaculate (Beach, 1956; Sachs, 1978). One of the difficulties in interpreting the effects of drugs on sexual behaviour in terms of alterations in appetitive (motivational/arousal), as opposed to consummatory (performance), mechanisms is that the latency measures which are taken to depend on the former are not independent of changes in peripheral mechanisms which directly affect the latter. For example, reduced latencies to mount following spinal treatments which facilitate erection (see below); clearly, this is not a direct effect of the treatments on central mechanisms underlying sexual arousal processes but an indirect - and no less interesting - effect of the drug (see Everitt, 1990a).

It is potentially of the greatest utility to design behavioural paradigms which can separate appetitive and consummatory mechanisms which together operate to generate the integrated pattern of sexual behaviour. Those that we have either developed or adapted for this purpose measure, in addition to parameters of copulatory behaviour and the sequence in which they occur, either arbitrary motor responses, including partner choice, which do not depend on the ability to make species-specific copulatory responses, and also utilise manipulations of sexual incentive stimuli which shift the emphasis of initiation of sexual contact onto the male.

BEHAVIOURAL PARADIGMS

Interactive sexual behaviour: there is no substitute for observations of sexual interaction. In addition to the familiar measures of masculine copulatory performance (mounts, intromissions, the rates at which they occur and latency measures, including the post-ejaculatory refractory period (PEI) - see Sachs & Barfield, 1976), it has proved both informative and useful to measure precopulatory responses such as olfactory investigation, pursuit and grooming (e.g. Everitt & Stacey, 1987; Hansen & Drake af Hagelsrum, 1984; Hughes et al., 1987, 1988) as well as paying particular attention to the sequence in which such responses occur (Stavy & Herbert, 1989). Giving males a choice between competing incentive stimuli, such as between a receptive female and a highly preferred ingestive reward (e.g. acesulfame) also provides a different perspective on the selection of motivated responses, particularly when a treatment selectively impairs or enhances the tendency to respond to one or other goal (Hawkins et al., 1989). Thus, ingestion of acesulfame by male rats is suppressed in the presence of a receptive female with which they interact sexually - except during the PEI, when sexual responses are inhibited and ingestion is more likely to occur (Hawkins et al., 1989). Castration of male rats is followed gradually by a decline in copulatory behaviour and a switch from sexual to ingestive responses (Hawkins et al., 1989). In this paradigm, then, the neuroendocrine mechanisms which determine the selection of responses towards biologically relevant goals which compete for the animal's attention may be studied.

Incentive manipulations: Sexual interaction between a male and female rat depends upon behaviour emitted by both animals. The male pursues and investigates the female, ultimately attempting to mount and intromit. The female is not the passive recipient of such attempts, but actively approaches and leaves the male, soliciting his interest and mounting attempts. In order to isolate more effectively the degree to which interaction is dependent on the male's interest and arousal, we have employed a procedure which changes the sexual incentive value of the female. Systemic injection of oestradiol-progesterone-primed female rats with low doses of the dopaminergic antagonist α-flupenthixol results in a preservation of receptive behaviour, in fact lordosis responses are elicited more readily and for longer durations, but the same treatment abolishes the active proceptive responses of hopping, darting and ear wiggling (Everitt et al., 1974; 1975). Thus, any copulatory sequence with this passively receptive female must be initiated by the male. Use of this procedure has provided insight into the nature of the processes disrupted by dopaminergic receptor blockade in the ventral striatum of male rats (Everitt, 1990a,b).

Partner preference: is the measurement of sexual partner preference. Here, the preference expressed by males for the opportunity to interact with a particular type of female is assessed by giving him the choice of an oestrous or anoestrous female in either side of the place preference apparatus. It is probably the case that the processes operating to determine such a choice are complex. However, this paradigm may indicate whether a male is able to classify the intrinsic, or primary, incentive stimuli associated with a female in a motivationally relevant way such that he interacts sexually with her.

Conditioned place preference: place preferences conditioned by the repeated exposure of males to a receptive, or an unreceptive, female in distinctive environments provides another perspective on the processes underlying sexual responses (see Everitt, 1990a for a description of the apparatus). This procedure allows the conditioned approach to environmental cues associated with prior sexual reinforcement to be measured, as with the second-order schedule of sexual reinforcement (see below), relatively independently of the ability to execute the consummatory acts of mounting and intromitting. The conditioned approach presumably reflects reward-related aspects of the prior sexual experience, since it is elicited by the conditioned incentive properties of the distinctive place that have accrued through their association with sexual interaction (see Carr et al., 1989; Everitt, 1990a; Phillips & Fibiger, 1985; Van der Kooy, 1985; White et al., 1985). The behavioural measures are probably related to those obtained using the second-order schedule, although there remains much to be established concerning the precise nature of such relationships. The advantage of this procedure is that it is relatively rapid and easy to establish and therefore provides a useful tool for investigating the effects of psychopharmacological and other neural and endocrine manipulations.

Instrumental responses with sexual reinforcement: the ability to quantify appetitive sexual responses in a way which is minimally affected by consummatory competence - in a sexual setting, the ability to mount, intromit and thrust - has long been acknowledged to be of great value in the study of motivational processes, but this has proved difficult to achieve in studies of sexual behaviour (see Everitt & Stacey, 1987; Everitt et al., 1987; 1990a,b). We have achieved this by utilizing a paradigm in which the instrumental behaviour of male rats is maintained by sexual reinforcement (the opportunity to copulate with a female in heat) presented under a second-order schedule. The method is described in detail elsewhere (Everitt et al., 1987). In the procedure we have adopted, males are first allowed to copulate with an oestrous female in an operant chamber in the presence of a discrete and arbitrary light stimulus (the conditioned stimulus, CS). Subsequently, when tested in the modified operant chamber in which they had previously copulated (see Everitt, 1990a for an illustration of the apparatus), they learn to respond at high rates on a lever to earn the presentation of the arbitrary CS which itself signals the ultimate presentation of the oestrous female at the end of the session. Thus, the male's instrumental behaviour is maintained by a secondary, or conditioned reinforcer, which has gained its (sexual) motivational significance through its prior, Pavlovian association with copulation with an oestrous female. The success of the method, in the sense that it generates high baseline levels of behaviour, is that responding is encouraged by the incentive motivational value of the CS which, presumably, evokes a central representation of the primary goal. Earlier methods were established by reinforcing responses through the presentation of the female herself which, because of the occurrence of mounts and intromissions, completely disrupted the ongoing appetitive response output (Everitt et al., 1987). Although acquisition of the behaviour takes a relatively long time, the stable baseline of responding is particularly suitable for the study of psychopharmacological and selected neural manipulations. The method certainly enables appetitive and consummatory (copulatory) elements of the male's sexual response repertoire to be dissociated and the effects of treatments on each to be assessed more precisely.

The utility of employing the above methods of studying sexual behaviour is apparent if, following a specific psychopharmacological treatment or neuroendocrine manipulation, clear dissociations between the effects in one or more of the behavioural paradigms is obtained, since this will potentially enable clearer definition of the psychological process(es) affected. Thus, the pattern of behavioural change across this range of methods is likely to be more informative than any single change within a procedure - and this is especially the case when attempting to understand, for example, why copulation is impaired or enhanced following a treatment if only interactive measures are taken (Everitt, 1990a). This point is illustrated below with respect to the effects of two kinds of psychopharmacological treatment: manipulation of opioid peptides within the hypothalamus and of dopamine systemically and within the ventral striatum.

OPIOID PEPTIDES AND MASCULINE SEXUAL BEHAVIOUR

Much of the interest in the effects of opioid peptides on sexual behaviour has been generated by the reports of the detrimental impact of chronic opiate use, especially heroin, on sexual function in both men and women (see Pfaus & Gorzalka, 1987). Furthermore, treatment of male heroin addicts with naltrexone during withdrawal has been reported to facilitate the reappearance of sexual desire (Hollister, 1975). Experimentally, morphine generally inhibits mounts, intromissions and ejaculations in male rats, while naloxone given systemically (often in high doses which are not selective for opiate receptors) tends to facilitate aspects of sexual behaviour, for example in sexually sluggish rats, or have no effect (see Pfaus & Gorzalka, 1987; Gessa et al., 1979; Hetta, 1977; Hughes et al., 1987, 1988; Myers & Baum, 1979, 1980). These effects of predominantly μ-receptor agonist and antagonist treatments have been studied using some of the above methods, initially by assessing the effects of naloxone given both systemically and into the medial preoptic/anterior hypothalamic area (POA) as well as the effects of the endogenous opioid peptide, ß-endorphin, infused within the preoptic area.

Beta-endorphin infused bilaterally into the POA dose-dependently inhibited sexual behaviour. Mounts and intromissions eventually ceased to occur and the latencies to mount, intromit and ejaculate were prolonged - eventually to the duration of the 15 minute test session or even longer (Hughes et al., 1987, 1988). What is the nature of this effect? Do males lose interest in females as a result of the treatment - i.e. do mount and intromission latencies lengthen as the result of a change in sexual arousal, or is the behavioural point of action of the drug more on the consummatory responses of mounting and intromitting? Careful observation of pre-copulatory, investigative responses revealed that ß-endorphin-infused males actively investigated and pursued females and, to some extent, made abortive mounting attempts (Hughes et al., 1987) - changes in behaviour which were seen in a more emphatic way following lesions of the POA (Everitt & Stacey, 1987). Analysis of the behavioural sequence revealed that, when a treated male and female first made contact, the pattern of interaction was not significantly different from that seen with control males, but that the sequence broke down at the point when investigative responses usually switched to the copulatory responses of mounting and intromitting; thus, ß-endorphin infused into the preoptic area appears not to affect sexual interest or arousal, but instead the transition between investigative and the species-specific, copulatory reflexes of mounting and intromitting (Hughes et al., 1987; Stavy & Herbert, 1989). These males also showed evidence of the thwarting of a motivated response tendency in that irrelevant behaviours, such as scratching and grooming, emerged at very high rates in ß-endorphin-treated males who could not copulate (Hughes et al., 1987; see Hansen & Drake af Hagelsrum, 1984). In addition, these sexually inhibited males now showed a release of the drinking of a preferred sweet solution which is normally suppressed in the presence of a receptive female with which the male would usually copulate (Hughes et al., 1987).

Infusion of ß-endorphin into the POA had no effect on instrumental responding for a receptive female presented under a second-order schedule of sexual reinforcement, nor on the expression of a place preference conditioned by sexual interaction with such a female (Everitt, 1990a; Everitt et al., 1990). However, the same treatment rapidly abolished a males preference for a receptive, over an unreceptive, female tethered in either side of the place preference apparatus.

An unexpected, further dimension to the analysis of the effects of ß-endorphin followed its infusion into the POA after an intromission, rather than before interaction with a female had begun. The peptide no longer had an inhibitory effect on mounting, intromission or ejaculation. Furthermore, the inhibitory effects of ß-endorphin were lost or reduced even if a delay of two hours or so was interposed between a single intromission and the infusion, provided the male was re-tested with the same female with which he had intromitted previously. If a different female was placed with the male following intromission, then the inhibitory effect on copulatory behaviour of ß-endorphin reappeared (Stavy & Herbert, 1989). These results clearly suggest that ß-endorphin does not simply act to prevent copulation, since if the male is allowed to begin interacting sexually with an individual female, the otherwise inhibited behaviour can be emitted normally and ejaculation will occur.

Taken together, the results of these experiments reveal the remarkable behavioural specificity of the effects of ß-endorphin infused into the POA. The peptide does not, apparently, influence appetitive aspects of sexual behaviour nor reward-related processes, so far as these are assessed in the place preference procedure. Instead, intra-hypothalamic ß-endorphin appears to prevent the display of the copulatory reflexes which together form the consummatory elements of the sexual response sequence. However, even here the effects of the peptide are not absolute, since if the male is allowed to engage sexually with a female, then the inhibitory effects of the peptide are themselves inhibited. The impaired preference for a receptive female which follows ß-endorphin infusion appears to be related to the failure to initiate the copulatory sequence - presumably only if the peptide is infused prior to an intromission, although this has not been tested explicitly by studying its effects on preference when infused after an intromission.

Therefore, a neural mechanism may exist in the POA which allows the appropriate behavioural responses of mounting and intromitting to be matched to a relevant sexual incentive stimulus and it is this which is impaired in the ß-endorphin-treated male rat. This may suggest a particular relationship between the intrinsic sensory properties of a receptive female and the species-specific motor output of copulatory reflexes, since acquired motor responses (e.g. bar-pressing or conditioned approach) for conditioned incentives, such as a light CS or the properties of the preferred place, are completely unaffected by the same ß-endorphin manipulation of the POA. The behaviour of a rat in the operant and place preference procedures may, therefore, be controlled by the same mechanisms that underlie unconditioned appetitive or preparatory responses and these appear not to reside within the POA, since lesions of the structure are also without effect on these parameters (Everitt, 1990a; Everitt & Stacey, 1987). Reduced preference, for an oestrous female after ß-endorphin infusions, however, may be related more to the switch from sexual to ingestive responses which is seen to follow manipulations of testosterone or hypothalamic ß-endorphin in male rats (Hawkins et al., 1989; Hughes et al., 1987).

The effects of naloxone on sexual behaviour have also been studied using a number of these behavioural procedures and the results also indicate the complexities underlying the superficially simple effects of the systemically administered opioid antagonist. Systemic naloxone may, in some circumstances, facilitate sexual behaviour in male rats (see above), but if given to intact, sexually active males these effects are vanishingly small (Hughes et al., 1987). However, the same treatment reduced instrumental responses for a female presented under a second-order schedule of reinforcement, promptly abolished a previously acquired conditioned place preference but had no effect on partner preference (Everitt, 1990a; Hughes et al., 1990). Infusing naloxone bilaterally into the POA resulted in a powerful facilitation of copulatory behaviour; males required fewer intromissions to ejaculate with a much reduced latency, yet had no effect on instrumental behaviour or a conditioned place preference. Partner preference was seen to be reduced following this treatment, but analysis of the data revealed this to be an epiphenomenon of the increased sexual activity, since males spent more of the test period in a state of refractoriness and therefore away from both females in the neutral compartment of the choice apparatus (Everitt, 1990a; Hughes et al., 1990).

It is clear from these results that opioid mechanisms within the POA are not involved with incentive motivational responses to sexual stimuli, but are more involved with consummatory sexual responses. However, more interesting, perhaps, is the indication that such incentive - or reward-related - responses are sensitive to systemic naloxone. In addition, it has been shown that this sensitivity is much greater in animals recently having been castrated (Mehrara & Baum, 1990; Miller & Baum, 1987). Since castration also profoundly affected a sexually conditioned place preference (Everitt, 1990a; Hughes et al., 1990) and instrumental sexual responses (Everitt & Stacey, 1987), while POA lesions, as well as intra-POA opioid manipulations, were without effect on these measures, opioid involvement in sexual reward-related processes may both be sex hormone-dependent and involve extrahypothalamic substrates (see Everitt,

1990a). The ventral tegmental area dopaminergic system innervating ventral striatal and limbic structures is an obvious focus for experiments investigating this problem (see Wise, 1989a,b).

DOPAMINE AND MASCULINE SEXUAL BEHAVIOUR

It has long been known that systemic treatment with neuroleptics impairs masculine sexual behaviour, while direct and indirect dopaminergic agonists can enhance sexual performance, especially in animals showing sub-optimal levels of sexual behaviour for some reason (see Bitran & Hull, 1987; Everitt, 1990a; Pfaus & Phillips, 1990). Systematic study of the effects of a variety of dopamine receptor agonists and antagonists infused into the POA by Hull and colleagues (Bitran & Hull, 1987; Bitran et al., 1988; Hull et al., 1986; Pehek et al., 1988) have made it clear that dopaminergic mechanisms within the POA influence the display of masculine copulatory behaviour, while spinal cord dopaminergic events may influence the display of penile reflexes (Pehek et al., 1990). However, these effects of altering dopaminergic transmission in the POA are not sufficient to explain the changes in appetitive aspects of sexual behaviour. For example, α-flupenthixol decreased instrumental responses for a receptive female presented under a second-order schedule of reinforcement at doses considerably lower than those which interfered with mounting and intromission (Everitt, 1990a; Robbins et al., 1989). In addition, the atypical neuroleptics, clozapine and sulpiride, said to influence ventral, rather than dorsal, striatal dopamine receptors had marked effects on the initiation of copulation, but had no effect at all on copulation itself. The typical neuroleptics, haloperidol and pimozide, not only delayed the initiation of copulation, but also affected copulatory measures as well, such as reducing the number of intromissions prior to ejaculation, whereas metaclopramide had no effect on initiation and only altered copulatory parameters (Pfaus & Phillips, 1989). While some of these actions may have been mediated by hypothalamic dopaminergic receptors (see above), it seems likely that actions on appetitive measures (e.g. the latency to initiate copulation) may reflect interference by the drugs with a more pervasive incentive motivational system involving the dopaminergic innervation of the ventral striatum (see Everitt, 1990a,b).

Infusion of d-amphetamine into the ventral striatum dose-dependently enhanced the rate of responding for a receptive female presented under a second-order schedule of reinforcement, shortened the latencies to mount and intromit once the female had been earned, but did not alter the latency to ejaculate or the numbers of mounts and intromissions preceding ejaculation (Everitt et al., 1989; Everitt, 1990a). Thus, selective manipulation of dopaminergic transmission in the ventral striatum primarily seems to affect appetitive, but not consummatory, elements of masculine sexual behaviour.

The subtle nature of the effects of dopaminergic receptor blockade in the ventral striatum was illustrated using the procedure in which proceptive/receptive female rats are treated systemically with α-flupenthixol. Normal males paired with such females show a small increase in the latencies to mount and intromit, but otherwise copulate normally (Everitt, 1990a). This is best interpreted as a consequence of removing one category of the incentive properties of the female (proceptive behaviour), such that responsibility for initiating interaction with the female rests solely with the male. Infusing the D_2-selective antagonist, raclopride, into the ventral striatum had little or no effect on the sexual behaviour of male rats when paired with normal females. However, when males so-treated were paired with the neuroleptic-treated females there was a very different result. Mount and intromission latencies were enormously prolonged such that, in some cases, treated males never initiated sexual interaction. However, replacing the treated female with an untreated one resulted in a fairly prompt initiation of copulation, emphasizing that the deficit was not simply one of motor performance and that consummatory sexual responses could be made quite effectively (Everitt, 1990a). Similar results followed 6-hydroxydopamine-induced dopaminergic depletions from the ventral, but not the dorsal, striatum (Everitt, 1990a). In the latter case, males were actually improved by pairing them with neuroleptic-treated females, suggesting that dopamine depletion from the caudate-putamen had mainly motoric effects, thereby interfering with the execution of consummatory, as well as appetitive, acts (Everitt, 1990a).

These results suggest that reducing dopaminergic transmission in the ventral striatum affects incentive motivational processes and renders animals so-treated very susceptible to devaluation of the goal object - in this case, preventing the display of proceptive responses by the receptive female. The results of experiments conducted by Pfaus, Phillips and colleagues (Phillips et al., 1990), using in vivo chronoamperometry and dialysis to measure extracellular dopamine in behaving rats, further illustrates this point. Placing an oestrous female behind a mesh screen resulted in a striking increase in the dopamine signal measured in the ventral, but not the dorsal, striatum of male rats. The dopamine signal in the ventral striatum remained elevated during the copulatory sequence which occured when the screen was removed, but tended to diminish progressively towards baseline as copulation progressed. In the dorsal striatum, extracellular dopamine drifted upwards gradually, but did not appear to be tightly coupled to discrete behavioural responses made by the males. Recrudescence of sexual interest during the PEI was actually preceded by a rise in the dopamine signal in the ventral striatum (Phillips et al., 1990; Pfaus & Phillips, 1989).

CONCLUSIONS

Taken in conjunction, this brief summary of experiments on the effects on masculine sexual behaviour of intracerebrally and systemically applied opioid and dopaminergic agonists and antagonists has, it is to be hoped, illustrated the utility of employing a variety of procedures to study drug-induced changes in behaviour. This utility has at least two facets: firstly, to facilitate analysis of the actions of a drug on such a complex behaviour, because these actions may be both complex and multidimensional. Secondly, to define more carefully the existence and operation of discrete neural, neurochemical and psychological processes which underlie the expression of the integrated sexual response pattern. Whether or not this approach will facilitate attempts to develop drugs with therapeutic potential for sexual disorders, or drugs to reduce sexual activity in, say, the treatment of sexual offenders, is difficult to say, but it seems unlikely at present. Of the neuroendocrine treatments which affect the display of sexual behaviour, only sex steroids would appear to be specific in their actions on this behavioural system. It is not surprising, therefore, to see that endocrine treatments, such as testosterone, can restore sexual activity in some hypogonadal men (Bancroft, 1989), while anti-androgens and progestational steroids have been used with partial success to treat certain categories of sexual offender (see Bancroft, 1989).

Aminergic and opioid treatments, on the other hand, have far reaching effects across motivational, as well as sensory and motor, systems and it seems unlikely that drugs interacting with these neurochemical systems could be developed so as to interfere selectively with the neural substrates of sexual behaviour. However, interactions between steroid hormones and naloxone-sensitive, as well as dopamine-sensitive, mechanisms which appear to underlie reward-related processes in a sexual context may prove to be an exception to this rather pessimistic view.

Acknowledgements: The research summarised here was supported by grants from the Medical Research Council. I thank Astra Alab, Södertälje, Sweden for supplying Raclopride.

REFERENCES

Ahlenius, S., Larsson, K. (1987) In: Pharmacology of central 5-HT1A receptors (C.T.Dourish, S.Ahlenius and P.H.Hutson, Eds), Ellis Horwood, Chichester, pp 185-198.
Ahlenius, S., Larsson, S. (1988) J. Psychopharmacol. 2, 47-53.
Bancroft, J. (1989) Human Sexuality and its Problems, 2nd ed. Churchill Livingstone, Edinburgh.
Brindley, G.S. (1986) Br. J. Psychiatry 149, 210-215.
Baum, M.J., Keverne, E.B., Everitt B.J., Herbert, J. (1977a) Arch. Sexual Behav. 6:173-192.
Baum, M.J., Keverne, E.B., Everitt B.J., Herbert, J. (1977b) Physiol. Behav. 18: 659-670.
Beaumont, G. (1973) J. Int. Med. Res. 1, 469-473.
Beach, F.A. (1956) In: Nebraska Symp. on Motivation (M.R.Jones, Ed), 4, Lincoln: Univ. Nebraska Press, pp 1-31.
Bermant, G., Davidson, J.M. (1974) Biological Bases of Sexual Behaviour, New York, Harper & Row.
Bitran, D., Hull, E.M. (1987) Neurosc. Biobehav. Rev. 11, 365-389.
Bitran, D., Hull, E.M., Holmes, G.M., Lokkingland, K.J. (1988) Brain Res. Bull. 20, 323-331.
Bowman, L.A., Dilley, S.R., Keverne, E.B. (1978) Nature 275, 56-58.
Carr, G.D., Fibiger, M.C., Phillips, A.G. (1989) In: The Neuropharmacological Basis of Reward (J.M.Liebman and S.J.Cooper, Eds), Oxford University Press, pp 264-319.
Eaton, H. (1973) J. Int. Med. Res. 1, 432-437.
Everitt, B.J. (1977) Postgraduate Med. J. 53, 202-210.
Everitt, B.J. (1979) In: Sex Hormones and Behaviour, Ciba Foundation Symp. 62. Elsevier North Holland, pp 329-358.
Everitt, B.J. (1990a) Neurosci. Biobehav. Rev. 14, 217-232.
Everitt, B.J. (1990b) In: Biological Psychology (T.Archer, S.Hansen, S.Ahlenius and P.Södersten, Eds), L. Erlbaum Associates. (In press).
Everitt, B.J., Cador, M., Robbins, T.W. (1989) Neuroscience 30: 63-75.
Everitt, B.J., Fray, P.J., Kostarczyk, E., Taylor, S., Stacey, P. (1987). J. Comp. Psychol. 101, 395-406.
Everitt, B.J., Fuxe, K., Hökfelt, T. (1974) Eur. J. Pharmacol. 29, 187-191.
Everitt, B.J., Fuxe, K., Hökfelt, T., Jonsson, G. (1975). J. Comp. Physiol. Psychol. 89, 556-572.
Everitt, B.J., Keverne, E.B. (1986) In: Neuroendocrinology (S.L.Lightman and B.J.Everitt, Eds), Blackwells Scientific Publications, Oxford. pp: 472-536.
Everitt, B.J., Stacey, P. (1987) J. Comp. Psychol. 101, 407-419.
Gessa, G.-L., Paglietti, E., Pellegrini-Quarantotti, B. (1979) Science 204, 203-205.
Hansen, S., Drake af Hagelsrum, L.J.K. (1984) Behav. Neurosci. 98, 868-883.
Hetta, J. (1977) Acta Pharmacol. Toxicol. (Copenh), Suppl. 4, 53-60.
Hollister, L. (1975) In: Sexual Behaviour: Pharmacology and Biochemistry (M.Sandler and G.-L.Gessa, Eds), Raven Press, New York, pp 85-92.
Hughes, A.M., Everitt, B.J., Herbert, J. (1988) Neuroscience 27, 689-698.

Hughes, A.M., Everitt, B.J., Herbert, J. (1987) Neuroscience 23, 1063-1073.
Hughes, A.M., Herbert, J.H., Everitt, B.J. (1990) Psychopharmacology (In press).
Hull, E.M., Bitran, D., Pehek, E.A., Warner, R.K., Band, L.C., Holmes, G.M. (1986) Brain Res. 370, 73-81.
Keverne, E.B. (1983) In: Mate Choice (P.Bateson, Ed), Cambridge University Press, pp 407-422.
Mehrara, B.J., Baum, M.J. (1990) Psychopharmacology 101, 118-125.
Miller, R.L., Baum, M.J. (1987) Pharmacol., Biochem. Behav. 26, 781-789.
Morales, A., Surridge, D.H.C., Marshall, P.G., Fenemore, I. (1982) J. Urol. 128, 45-47.
Myers, B., Baum, M. J. (1979) Pharmacol., Biochem. Behav. 10, 615-618.
Myers, B., Baum, M. J. (1980) Pharmacol., Biochem. Behav. 12, 365-370.
Pehek, E.A., Warner, R.K., Bazzett, T.J., Bitran, D., Band, L.C., Eaton, R.C., Hull, E.M. (1988) Brain Res. 443, 70-76.
Pehek, E.A., Thompson, J.T., Eaton, R.C., Bazzett, T.J., Hull, E.M. (1990) Pharmacol., Biochem. Behav. (In press).
Pfaus, J. and Gorzalka, B. (1987) Neurosci. Biobehav. Rev. 11, 1-34.
Pfaus, J.G., Phillips, A.G. (1989) Psychopharmacology 98: 363-368.
Phillips, A.G., Blaha, C.D., Fibiger, H.C. (1989) Neurosci. Biobehav. Rev. 13, 99-104.
Phillips, A.G., Pfaus, J.G., Blaha, C.D., Fibiger, H.C. (1990) In: The Mesocorticolimbic Dopamine System: From Motivation to Action (P.Willner and J.Scheel-Kruger, Eds), (In press).
Phillips, A.G., Fibiger, H.C. (1985) In: Methods of assessing the reinforcing properties of abused drugs (M.A.Bozarth, Ed), New York: Springer Verlag, pp 275-290.
Phillips, A.G., Pfaus, J.G., Blaha, C.D., Fibiger, H.C. (1989) Behav. Pharmacol. 1: (Suppl. 1) 275.
Reid, K., Surridge, D.H.C., Morales, A. (1987) Lancet ii, 421-423.
Robbins, T.W., Cador, M., Taylor, J.R., Everitt, B.J. (1989) Neurosci. Biobehav. Rev. 13, 155-162.
Sachs, B.D. (1978) In: Sex and Behaviour. (T.E.McGill, D.A.Dewsbury, B.D.Sachs, Eds), New York: Plenum Press, pp 267-295.
Sachs, B.D., Barfield, R.J. (1976) In: Advances in the Study of Behaviour (J.S.Rosenblatt, R.A.Hinde, E.Shaw, C.Beer, Eds), vol 7. Orlando FL, Academic Press, pp 91-154.
Sachs, B.D., Meisel, R.L. (1988) In: The Physiology of Reproduction (E.Knobil, J.Neill, Eds), New York: Raven Press, pp 1393-1485.
Stavy, M., Herbert, J. (1989) Neuroscience 30, 433-442.
Wise, R.A. (1989a) In: Neuropharmacological basis of Reward (J.M.Liebman, S.J.Cooper, Eds), Oxford, Clarendon Press, pp 377-424.
Wise, R.A. (1989b) Neurosci. Biobehav. Rev. 13, pp 129-134.

Animal Models in Psychopharmacology
Advances in Pharmacological Sciences

ADULT PARTNER PREFERENCE OF MALE RATS 'ORGANIZED' PERINATALLY

T. Brand[1], J. Kroonen[1], J. Mos[2] and A.K. Slob[1]

[1]) Department of Endocrinology and Reproduction, Faculty of Medicine and Health Sciences, Erasmus University, P.O.Box 1738, 3000 DR Rotterdam, The Netherlands,
[2]) Department of Pharmacology, Duphar B.V. P.O.Box 900, 1380 DA Weesp, The Netherlands

SUMMARY: Intact adult male rats, in which aromatization of testosterone to estradiol was pre- or perinatally prevented by ATD (1,4,6-andro-statriene-3,17-dione), were repeatedly tested for partner preference behaviour (choice: estrous female vs active male. Perinatally ATD treated males had significantly lower preference scores for an estrous female than prenatally ATD treated and control males. It seems that this behaviour is, at least partially, organized by estradiol neonatally. 8-OH-DPAT, a serotonin 5-HT_{1A}-receptor agonist, enhanced preference for an estrous female in all groups, albeit that perinatally ATD treated males still had lower preference scores.

It has been shown that the estradiol (E_2) metabolite of testosterone (T) is essential during a critical perinatal period for masculinization and defeminization of sexual behaviour in the male rat (e.g. Baum, 1979). We have recently found that partner preference behaviour is affected by neonatal castration (Brand & Slob, 1991). This suggests that T may play a role in the organization of this behaviour. Whether it is also the estradiol metabolite of T that plays a role in this process is not clear. Davis et al. (1979) administered (silastic implant) the aromatase inhibitor ATD to male rats neonatally (days 2-10). In adulthood there were no differences in partner preference behaviour (estrous female vs active male) between ATD and control males. It is possible, however, that T or its E_2 metabolite still could have organized the behaviour during the first 24 hrs following birth, when T levels are high (Baum et al., 1988). We have, therefore, tested adult male rats for partner preference behaviour, which were treated with ATD before and/or immediately after birth. Also, the effects on partner preference and sexual behaviour were studied of 8-OH-DPAT, a 5-HT_{1A}-agonist with "aphrodisaical" properties in male rats (Ahlenius et al., 1981; Schnur et al., 1989).

MATERIALS AND METHODS

Animals and treatments: Pregnant Wistar albino rats were treated with ATD (5 mg/day) or propylene glycol (0.1 ml/day; solvent) from day 10 to 22 of pregnancy. Within 9 h after birth a silastic implant (i.d. 1.5 mm; o.d. 2.1 mm; length 5 mm) was placed s.c. in the back of the male under ice anaesthesia. The implant contained ATD (approx. 8 mg) or was left empty. At 21 days of age the animals were weaned and their implant removed. The following groups were formed: control (n=14), prenatal ATD (n=12) and pre- and neonatal ATD (n=18). The males were housed 2-3 of same treatment to a cage, with water and food freely available and left undisturbed until start of behavioural testing at 11 weeks of age. Lighting in the animal room was artificial only for a 14 h day (lights on: 5.30 pm to 7.30 am). Stimulus animals were F1 hybrids of two inbred Wistar strains (R x U). Ovariectomized females were brought into behavioural estrus by injecting 30 μg estradiol benzoate 24-48 h, followed by 2.5 mg progesterone 3-4 h prior to testing.

Behavioural testing: Thirteen consecutive weekly 15 min tests were carried out in a test box with 3 adjacent compartments (60 x 30 x 40 cm each) with a small opening (13 x 12 cm) in both partitions near the front window (Slob et al., 1987). These openings could be closed by a guillotine door. Before testing stimulus animals were put in the lateral compartments and the experimental male in the central compartment (sliding doors closed) for 15-20 min adaptation. The incentives, an estrous female and an active male, were placed behind a wire mesh separation halfway the lateral compartment (tests 1-7) or weared a leather harness which was attached to the rear of the compartment (tests 8-13). At the beginning of each test the sliding doors were opened and the time spent by the experimental male in each compartment was recorded. To quantify partner preference, a preference score was calculated in which time spent with the male was subtracted from time spent with the female; thus positive score=preference for female; negative score=preference for male. In tests 8-13, in which social interaction was possible, various sexual behaviours were scored.

Thirty min prior to test nrs. 6, 10 and 12 the males were s.c. injected with the 5-HT_{1A} receptor agonist 8-OH-DPAT (tests 6 and 10: 0.2 mg/kg; test 12: 0.4 mg/kg). In tests 4, 5, 9, 11 and 13 saline (2 ml/kg) was injected 30 min prior to testing.

RESULTS

Partner preference behaviour

The data of tests 1-7 were analysed separately from tests 8-13. In the very first test, most control males (10/14), half of the prenatally ATD treated males (6/12) and only a minority of perinatally ATD treated males (5/18) preferred the estrous female. Two-way ANOVA and further LSD-analysis both on test-series 1-7 and 8-13 showed gradually increasing preference scores for the estrous female in all groups (in test 13 all males preferred the female), albeit that perinatally ATD treated males had significantly lower preference scores than prenatally ATD treated and control males; the latter two did not differ. 8-OH-DPAT given in test 6 had no effect; when given in tests 10 and 12 the preference for the estrous female significantly increased in all groups, but the perinatally ATD males still remained the lowest; pre-ATD and control groups did not differ.

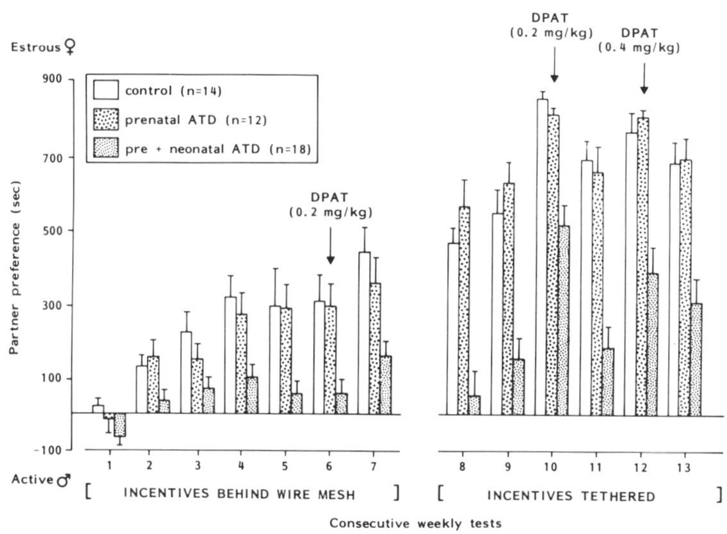

Fig. 1 Mean (±SEM) preference (seconds) for an estrous female over a sexually active male or intact male rat after perinatal ATD or control treatment. In tests 8-13 sexual interaction with tethered incentives was possible; in tests 1-7 sexual interaction was prevented by wire mesh. See text for more details.

Sexual behaviour with tethered female incentive

Two-way ANOVA showed that prenatal ATD-males ejaculated less than controls. Perinatally ATD treated males almost never ejaculated except after 8-OH-DPAT treatment. The latter treatment also decreased the number of intromissions prior to the first ejaculation, resulting in 'premature ejaculation' (i.e. ejaculation with the first intromission): with the low dose (0.2 mg/kg) in 21% of the males (8/38); with the high dose (0.4 mg/kg) in 81% of the males (35/43).

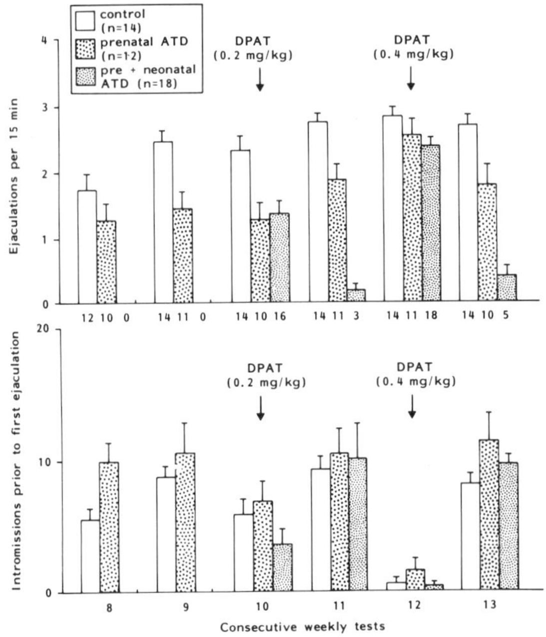

Fig. 2 Top. Mean (± SEM) ejaculation frequency during partner preference tests 8-13. The digits below the bars indicate the number of males per group that ejaculated at least once. Bottom. Mean (± SEM) number of intromissions prior to the first ejaculation.

DISCUSSION

The present data show that adult partner preference behaviour can be added to the list of behaviours in the male rat that are (at least partially) organized during a critical perinatal period by testosterone, through its estrogenic metabolite. The data are not in line with earlier findings by Davis et al. (1979) who could not find changes in adult partner preference following ATD treatment during postnatal days 2-10. We assume that Davis et al. started their ATD treatment too late. High levels of testosterone are circulating during day 1 (Baum et al., 1988) when ATD treatment was started in the present experiment. Prenatal ATD treatment had no effect on adult partner preference behaviour, although it was not without behavioural effect: sexual behaviour was significantly inhibited compared to control males. The perinatally ATD treated males were severely inhibited in their adult ejaculatory behaviour. The administration of the 5-HT_{1A} receptor agonist 8-OH-DPAT, a known stimulator of male sexual behaviour (Ahlenius et al., 1981), improved sexual behaviour to the level of the controls. Apparently, perinatally ATD treated males are still able to perform sexual behaviour and show comparable sexual behaviour stimulating effects in response to 5-HT_{1A}-compounds like normal naive males.

In conclusion, the availability of neonatal estrogen (derived from T) organizes, at least partially, male rats to display a preference for an estrous female in adulthood. The lack of neonatal estrogen causes males to be less masculinized: both in partner preference behaviour and ejaculatory behaviour. Treatment with a serotonin agonist abolishes, at least partially, the perinatal ATD effects on adult partner preference behaviour and on ejaculatory behaviour.

REFERENCES

Ahlenius, S., Larsson, K., Svensson, L. (1981) Pharmacol. Biochem. Behav. 15, 785-792.
Baum, M.J. (1979) Neurosci. Biobehav. Rev. 3, 265-284.
Baum, M.J., Brand, T., Ooms, M., Vreeburg, J.T.M., Slob, A.K. (1988) Biol. Reprod. 38, 980-986.
Brand, T., Slob, A.K. (1990) Physiol. Behav. 49, (in press).
Davis, P.G., Chaptal, C.V., McEwen, B.S. (1979) Horm. Behav. 12, 12-19.
Slob, A.K., de Klerk, L.W.L., Brand, T. (1987) Physiol. Behav. 41, 571-576.

Animal Models in Psychopharmacology
Advances in Pharmacological Sciences

EXPERIMENTAL SOCIAL DIFFERENTIATION IN THE RAT: BEHAVIOURAL CHARACTERISTICS IN WISTAR AND LONG-EVANS STRAINS.

D. Desor*), C. Colin*), F. Thuillier*), J. Mos**) and B. Krafft*)

*) Laboratoire de Biologie du Comportement, URA C.N.R.S. 1293, B.P. 239, 54506 Vandoeuvre les Nancy CEDEX, France
**)Duphar B.V., P.O.Box 900, 1380 Weesp, The Netherlands

SUMMARY: When grouped Wistar rats progressively have to dive and swim to reach food, a behavioural differentiation into 2 main types appears: "Carrier" rats which display this behaviour, and Non-Carrier rats which get their food only by agonistic encounters. It is possible to split these types into sub-types; (Resistant and Supplier Carrier rats, Efficient and Non-efficient Non-carrier rats). Various arguments indicate that this differentiation is based upon social interactions. Long-Evans hooded rats basically act as Wistar ones when placed in this situation, but hoarding behaviour is observed, which modifies the agonistic interactions.

The present study was conducted in order to determine the reaction of a group of rats faced with a problem of food supplying. The basic question is how interindividual behavioural variability may lead to stable role differentiation in the groups.

Groups of 6 rats are installed in experimental devices composed of a cage (40x40x40 cm) and a tunnel which allows the rats access to an aquarium (150x25x40 cm). A feeder is disposed at the extremity of the aquarium. From the beginning of the experiment, the access to food is restricted to a daily 3 hours period. On days 1 and 2, there is no water in the aquarium, so the rats can easily learn the spatial characteristics of their new environment. From day 3 to 7, the level of water is progressively increased and consequently, rats have to walk in water, then they have to swim, and finally when the water has reached its maximum level, they have to dive without breathing to go to the feeder and bring back the pellet to the cage in order to eat it. In the Wistar strain, this experimental schedule leads to behavioural differentiations among the rats:

- About 50% of the rats dive and bring back food pellets to the cage (Carrier rats = C. rats). The defence of the food in the cage is insured by various behavioural items, mainly: flying, dodging, kicking with hind legs. C. rats differ in their abilities to defend food: Cluster analysis differentiates "Resistant" carrier rats, which are autonomous in the sense that they can hold the food they got by diving, and "Supplier" carrier rats which have to feed various numbers of non-carrier rats before eating.

- The other 50% rats never dive to bring back food (Non-Carrier-N.C. rats). Their only way to eat is to steal food from other rats. The efficicacy of N.C. rats in stealing food are not equal: some are very efficient and display brief and violent attacks in the cage or at the tunnel. Some other rats spend considerable time in harassing food possessors. Finally, some rats scrounge the pellets lost in the aquarium or under the floor of the cage and show a strong reduction of their social interactions in the alimentary context.

This behavioural differentiation rapidly occurs (lst week) when the rats are placed into the experimental devices, and it remains stable for long periods: 6 groups were observed during 8 months without notable variations in their compositions.

The determinants of individual roles in this situation are not exactly known at present, but 3 facts indicate the importance of social factors: 1. when they are individually trained in the device, about 100% of the rats are able to dive in order to get food, whereas only 50% evolve to this type in the groups of 6. 2. making new groups composed of previously differentiated Carrier or Non-Carrier rats leads to new differentiations in the two cases. 3. incentive behaviour is often observed. At the tunnel, N.C. rats can push the Carriers with their head or their forelegs, or they can bite them. As a result, the time of hesitation of Carrier rats before diving is considerably reduced.

In hooded (Long-Evans) rats, this basic behavioural differentiation is present too. But generally, from the 2nd week of the experiment, the Carrier hooded rats carry out hoarding, and from the time pellets are left on the ground of the cage, N.C. rats collect them and stop attacking the Carriers. It seems interesting to note that the latency for first hoarding decreases with time, and about the 3rd week, even the first pellet is left on the ground of the cage. Consequently, the level of aggression drops down to very low values in these groups.

This model seems to be relevant in order to study the effects of various drugs in a complex social situation.

Animal Models in Psychopharmacology
Advances in Pharmacological Sciences

EFFECT OF THE SEX OF THE INTRUDER ON TERRITORIAL AGGRESSION OF WILD RATS

Aldo B. Lucion, Alberto A. Rasia Filho and Rosa M.M. de Almeida

Department of Physiology, Pharmacology and Biophysics, Institute of Biological Sciences, Federal University of Rio Grande do Sul, Porto Alegre, RS, 90.049, Brazil

SUMMARY: The aggressive behaviour of 18 colonies (2 male and 2 females each) of wild rats against castrated albino males was compared to that towards sexual receptive albino females. The number of deaths and the latency of investigate and attack the castrated male were the same as for normal males. When females were intruders, the number of deaths was not different from randomness, but significantly lower than that of normal or castrated males. The latency of investigating females was the same as males. Sexual behaviour was observed. The results were discussed with reference to the influence of stimuli which trigger sexual behaviour versus territorial aggression.

INTRODUCTION

The presence of a conspecific intruder in a rat colony triggers aggressive behaviour (Blanchard & Blanchard, 1977; Blanchard et al., 1977; Boice, 1972; Koolhaas et al., 1980; Takahashi & Blanchard, 1982). The characteristics of the intruders' age, sex, and level of sexual hormones are important variables in eliciting this aggression (Adams & Boice, 1983; Albert et al., 1986; Blanchard et al. 1984; Leshner, 1975; Takahashi & Lore, 1982).

Beilharz (1975) described the aggressive behaviour of various strains of mice toward different social stimuli and reported that the type of stimulus seems to be less important than the nature of the subjects. On the other hand, Flannelly and Thor (1978) argued that the stimulus which provokes aggressive behaviour is of gonadal origin and that castration of male intruders reduces this behaviour.

The present study continues the analysis of the stimuli which elicit the aggressive behaviour of territorial defense, checking the role of sexual hormones of the intruders.

We intended to test the hypothesis that the absence of the male sexual hormone (castrated intruder) or the presence of receptive females inhibited aggression by the resident male. We developed this study with wild rats colonies which present clearly defined aggressive behaviour.

MATERIAL AND METHODS

Eighteen wild rats (Rattus norvegicus) colonies were studied (2 males and 2 females, non-siblings), laboratory bred from ancestors captured in the city of Porto Alegre (Brazil) in 1986. These 18 colonies were part of a set of 23, which had already been tested with albino and normal wild male intruders. The age of resident males was between 111 and 238 days.

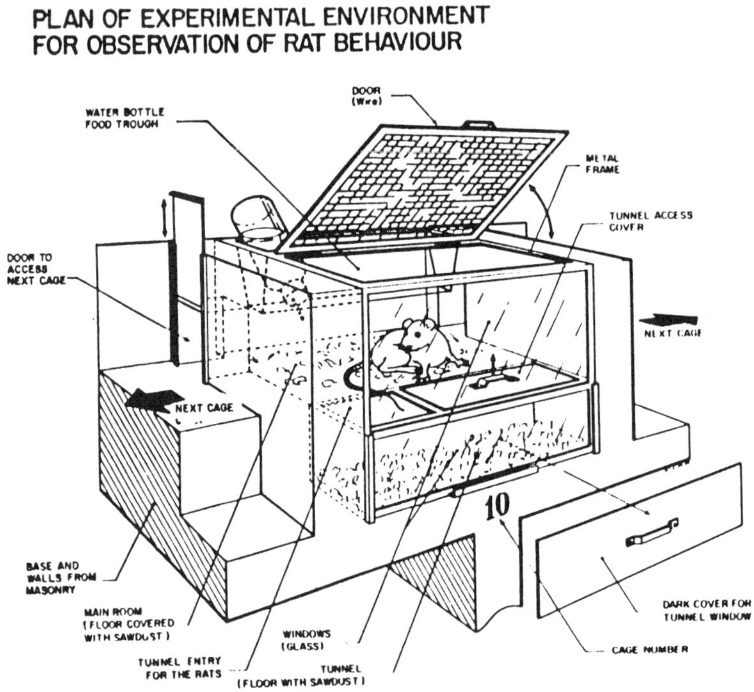

Fig. 1. Observation cage with its dimensions.

The subjects in the colonies were in specially constructed cages (Fig. 1) in a room where the dark/light cycle was controlled (12:12 hours, with the dark phase beginning at 5 p.m.) as well as temperature and noise. Water and food were given ad libitum.

The effect of the introduction of receptive albino females and castrated albino males was tested, according to a random distribution of the testing order. The females, 70 to 90 days old, were induced to sexual receptiveness by injecting 0.125 mg of estradiol benzoate intramuscularly 24 hours before the test. The intruder males, with weight and age approximately equal to that of the residents, were castrated 20 days before the test. These intruders were placed in the colonies at the beginning of the dark cycle, and stayed there for 15 hours, then being removed and the number of deaths computed. During the first 10 minutes the social interaction between the intruder and the resident was observed, and latency of the investigative and fighting behaviour of the resident male toward the intruder was recorded in seconds from the beginning of the test. Two tests were performed introducing castrated males and receptive females in each colony. The frequency of deaths was compared using the exact test of Fisher (Siegel, 1956) both regarding receptive females and castrated males, with random distribution and frequency of normal albino males deaths in these same colonies. The mean latency of the investigative and fighting behaviour was compared to that of the normal albino male using the Student t test (Dixon & Massey, 1969).

RESULTS

Table 1 presents the number of deaths of receptive females, castrated males and normal males. For a same type of intruder, very close values were observed in both tests performed. The frequency of deaths among the receptive females was no different than would be expected at random, but it was significantly lower than that of the castrated and normal males. On the other hand, the number of deaths of castrated males was equal to that of the normal males. The receptive female or castrated male's investigative latency and the attack latency against the latter male did not differ from that found against the normal male (Table 2).

Direct observation showed that the male resident seemed to behave toward the castrated male in the same way as against the normal male. As to the females, the behaviour of sniffing the genitals followed immediately upon that of investigating the body, which did not occur with the male intruder. In this case at the same time the resident male attacked and mounted the receptive female. Strangely enough both males mounted the female, and the resident females attacked and bit the female intruder. This neither happened in relation to the normal male nor to the castrated one; a single male attacked the intruder, and the others did not even leave the tunnel.

Table 1 Number of deaths of albino intruders: normal male (N=23) castrated male (N=18) and receptive female (N=18).

		intruders			
		Normal male (N=23)	Castrated male (N=18)		Receptive female (N=18)	
		3rd test	1st test	2nd test	1st test	2nd test
Number of deaths		23	17	18	8	10
Number of deaths expected by randomness		12	9	9	9	9
Fisher exact probability	$p1^{+}$	0.0001	0.0036	0.0005	O.5000	0.5000
	$p2^{*}$	-	0.4390	1.0000	0.00003	0.0004

+; p1 = comparison to randomness
* ; p2 = comparison to the number of deaths of normal albino male intruder in the third test

Table 2 Mean latency in seconds (s) of investigate and attack behaviour of female (n=18), normal (n=23) and castrated (n=18) male intruders

Intruder	Mean latency (s) *	
	investigate	attack
Normale male	11.14	36.86
Castrated male	13.64	46.50
Receptive female **	19.66	-

* Student t test not significant ($\alpha < 0.05$)
** Latency of attack against female intruder was not computed

DISCUSSION

The results showed that the receptive female was the only stimulus able to inhibit the aggression of the resident males. The sexual stimulation produced by her seems to have counterbalanced her role as an intruder, and so the significant reduction in the aggressive territorial defense behaviour can be explained. Thus, there seems to have been a competition between the aggressive and sexual behaviours. Half the intruder females were killed, i.e., there was no single, homogeneous behaviour on the part of the resident male. Female hormonal stimulation plays a specific role. The absence of male sexual hormones in the castrated intruder did not inhibit aggression.

Flannelly and Thor (1978) showed a reduction in the aggressiveness of albino males against castrated intruder males. We were unable to replicate these results probably because the rats were of different types, the wild one is more aggressive than the albino (Boice, 1972; Koolhaas et al., 1980; Price, 1978). Besides, the intruders were placed in colonies of males and females, whereas Flannelly and Thor (1978) recorded the behaviour of an isolated male at the time of the test. The presence of females in the colony seems to increase the aggressiveness of the resident males (Barnett et al., 1968; Flannelly & Lore, 1977).

The observation of the colony's behaviour showed that the resident females attacked the intruder femaleas already observed (Blanchard et al. 1984). Therefore, we cannot state with certainly that the intruder females were killed by the larger male of the colony, who, normally defends the territory against other intruders. However, injuries produced by females are relatively inoffensive (Blanchard et al., 1984).

Our results are in agreement with Beilharz (1975) in which wild mice attack and kill intruder males, but they change this pattern against the females, with diminished aggressiveness. Thus, wild rats in small colonies seem to have a rigid aggressive nature but they can discriminate sexual stimuli which obviously plays an imrortant role in preserving the specie.

Acknowledgements: Special thanks to Prof. W.A. Dal Zot for assistance in statistical analisys of the data. This research was supported by grants from FINEP and FAPERGS. R.M.M.A.and A.A.R.F. were pre-doctoral fellowships from CNPq.

REFERENCES

Adams, N., Boice, R. (1983) J. Comp. Psychol. 97, (l): 24-33.

Albert, D.J., Walsh, M.L., Gorzalka, B.B., Siemelis, Y., Louie, H. (1986) Physiol. Behav. 36, 401-407.

Barnett, S.A., Evans, C.S., Stoddart, R.C. (1968) J. Zool. (Lond.) 154, 391-396.

Beilharz, R.G. (1975) Z. Tierpsychol. 39, 141-149.

Blanchard, R.J., Blanchard, D.C. (1977) Behav. Biol. 21, 197-224.

Blanchard, R.J., Takahashi, L.K., Blanchard, D.C. (1977) Anim. Learn. Behav. 4, 365-369.

Blanchard, D.C., Fukunaga-Stinson, C., Takahashi, L.K., Flannelly, K.J., Blanchard, R.J. (1984) Behav. Proc. 9, 31-48.

Boice, R. (1972) Behaviour 42, 198-231.

Dixon, W.J., Massey, Jr., F.J. (1969) International Student Edition. McGraw Hill Kogakusha, Ltd.
Flannelly, K., Lore, R. (1977) Anim. Behav. 25, 654-659.
Flannelly, K.J., Thor, D.H. (1978) Physiol. Behav. 20, 785-789.
Koolhaas, J.M., Schuurman, T., Wiepkema, P.R. (1980) Progress in Neurobiology, 15, 247-268.
Leshner, A.I. (1975) Physiol. Behav. 15, 225-235.
Price, E.O. (1978) Behaviour 64, 340-353.
Siegel, S. (1956) Nonparametric Statistics for the Behaviour Sciences. McGraw-Hill, New York.
Takahashi, L.K., Lore, R.K. (1982) Physiol. Behav. 29, 1013-1018.
Takahashi, L.K., Blanchard, R.J. (1982) Behav. Proc. 7, 49-62.

Animal Models in Psychopharmacology
Advances in Pharmacological Sciences

ROLE OF THE INTRUDER IN THE AGGRESSIVE BEHAVIOUR OF COLONIES OF WILD RATS (RATTUS NORVEGICUS)

Aldo B. Lucion and Rosa M.M. de Almeida

Department of Physiology, Pharmacology and Biophysics, Institute of Biological Sciences, Federal University of Rio Grande do Sul, Rua Sarmento Leite, 500, Porto Alegre, RS, 90.049, Brazil

SUMMARY: Wild rats (Rattus norvegicus) formed colonies with 2 males and 2 females, in which albino male (N=23) and wild male intruders (N=10) were placed. The results showed that 34% of the albino intruders were killed during the first test; that a progressive increase in the number of deaths occurred over time; and that the wild intruders presented a lower rate of deaths than the albinos. In experiment II, the intruders, both albino (N=12) and wild (N=9) were placed in the same colonies as previously, under anesthesia. It was found that the latency to investigate and attack the anesthetized intruder was the same to albino and wild, but was higher than when they were awake. It may be concluded that the intruder's behaviour has a decisive influence on the aggressiveness of the resident, and that visual stimuli are important during aggressive interactions, but that olfactory sensation seems to be required to trigger it.

INTRODUCTION

The aggressive behaviour of the resident male in a colony against an intruder has been reported in various studies (Barnett, 1963; Blanchard & Blanchard, 1977; Eibl-Eibesfeldt, 1961; Flannelly & Lore, 1977a; Grant, 1963; Koolhaas et al., 1980; Lore & Flannelly, 1977; Robitaille & Bovet, 1976; Scott, 1966; Seward, 1945). This behavioural pattern has been described as territorial defense, and seems to play an important role in social dominance (Blanchard et al., 1977, 1984, 1988; Boice, 1972). Descriptions have reported that this behaviour only arises after 100 days of age (Koolhaas et al., 1980; Takahashi & Blanchard, 1982); that it is the males who react against the intruder (Barnett, 1963; Blanchard et al., 1984); that the females seem to play an important role in triggering the aggressiveness of the male (Albert et al., 1988;

Barnett et al., 1968; Flannelly & Lore, 1977b); that this behaviour depends on male sexual hormones (Albert et al., 1986; Flannelly & Thor, 1978; Leshner, 1975); and that social relations at an early age seem to influence its development (Meany & Stewart, 1981; Panksepp, 1981; Takahashi, 1986, Taylor, 1980).

Most of these results were obtained by studying colonies of albino rats, but few have studied the behaviour of wild Norway rats. It was demonstrated that albino rats show aggressive behaviour towards intruders, in the same way as wild ones, but that the latter demonstrate their aggressiveness more obviously than the albino variety (Barnett, 1963; Boice, 1972; Price, 1978; Takahashi & Blanchard, 1982).

The study of a given behaviour involves the analysis of three main factors: the external stimuli which trigger it; the variables of the endocrine and central nervous systems of the animal involved in creating it; and the various forms in which this behaviour can manifest itself in the environment (Ewert, 1985).

In this paper, we analyze the stimuli which trigger the aggressive behaviour of territorial defense in stable colonies of laboratory bred wild rats especially the relative importance of olfactory sensorial and visual inputs.

The study was divided into two experiments: experiment I analyzed the effect of introducing wild male intruders and albino males in stable colonies of wild rats. Experiment II compared the effect of the introduction of wild male intruders, under anesthesia, to that of anesthetized albino males, in the same colonies as experiment I.

MATERIAL AND METHODS

Animals and experimental environment: We studied the lst, 2nd and 3rd generation of wild Rattus norvegicus rats, laboratory bred from ancestors captured in the city of Porto Alegre (Brazil), in 1986. Each colony was composed of two males and two females, which were not siblings, and began to live together from the time of weaning until the end of the experiment. They received water and food ad libitum.

The animals lived throughout their lives in two environments. They were born in standard plastic cages, and stayed there with their mothers until weaning which occurred between the 21st and 31st day of age. After this they were placed in specially built observation cages. These cages were made of masonry with a wire screen roof, which could be moved and allowed access into the cage. On this, food and water was placed. Inside the cage, in the forward section of the floor, a tunnel was built, partly covered by a metal cover, where the animals could hide. The rats spread sawdust in this tunnel in a way which much resembles holes made in the ground (Lore & Flannelly, 1977). The front wall of the cage including that of this tunnel, was made of transparent glass and allowed plenty of light and a good view of the inside, but the part which corresponded to the tunnel was normally covered by a non-transparent acrylic plaque. The cage floor was covered by sawdust which was changed every week when the animals had to be manipulated.

There were a total of 19 cages, one next to the other, on a concrete platform 110 cm of the ground, disposed in an U-shape inside a room. One meter away from the cages, a wall, which was also U-shaped was built, with windows through which it was possible to see inside the cages. The 12:12 hour light-dark cycle, with the dark period beginning at 17:00 hours, was caused by fluorescent lamps of 4OW each, placed on the ceiling of the room. In front of, and above the cages there were incandescent red lamps of 15W each, which were only switched on for a few minutes before recording and then switched off. Room temperature was kept between 22 and 26°C, by an air conditioning system and the room was ventilated by two exhausts. The outside noise was controlled by acoustic insulation.

Experiment I

Procedure: Analysis

From 90 days of age onwards, colonies were placed in contact with male intruders who stayed there for 14 hours from the beginning of the dark cycle. The intruders were aged from 80 to 120 days, and their size was smaller than or equal to the largest resident male.

Initially male albino intruders were placed and later wild males. Therefore, the first contact of the colony with intruders was always with albino males. The albino intruders were of the Wistar variety and came from isosexual colonies, obtained at the animal breeding section of the Institute of Biological Sciences, at the Federal University of Rio Grande do Sul (UFRGS), whereas the wild ones came from the same ancestor as the colony members, but were not siblings to the resident, and before being placed in the colonies as intruders lived in mixed standard colonies (2 males and 2 females).

The effects of the introduction of male albino intruders were studied in 23 colonies, and of male wild intruders in 10 of these same colonies. Each colony was tested 3 times against albino intruders and twice against wild ones, and in the first situation 3 test blocks with 23 recordings and in the second one, 2 blocks of ten were generated.

After the entry of the intruder who was placed directly in the center of the cage, the behaviour of the colony was observed during 10 minutes and the latency in seconds of the first social investigation and of the first attack on the intruder was recorded. Social investigation was understood as sniffing any part of the intruder's body. Attacking includes investing and biting the intruder. There was no distinction between the 2 resident males, and these behaviours were ascribed to the first one to execute them. Usually it was the same one: the largest male. There were colonies in which the 2 males were practically the same size, but only one of them attacked the intruder. After the test had lasted 14 hours, the intruder was removed, the location of the injuries was recorded, and the number of deaths was calculated, compared both to a random distribution, and among the various introductions, using the exact test of Fisher (Siegel, 1956). The meams of the investigation and attack latencies among the subjects themselves were also calculated and compared using a Student t test (Dixon & Massey, 1969).

RESULTS

The results (Table 1) showed that the number of deaths of albino intruders during the first test was no different than was to be expected from a random distribution. Already at the third test there was clearly a higher percentage of deaths as related to chance. Thus, an increase in the effectiveness of aggressive behaviour of the residents after increased experience was noted. After the third test, the percentage of deaths was maintained around 100%.

The number of deaths of wild intruders was no different from the random one, and was less than that of albinos. The latency of investigation and attack on the wild intruder was greater than that of the albino (Table 2). Therefore, the wild intruder was attacked less than the albino.

Direct observations showed that the behaviour of the albino intruder was different from the wild one. The albino, when it was in the colony, immediately began to explore the environment, walking and sniffing the whole cage. But it rarely entered the tunnel of the residents. On the other hand, the resident would leave its "house", approach and sniff the intruder who froze. The resident would walk slowly in an aggressive display, around the intruder who sometimes fled, and was pursued, and a sequence of pursuit, fighting and pursuit was set off. After some time, the resident would dominate the intruder and biting him harder.

Table 1 Number of deaths of male albino (N=23) and wild (N=10) intruders in wild rats colonies. Fisher exact probability (p) compares the number of deaths to randomness.

	Intruders				
	albino male (N=23)			wild male (N=10)	
	1st test	2nd test	3rd test	1st test	2nd test
Number of deaths observed	8	18	23	6	5
Number of deaths expected by randomness *	12	12	12	5	5
Fisher exact probability (p)	0.19	0.06	0.0001	0.5	0.67

We used 12 as half of 23. If it were 11, "p" would be 0.2749; 0.0327 and 0.0000347 respectively

Table 2 Mean latency in seconds (s) of investigate and attack behaviours of the wild resident male towards albino (N=23) and wild (N=10) intruders.

	Mean Latency (s)	
Intruder	Investigate	Attack
Normal albino male	11.1	36.9
Normal wild male	22.2	210.9 *

*= Student t test significant ($\alpha < 0.05$) comparing wild and albino intruders

The wild intruder, upon entering the colony, did not explore the environment. It would take on an attitude of avoidance. The resident on its turn, presented the same behaviours as in relation to an albino. When it approached the intruder the latter would flee, and often hide inside the tunnel, mixing with the other members of the colony. There were fights with aggressive display, piloerection, investing and attack, but the wild intruder was more skilled in avoiding the resident's attack.

The intruders, both albino and wild, found dead after 14 hours, presented deep injuries in the dorsal cervical region and/or in the ear. When they were alive the injuries were distributed mainly over the posterior dorsal region and the genital region.

DISCUSSION

In discussing the results obtained, it must be taken into account that the members of the colonies were wild rats and the male intruders were Wistar albino and wild; and that the first contact of the colony was always with an albino.

The first evidence to be discussed is that the colony can kill the intruder already at the first encounter, i.e, territoriality is complete without any previous experience. This data is different from that obtained with albino residents (Blanchard et al., 1977), in which no deaths of intruders were recorded, and aggression occurred only after several contacts, increasing progressively. The stronger aggressiveness described in our results is probably due first of all to the nature of the wild rat which is more aggressive, and secondly, the intruder was an albino, which seems not to present the same behaviour as a wild intruder, rendering aggression by the resident easier. This difference in defensive behaviour between the wild and the albino intruder had already been perceived (Takahashi & Blanchard, 1982).

However, after some time, it was observed that effective aggression against the intruder increased, as measured by the larger number of deaths at the second and third tests. This is perfectly in accordance with the findings of Blanchard et al. (1977), that the experience with intruders, leads in later contacts to high levels of attack against the albino rats (Blanchard et al., 1977).

Thus, it may be concluded that for the wild rats the aggressive behaviour of territorial defense only requires some experience or no experience at all to be demonstrated completely. At the age studied, this pattern seems to be ready to be uncovered and carried out.

The second result to be analyzed was that the aggression to the wild intruder was less effective than upon the albino. There was a lesser number of deaths, as compared to the third test in which the male albino was introduced, and this number was no different from what might have been expected at random. The latency for investigation and attack of the wild intruder was significantly longer. This may have been due to the behaviour of the intruder who efficiently avoided interaction with the resident (Takahashi & Blanchard, 1982). The latency of attack of the wild resident against the wild intruder, mentioned by Takahashi & Blanchard (1982) was 369.3 seconds, whereas we record 210.9 seconds. We believe that this difference may also be explained by the behaviour of the intruder which, in their study was captured and thus might be more skilled at escaping contact with the resident. Taking these results as a whole, it can be concluded that apparently the behaviour of the intruder defines the development of the fight.

Experiment II

It is possible that the lesser aggressiveness upon confronting the wild intruder in Experiment I might be due to color, to some scent, or to different display behaviours, which all may influence the aggressive behaviour of the resident. That is, that the albino intruder was so deviant with regard color, scent and behaviour, that it would be immediately recognized as an intruder. Thus, we tested, in Experiment II, the role of the intruder's behaviour, since we had observed that the behaviour of the wild intruder was very different from that of the albino intruder.

In this experiment we compared the effect of introduction of anesthetized male albino and anesthetized male wild rats. In this way, we eliminate the behaviours displayed by the intruder, leaving only the scent and physical presence to the resident.

METHODS

The colonies and the experimental environment utilized were the same as in experiment I. The subjects already had previous experience with albino and wild intruders. The intruders had the same weight and age as those in experiment I, but 30 minutes before each recording they were deeply anesthetized with 35 mg/kg i.p. pentobarbital. They were placed in right lateral decubitus in the center of the cage and removed after 2 hours. Twelve colonies were studied in the presence of anesthetized albinos and 9 colonies with anesthetized wild rats. The latency, in seconds, of the behaviours of investigation and attack by the larger male resident on the intruder's body, was computed from the begining of the test. After its removal, the sites of injuries were checked. The latencies of these behaviours among the albino and wild anesthetized intruder were compared as in Experiment I.

RESULTS

The latency of the behaviours of investigating and attacking performed by the resident male were the same, both against the albino intruder and against the anesthetized wild rat (Table 3). When the anesthetized albino intruder was compared to the one who was awake, it was found that both the investigative and attack behaviour had a greater latency when the intruder was anesthetized. For the anesthetized wild intruder, only the investigative behaviour had a greater latency than for the normal wild one, while the attack behaviour was not significantly different.

Table 3 Mean latency in seconds (s) of investigate and attack behaviour of the wild resident male towards albino (N=17) and wild (N=9) anesthetized intruders.

	Mean Latency (s) +	
	Investigate	Attack
Anesthetized Albino male	64.23	148.62
Anesthetized wild male	54.77	150.60

+ Student t test not significant ($\alpha < 0.05$) comparing albino and wild anesthetized intruders.

It was observed that the interaction between the resident male and the anesthetized intruder, both albino and wild, was different from the relations with the intruder when awake. In the first case the resident left his home "cautiously" and stood there for a few seconds staring at the intruder, and only then would walk slowly in its direction, beginning to sniff it, especially the genitals, and walking around its body, sometimes over it. Frequently it would throw sawdust over the intruder, burying it, or then drag it from one side to the other of the cage, without biting it. Sometimes it lay for long periods in ventral decubitus near the intruder, facing it as though "controlling" it. Piloerection occurred after sniffing the genitals, but not the aggressive display. Usually only the large male presented this behaviour. On the other hand, when the intruder was awake, the behaviours displayed before attack were of a different nature. They were more "direct", there was not so much "cautiousness". It is important to note that the injuries observed in the anesthetized intruder were all located either on the head, more specifically in the ear, or slightly more caudally in the cervical dorsal region.

Therefore, the behaviours which preceded the attack explain the main differences observed between the social relations among the resident and the two types of intruders, anesthetized and awake.

DISCUSSION

Experiment II analyzed the possible reasons why normal wild intruders were less injured than albinos in Experiment I. Apparently, it is not due to its color. And there does not seem to be an olfactory difference between the wild intruder and the albino which could explain the greater effectiveness of the aggression to the latter.

The results of this experiment showed that the resident male investigates and attacks anesthetized intruders, both albino and wild, with the same latency. Thus it can be concluded, for Experiment I, that the wild intruder was capable, by behavioural means of avoiding physical aggression. In Experiment II, the anesthetic, by preventing the behavioural reaction, abolislied the differences in latencies recorded previously among the albino and wild rats.

Another fact found was that the resident seemed to expect a behavioural pattern from the intruder compatible with its role, and if he did not fulfill this role due to the anesthetic, the resident tried to evoke normal behaviour from the intruder. As previously observed (Takahashi & Blanchard, 1982) the resident dragged the intruder from one side to the other, movements which would be the expected behaviour.

From these observations it may be concluded that the agonistic behavioural pattern performed in aggressive relationships apparently is quite rigid. The behavioural sequence of the resident seems to be in phase with that of the intruder, each of them depending on the other. When the intruder is anesthetized, a response may occur similar to a vacuum reaction (Hinde, 1970, 1971).

We may thus conclude that the sensorial route of vision plays an important role in this relationship. The subjects in a fight seem to be monitoring the behaviour of their opponent at all times and, with this information they are capable of acting appropriately in a mutual communication process (Enquist, 1985).

The auditory sensorial route, besides the visual and olfactory ones, may also be involved (Brown, 1976). It is interesting to observe that when the intruder was placed in a given colony, the resident would leave the tunnel without having it scented or seen; and the males of other nearby colonies also left their "homes", and stayed alert, sniffing the air. This would happen without any fight or perceptible noise having occurred in the colonies.

The complete aggressive stimulus, i.e., the intruder awake and displaying behaviours, triggers the complete aggressive sequence of the resident. However, incomplete stimulation, i.e., the absence of auditory stimuli and the distortion of visual information regarding the anesthetized intruder, generate an aggressive behaviour which is also incomplete, in which display behaviours of piloerection and aggressive posture are missing. But possibly olfactory stimuli are sufficient to trigger the attack.

Acknowledgements: Special thanks to Prof. W.A. Dal Zot for assistance in statistical analysis of the data. This research was supported by grants from FINEP and FAPERGS. R.M.M.A.was a pre-doctoral fellowship from CNPQ.

REFERENCES

Albert, D.J., Walsh, M.L., Gorzalka, B.B., Siemens, Y., Louie, H. (1986) Physiol. Behav. 36, 401-407.

Albert, D.J., Dyson, E.M., Walsh, M.L., Petrovic, D.M. (1988) Physiol. Behav. 44, 735-740.

Barnett, S.A. (1963) The rat: a study in behaviour. Aldine, Chicago, 1963.

Barnett, S.A. (1967) Rats. Sci. Am. 216, 78-87.

Barnett, S.A., Evans, C.S., Stoddart, R.C. (1968) J. Zool. (Lond) 154, 391-396.

Blanchard, R.J., Takahashi, L.K., Blanchard, D.C. (1977) Anim. Behav. 5, 365-369.

Blanchard, R.J., Blanchard, D.C. (1977) Behav. Biol. 21, 197-224.

Blanchard, D.C., Stinson, C.F., Takahashi, L.K., Flannelly, K.J., Blanchard, R.J. (1984) Behav. Proc. 9, 31-48.

Blanchard, R.J., Flannelly, K.J., Blanchard, D.C. (1988) Physiol.Behav. 43, 1-7.

Boice, R. (1972) Behaviour 42, 198-231.

Brown, A.M. (1976) Biochem. Physiol. 53, 313-317.

Dixon, W.J., Massey, F.J. (1969) Introduction to statistical analysis. International Student Edition, McGraw-Hill Kogakusha, Ltd.

Eibl-Eibesfeldt, I. (1961) Sci. Am. 205, (6): 112-122.

Enquist, M. (1985) Anim. Behav. 33, 1152-1161.

Ewert, J.P. (1985) Anim. Behav. 33, 1-29.

Flannelly, K., Lore, R. (1977a) Phychol. Rec. 2, 315-329.

Flannelly, K., Lore, R. (1977b) Anim. Behav. 25, 654-659.

Flannelly, K.J., Thor, D.H. (1978) Physiol. Behav. 20, 785-789.

Grant, E.C. (1963) Behaviour 21, 260-281.

Hinde, R.A. (1969) Psychosom. Res. 13, 213-219.

Hinde, R.A. (1970) Animal Behaviour. McGraw-Hill Book, Company.

Koolhaas, J.M., Schuurman, T., Wiepkema, P.R. (1980) Progr. Neurobiol. 15, 247-268.

Leshner, A.I. (1975) Physiol. Behav. 15, 225-235.

Lore, R., Flannelly, K. (1977) Sci. Am. 236, 106-116.

Meaney, M.J., Stewart, J. (1981) Anim. Behav. 29, 34-45.

Panksepp, J. (1981) Psychobiology 14, 327-332.

Price, E.O. (1978) Behaviour 64, 340-352.

Robitaille, J.A., Bovet, J. (1976) Biol. Behav. 1, 289-308.

Scott, J.P. (1966) Am. Zool. 6, 683-701.

Seward, J.P. (1945) J. Comp. Psychol. 38, 175-179.

Siegel, S. (1956) Nonparametric statistics for the Behavioural Sciences. New York, McGraw-Hill.

Takahashi, L.K., Blanchard, R.J. (1982) Behav. Proc. 7, 49-62.

Takahashi, L.K., Lore, R.K. (1982) Physiol. Behav. 29, 1013-1018.

Takahashi, L.K. (1986) Behav. Proc. 12, 237-260.

Taylor, G.T. (1980) J. Comp. Physiol. Psychol. 94, 953-961.

part 5:

DRUG DISCRIMINATION

Animal Models in Psychopharmacology
Advances in Pharmacological Sciences

DRUG DISCRIMINATION AND ANIMAL MODELS

J.L. Slangen

University of Utrecht, Faculty of Pharmacy, Section Psychopharmacology, Sorbonnelaan 16, 3584 CA, Utrecht, The Netherlands

INTRODUCTION

Animal models can be classified into three broad categories: homologous, physiological and empirical models. The use of animal models is dictated by our limited current knowledge of the aetiology and basis of psychiatric illnesses. It is often impossible to simply describe the human pathology, point to certain brain structures and neurochemical substrates involved, and start to look for all the determining factors while remaining at the human level all the way. Especially in the search for therapeutically effective drugs one has to resort to animal models. A model is here defined as a behavioural or physiological index of drug action that predicts a desirable clinical drug action.

Empirical models

Given a clearly defined therapeutic target it is not too difficult to assess the clinical effectiveness of a new drug, provided there is agreement concerning outcome measures. Once the clinical effectiveness of a drug has been established, the construction of an appropriate animal model can proceed in anempirical way, because the characteristics of the clinical drug response provide the criteria to evaluate the model. Most of the animal models used in pharmaceutical companies are of the empirical type. The quality of an empirical model can be judged by the criteria extensively discussed by Carlton (1983):

- Only drugs that affect the clinical drug response (CDR) affect the response used in the model (no false positives and false negatives);
- rank order of potency of drugs is the same for CDR and the response used model;
- latency of onset of action is similar for CDR and model;
- tolerance in the CDR is mimicked in the model;
- drugs that antagonize (or augment) the CDR do so in the model.

It must be noted however that many of the empirical models in use today have not been systematically judged according to all of the aforementioned criteria.

Physiological models

Physiological models find their origin in a hypothesis concerning a pathology. Ususally it is assumed that a particular class of brain neurons is involved in the pathological behaviour for which a treatment is sought. Any response, be it behavioural, biochemical or neurophysiological, indexing the activity of the aforementioned neurons, can be a model either to drugs that may change the pathological behaviour or to further investigate the determinants of the pathology. Evidently the success of this approach depends on the extent to which the chosen response is a correct index of the activity of the neurons involved and the extent to which changes in the activity of the chosen class of neurons indeed are essential for the pathological behaviour.

Physiological models may be regarded as a subset of homologous models. This will become clear in the next paragraph.

Homologous models

In homologous models it is hypothesized that the response in the model is caused by a process similar or equal to the process that is causing the human response to be treated clinically. These models are based on apparent similarities between the processes or causes that underlie the human condition and animal test response, and therefore easily subjected to the pitfalls of face validity. In developing homologous animal models of psychopathology one is confronted with the difficulty of identifying in animals the range of autonomic, endocrine, emotional and cognitive effects that is at the core of conditions such as anxiety, depression and schizophrenia. Often the measurement of the whole range of effects is reduced for practical and financial reasons to the measurement of one or two effects, thereby blurring the distinction between homologous and physiological models. The advantage of both types of models is that they do not require for their development the presence of a reference compound that is therapeutically effective.

Drug Discrimination

Drug discrimination has not been developed as a model and cannot simply be fitted into the types of models described sofar. It has characteristics however that may link it to the physiological models. It also has been used as an empirical model. In the following paragraphs its characteristics will be described and examples will be given of its use. Finally its use as a model will be evaluated.

DRUG DISCRIMINATION LEARNING AND STIMULUS GENERALIZATION TESTS

In a typical drug discrimination (DD) experiment subjects are presented with a simple choice task. For example, subjects might be differentially reinforced for the direction that they run in a T-maze or for the pressing of one of two levers in an operant chamber. The reinforcer can be anything that maintains responding. Usually this involves appetitive (food or water) or aversive (electric shock) stimuli. To train the discrimination, either vehicle or a dose of drug is injected prior to training, and subjects are allowed to respond while the drug is producing its effect. If experimental contingencies are arranged such that one of the responses will only be reinforced in the presence of the drug effect, while the other response is only reinforced in the absence of the drug effect, then, if the drug is capable of serving as a discriminative stimulus (DS), its presence or absence will differentially control responding. Drug stimulus control means that the subject chooses the response which is appropriate for the drug condition only in the presence of the drug effect, and prefers the alternative response in the absence of the drug effect. Learning the discrimination generally requires between 30 to 60 training sessions. Stimulus control develops more readily when stimuli have markedly different characterics. Therefore speed of learning has been used as an index for the discriminability of the stimulus effect generated by the training-drug (Overton, 1971).

In substitution tests with other doses of the training-drug it is observed that the subject's tendency to respond on the "drug-lever" (DL) diminishes when the dose is lowered. A dose-response curve can be constructed. This curve may be equated with the generalization gradient known from experimental psychology. It is considered to demonstrate the behavioural control exerted by the drug-stimulus. The lawful relation between dose and behaviour indicates that it is a drug effect that acts as the DS and nothing else. The gradients allow the investigator to determine slope-functions and potencies. Doses higher than the training dose usually do not lead to a decrease in stimulus control, unless the higher doses cause sedation and disruption of behaviour.

When subjects trained in a two-choice task using drug versus vehicle are tested with drugs other than the training-drug, it happens with some of these drugs that the subjects tend to respond to the test-drugs in much the same way they had previously responded to the training-drug. It is then said that the training-drug stimulus generalizes to the stimulus effect of the test-drug. or that the test-drug substitutes for the training-drug.

In DD the control of the drug stimulus over behaviour is specific, in the sense that there are many test-drugs that do not substitute for a particular training-drug, whereas some do. The ones that do must produce a stimulus that is similar in one way or another to the training-drug stimulus. The ones that do not substitute may or may not produce a stimulus. There certainly are drugs without stimulus effects. However, it has been demonstrated that many of the psychoactive drugs which fail to substitute for one training-drug, do substitute for another training-drug, and do produce stimulus effects when used as training-drug themselves. Zero generalization therefore can be a genuine outcome of a generalization test indicating that the effects induced by the test drugs differ so markedly from the training stimulus that they cannot assume any of the differential control exerted by the training stimulus. This outcome has led to theoretical considerations about the nature of the stimulus complex that is effective in DD learning (Jarbe and Swedberg, 1982).

Because the drug discrimination procedure is not complete without generalization tests, DD research almost automatically leads to pharmacological comparisons of drug effects. Dose-response curves can be constructed, slopes can be compared and differences in potencies of drugs can be established. The fact that drug stimulus control is specifich as led to a great number of studies investigating how drugs that produce an internal stimulus fall into classes. This however has not been the sole purpose of DD.

We now will describe the different purposes for which DD has been used,viz. in:

A. methodological research elucidating determining factors
B. questions concerning drug abuse and drug dependence
C. drug screening
D. the analysis of the neurochemical mode of action of drugs.

METHODOLOGICAL RESEARCH

Many studies have addressed questions concerning the factors that determine the outcome of the DD procedure (Colpaert and Balster, 1988; Colpaert and Rosecrans, 1978; Colpaert and Slangen, 1982; Ho et al., 1978; Lal, 1977). It comes as no surprise that organismic and environmental variables, e.g. level of deprivation and schedule of reinforcement, that influence the discriminative performance of subjects in experiments with exteroceptive stimuli, likewise determine the performance in experiments with interoceptive stimuli. Also the factors species, sex, heridity, psychological and physiological state play a role.

As an example of the role of the physiological state of an organism in DD learning, consider the following. When given in non-dependent rats, pure opioid antagonists have no or very low potency to act as a DS. In contrast, if subjects are made dependent on opioids, the antagonists are very well able to function as DS. Drug stimulus control is an aspect of operant conditioning. For this reason it is to be expected that conditioning history and contextual cues codetermine the final performance of an animal in a DD experiment. In addition there may be an interplay between extero- and interoceptive stimuli in DD experiments in that both drug stimuli and sensory cues may be instrumental in controlling behaviour in a DD experiment. Basic research also has looked into the relation between DS effects and other properties of drugs such as reinforcing and aversive properties, and into the relative merits of different experimental situations in which DD can be measured (for example, two-choice versus three-choice discrimination). Also a number of other interesting topics have been studied with the DD procudure e.g., the establishment of DD in humans and the DS properties of drug mixtures.

Much attention has been given to the measurement of DD behaviour. DD is measured relative to the asymptotic performance displayed after many administrations of a particular dose of the training drug and an equal number of vehicle administrations. The discriminative performance usually is described in terms of drug appropriate responses. When under drug condition only the drug lever is chosen and under vehicle condition exclusively the vehicle lever is chosen, it may be stated that there is a 100% discrimination. But how to interprete intermediate results? For example, after administration of a lower dose than the training dose, animals may emit 60 responses on the drug lever and 40 responses on the vehicle lever. Is this an indication that the teststimulus is less intense than the training stimulus and that the intensity ratio of training and test stimulus is somewhat like 10:6? Or does this hypothetical result reflect variations in threshold for detecting the discriminative stimulus? If the latter is the case then the experimenter must decide on an upper and lower limit, for example 80% and 20% DL responses, indicating the presence and absence of the drug stimulus respectively. In this conception the individual subject can only detect or not detect the DS and the detection is expressed by what the experimenter defines as the selection of one or the other response. Because there are differences between subjects in sensitivity, the experimenter must make a second decision on where to draw the line between a result that indicates the presence of the DS in a group of animals and a result indicating its absence. Again he may choose the 80% level for determining the criterion response. In this approach it is assumed that DS effects cannot be measured as graded responses on a continuous scale. The drug stimulus occurs or does not occur and is considered a quantal response (Mathis et al., 1987). The values and pitfalls of this approach are extensively discussed in a recent paper (Mathis and Emmett-Oglesby, 1990).

It must be realized that the relationship between dose and quantal response does not give information on the magnitude of effect as related to concentration of drug but on the number of subjects either responding or not responding to each given dose, i.e. the frequency with which any dose of a drug produces an all-or-none pharmacological effect. A quantal dose-response curve provides a measure of the variation in threshold dose needed to produce a stated effect in a sufficiently large group of subjects.

When used in this context partial generalization means that a group of trained subjects reacts differentially to a testcompound: some select the lever associated with training drug and some select the saline-associated lever. One can speculate on the reasons for this, but once this approach is chosen in DD research the results can tell us two things: a. whether a training compound is likely to produce a DS effect at or above a particular dose, b. whether at or above a particular dose a test compound is likely to exert the same stimulus control that is shown by the training drug, thus pointing to a similarity between the stimulus effects of training and test drug.

Similarly, when this approach is chosen in antagonism tests the outcome can only tell that a particular dose of an antagonist is effective in blocking the stimulus effect of the training drug or not.

It is important to realize that the concepts of partial generalization and partial antagonism have a different meaning when it is assumed that the response of the subject increases measurably with increasing concentration of drug. In the graded type of dose-effect relationship the training drug may be said to exert a 100% effect. When after administration of a lower dose of the training drug or after pretreatment with an antagonist the drug lever is chosen only 60%, it is concluded that this diminished responding is caused by a decrease in stimulus intensity. The behavioural effect is considered to be graded as a function of dose (or as a function of number of receptors that is activated). How appropriate the gradation is measured in the DD-techniques currently in use is still a matter of discussion.

Is has been assumed that partial generalization indicates lack of selectivity of the testdrug for the receptors underlying the discrimination. However, it also can be argued that partial generalization does not necessarily refute selectivity and does not necessarily indicate that the testcompound acts at different binding sites. Given a training compound that acts at one receptor site in particular, subsequent substitution tests with various agents might reveal different degrees of intrinsic activity for that receptor site. In other words: partial generalization might be an index for true partial agonism of compounds with low efficacy at the receptor that underlies the DS effect of the training-drug.

For example, the sigma agonist cyclazocine partially substitutes for the μ agonist morphine. Evidence has been offered for the interpretation that this is not an unselective effect of cyclazocine, but an indication that cyclazocine has low intrinsic activity at the μ-receptor (Colpaert and Janssen, 1984). Koek has summarized the predictions that can be derived from the hypothesis that partial generalization is based on partial agonist properties at the receptors that underly the DS effects of the training-drug (Koek and Colpaert, 1989; Koek and Woods, 1989):

1. The generalization gradient of a partial agonist is less steep than the generalization gradient of a full agonist.
2. The partial generalization is blocked by a full antagonist of the DS effects of the training-drug.
3. A partial agonist antagonizes to some extent the DS effect of the training-drug.
4. A partial agonist produces less generalization and more antagonism of the effects of the training-drug as the training dose is increased, and produces more generalization and less antagonism of the training-drug as the training dose is decreased (a low-efficacy drug can be a full agonist, a partial agonist, or an antagonist, depending on the sensitivity of the test system).
5. The DS effects of a partial agonist can be blocked by another partial agonist that has lower intrinsic activity.

Not yet all of these predictions have been tested.

QUESTIONS CONCERNING DRUG ABUSE AND DRUG DEPENDENCE

A relationship between DS properties and abuse liability has been suggested. It is hypothesized that drugs which show stimulus properties may also show abuse potential and reinforcing properties. However, discriminability does not correlate well with the known abuse liability of drugs (Overton and Batta, 1977). And further research has shown that the degree of discriminability of drugs is not predictive of abuse potential. However, the reverse may be true: drugs that cannot be discriminated are unlikely to have abuse potential.

Also drug similarity seems not a direct index of abuse liability. When drugs of abuse are used as training-drug in a DD procedure and it is observed that the training drug generalizes to a new compound, this compound may be regarded as a potential drug of abuse. The outcome of such generalization tests however has not been found to be predictive of the actual abuse by humans. A similar remark can be made concerning the use of hallucinogenic drugs in DD research in order to identify drugs with hallucinogenic activity in humans. The DS produced by hallocinogens in animals trained in a DD procedure could be hypothesized to be linked with the hallocinogenic property of the training drug and therefore be used in the detection of hallocinogenic properties of unknown drugs. Not all drugs however that substitute for a hallucinogenic training drug have hallucinogenic effects in humans.

When DD is used as a strategy to study animal analogs of symptoms of withdrawal, subjects are trained to emit one set of responses under a withdrawal condition, and to emit a second set of responses under the baseline condition of chronic drug administration (Emmet-Oglesby et al., 1990). This procedure has been applied for testing the putative therapeutic effect of drugs in the treatment of withdrawal symptoms.

DRUG SCREENING

Incidently DD has been used for classifying drugs with presumed similarity of action. The usual classification for psychoactive drugs is based on a mixture of properties e.g., their chemical structure, their major pharmacological laboratory effects and their clinical uses. When drugs differ widely in their actions, and are studied with very diverse procedures, it becomes difficult to classify them and to assess their pharmacological specificity. Drug discrimination behaviour offers a technology for the classification of drugs and the assessment of their pharmacological specificity, because the choice behaviour in the DD test has been found to be very robust, sensitive and specific. Once an animal has been trained to discriminate the effects of one particular drug from the non-drug condition, it is the rule that only drugs of the same pharmacological class are found to generalize with it. The fact that the discriminative stimulus effect of a training-drug is extremely specific for a class of drugs is considered a very useful property of the DD test. Consider rats trained to discriminate morphine from saline. When other drugs are administered in generalization tests, only compounds with morphine-like activity elicit drug-appropriate responding. All other compounds produce saline-appropriate bar-pressing. Therefore the test may be used to detect new compounds with a morphine-like action. In addition the relative potencies of drugs may be assessed. Such use of DD is comparable to the use of a behavioural profile. If the profile of a new drug is similar to that of an established clinically effective drug, e.g. a neuroleptic, this may be taken as an initial indication of neuroleptic activity. As an example we refer to the research of Holtzman. It clearly demonstrates how the DD procedure can be used for classifying drugs as opioids and for the determination of the relative potencies of opioid drugs.

If classifications of drugs according to their stimulus properties yield much the same categories as would be arrived at without consideration of stimulus effects, the value of DD would not be great. An added value comes from the fact that DS effects may be linked with particular properties of a drug, i.e. the analgesic property of a drug may be related with DS-property A, whereas the reinforcing property of that drug may be related to its DS-property B. The specific property A then may be used in the search for drugs with a similar specific action. Such screening for specific actions is encouraged by DD research because in drug-drug discrimination, animals may be trained to discriminate a drug with two properties, say anxiolytic and depressant, against a drug with only a depressant action. If the resulting discrimination would be based on what the two drugs not have in common namely the anxiolytic property, then possibly this DS property could be used in screening drugs with a rather specific therapeutic action.

THE ANALYSIS OF THE NEUROCHEMICAL MODE OF ACTION OF DRUGS

A second additional value of the DD procedure is to be found in its potential for advancing the classification of drugs in terms of their mechanism of action. For produced a further classification in mu, kappa and sigma agonist action. This points to the possibility of using DD as a behavioural bioassay. For example, it has been shown extensively that the discriminative stimulus properties of d-amphetamine are mediated by the dopamine receptor. Thus rats trained to discriminate the effects of d-amphetamine can be used to answer questions like: does this new drug have effects similar to d-amphetamine, does it block the d-amphetamine cue, or does it neither? If the new drug substitutes for d-amphetamine, or if it blocks the d-amphetamine cue, it is thought that the new drug's mode of action is through dopamine receptor sites (D'Mello and Stolerman, 1977). In this approach the training drug always is a compound whose neurochemical effects are known and the conclusions with regard to the test drugs is always in reference to the neurochemical effect of the training drug.

DD not only has been used for detecting the underlying neurochemical effect of a compound, it also has been used as a behavioural assay for detecting intrinsic activity. It must be realized that the drug produced DS as much is a specification of the drug under study as any other biological response it produces. The stimulus action of a drug adds to the various other actions of the drug. Sometimes this intrinsic action is unsuspected. For example in the case of the benzodiazepine antagonist flumazenil it was found that this drug was unable to completely antagonize the DS of chlordiazepoxide. Further research demonstrated that flumazenil can produce a DS effect and therefore must be considered to have mixed agonistic-antagonistic actions (De Vry and Slangen, 1985).

DD also has been used in the study of structure activity relations (SAR). There the DS effect of a drug can be used like any other drug response. The choice of the DS effect may be dictated by a specific link between the DS effect and e.g. the binding to a particular receptor type. The link may be less specific for other effects of that drug. The research of Glennon and coworkers into central 5-HT receptor agonists and antagonists is an example in case.

In connection with the topics mentioned research has been done concerning the onset of action, duration of action and involvement of metabolites of the training drug with regard to its stimulus effects.

DEVELOPMENTS TOWARDS THE USE OF DD AS A MODEL

The DS as a model of a state of anxiety

Often it is hypothesized that the DS produced by a drug may be equated with a subjectively experienced state. For example the DS induced by a subconvulsive dose of pentylenetetrazole (PTZ) in rats has been related to the subjective state of anxiety. First because the autonomic signs induced by subconvulsant doses of PTZ resemble those of anxiety. Second, because the DS induced by PTZ is antagonized by compounds with known anxiolytic effects and third, because compounds known to induce anxiogenic effects substitute for PTZ in generalization tests. Since PTZ causes feelings of anxiety in humans, it is tempting to interprete the PTZ cue in animals as representing a state of anxiety and therefore useful as a model of a distress-state. Lal and Fielding (1984) concluded that the antagonism of the PTZ stimulus may be a reliable procedure to identify anxiolytic drugs. However the model may be relevant only for anxiolytics and anxiogenics that are related to the anti- and proconvulsant activity of PTZ. All of the substances reported to antagonize the PTZ cue are anticonvulsants and many of the substances that substitute for PTZ are convulsants or have proconvulsant effects. Andrews et al. (1989) have provided evidence that the PTZ-cue-model does not adequately separate the apparent anxiolytic and anticonvulsant properties of test-drugs. Note that this model may be considered an empirical model.

The DS as a model of drug withdrawal symptoms

The reasoning behind this DS-model is that anxiety always is a major component of withdrawal in humans. The anxiety stimulus caused by withdrawal may be equated with the stimulus produced by PTZ. And therefore procedures using discrimination of the anxiogenic drug may be useful for detecting withdrawal in animals. This approach is extensively discussed in a recent review (Emmet-Oglesby et al., 1990). It is implied that a drug which blocks the generalization to PTZ might be a therapeutically effective drug in relieving withdrawal symptoms. The procedure is first to train animals to discriminate PTZ from saline, then to treat them chronically with high doses of benzodiazepines and finally to test the animals under conditions of withdrawal, induced by the administration of a benzodiazepine antagonist. During this acute withdrawal experience animals will select the PTZ-associated lever, supposedly because withdrawal is accompanied with anxiety, and the withdrawal induced anxiety substitutes for the anxiogenic PTZ cue.

To the extent that the aforementioned models seeks to relate withdrawal symptoms and states of anxiety in humans with stimulus symptoms in animals, they are homologous model. To the extent that these models are used for finding drugs that alleviate symptoms of withdrawal and anxiety they are empirical models.

The DS as an index of one particular effect of a compound with several psychoactive effects

When drugs are classified according to their discriminative stimulus attributes they generally fall into the traditional classes of opiates, CNS stimulants, hypnotic/sedatives, anaesthetics etc.. However members of a group may be found to differ and this may lead to further classification. For example in DD research with depressant drugs it has been observed that barbiturates and benzodiazepines generalize to each other, suggesting that these compounds belong to the same class. Appropriate doses of the benzodiazepine antagonist Ro15-1788 will block the stimulus effects of benzodiazepine agonists, and not the stimulus effects of barbiturates (Herling and Shannon, 1982). This points to differences in the mechanisms through which barbiturates and benzodiazepines produce the DS. It not necessarily points to dissimilarities in the stimulus effects of the two types of compounds that cross-generalize to each other. As far as their stimulus effects are concerned the two types of drugs may have effects that belong in the same class, and therefore point to a similar therapeutic effect. Further research however has clearly shown that there are conditions in which differences between the stimulus properties of benzodiazepines and barbiturates are indicated, because asymmetrical cross-generalization is found. For example, pentobarbital-trained rats generalize to lorazepam, but lorazepam-trained rats do not generalize to pentobarbital (Ator and Griffiths, 1985). Ethanol-trained rats generalize to pentobarbital, but pentobarbital-trained rats do not generalize to ethanol (De Vry and Slangen, 1986). In view of these and other results it is now suggested that members of the class of depressant drugs produce stimuli that may be differentiated from each other. However, the current status of data on mechanisms does not explain why these differences occur. To date they have not been linked to differences in neurochemical action or differences in behavioural action. It is tempting to hypothesize that results from drug discrimination studies might indicate relations between particular actions of a drug and its stimulus properties. The depressants are a case in point. They generalize with anaesthetic, sedative/hypnotics, muscle relaxant and anxiolytic drugs, probably because they produce a rather broad stimulus complex. It could be however that the different actions of these compounds are linked with different neurochemical actions and perhaps with different receptors or receptor subtypes. In turn, these receptors or receptor subtypes may be linked with particular components of the broad stimulus complex. An individual component may function in isolation whenever a high level of stimulus specificity is obtained. And such specificity can be obtained by experimentally promoting a differential attention to certain aspects of a stimulus effect. In this way stimulus specificity may help to identify behaviourally relevant receptor types and subtypes. In all this it is assumed however that there is a one to one relation between the stimulus effect of a drug and one of its major other effects, and that when a single drug functions as a compound stimulus, each component stimulus effect is mediated by a distinct pharmacological action at a particular receptor. Unequivocal evidence for these assumptions has not been provided yet.

The DS as a tool in elucidating mechanisms of action and as a tool in seeking new therapeutic agents

The accuracy and selectivity of the DD procedure can be used profitably in the search for drugs with particular characteristics. For example it has been observed that neuroleptics which antagonize central dopaminergic effects are therapeutically effective against the so called positive symptoms of schizophrenia but in addition induce extrapyramidal side effects, whereas the combined administration of D_2 and 5-HT_2 antagonists is therapeutically active against both the positive and negative symptoms of schizophrenia and reduces the incidence of untoward side effects. It therefore seemed attractive to search for drugs that selectively block 5-HT_2 and D_2 receptors alike. Janssen Pharmaceuticals developed risperidone, a potent 5-HT_2 and D_2 antagonist. However the serotonergic activity of this compound is greater than its dopaminergic activity (Janssen et al., 1988). A new compound (R 79598) was developed with presumably more equipotent 5-HT_2 and D_2 antagonism. In order to test for the combined serotonergic and dopaminergic action the experimental compound R 79598 was compared in a DD procedure with risperidone, with another 5-HT_2 antagonist ritanserin, and with haloperidol, the classical D_2 antagonist. In previous experiments risperidone and ritanserin had been compared as to their antagonistic effect on the dopaminergically mediated d-amphetamine cue. It was found that risperidone, but not ritanserin affected the stimulus effects of d-amphetamine. This is in agreement with studies showing that ritanserin has little affinity for D_2 binding sites, whereas risperidone has a strong affinity. (Meert et al., 1989, 1990).

All four test drugs listed in table 1 were tested in different groups of animals, trained to discriminate saline from respectively xylazine, d-amphetamine, d-LSD and DOM.

None of the training drugs generalized to any of the test drugs, indicating that the antagonists were not producing the stimulus effect of the training drugs. When antagonism studies were performed it turned out that, except for ritanserin, all test drugs antagonized the D_2 mediated d-amphetamine cue. R79598 clearly was the most potent compound. None of the testdrugs had an effect on the $alpha_2$ mediated xylazine cue. Only ritanserin, risperidone and R79598 antagonized the serotonergically mediated LSD and DOM stimulus, R79598 again being the more potent compound. Taken together these results suggest that the experimental drug R79598 is specifically active at D_2 and 5-HT_2 receptor sites. They demonstrate how the DD technique can be used in developing new therapeutic agents. Moreover this approach shows that the in vivo involvement of particular receptor systems in drug effects can be evaluated rather precisely with the DD procedure.

Table 1

Training drug	Receptor	Test drug	Receptor
DOM	5-HT_2	Risperidone	5-HT_2
LSD	5-HT_2	Ritanserine	5-HT_2
d-Amphetamine	D_2	Haloperidol	D_2
Xylazine	$alpha_2$	R79598	?

CONCLUDING REMARKS

There is no such thing as "the" DS-effect. Discriminative stimuli are not invariable and do not operate in isolation. Properties of drug stimuli like those of other stimuli depend on a multitude of factors, some inherent and some acquired. This is shown by research employing animals having genetically different characteristics, as well as by animals subjected to different forms of conditioning. Or by employing animals brought into different states by pharmacological and surgical means. Conditioning history, prevailing circumstances and contextual cues interact. Therefore,in using the DD technique it must be realized that there are no principal differences between the complexities as to what drug stimuli and conventional, exteroceptive stimuli do to the organism.

Considerable time, skill and expense is involved in training animals for DD experiments. Therefore application of the technique is best at later stages in drug development. Adequately applied however it seems a remarkably sensitive technique for in vivo measurement of selective drug effects, especially when questions concerning the mechanisms of action and the involvement of different neurochemical systems are prominent. DD can help to elucidate the role of behaviourally relevant receptor subtypes, and perhaps can help to disentangle the different actions of a drug by relating its different behavioural effects to different stimulus effects, in those cases in which each stimulus effect is mediated through a different mechanism.

REFERENCES

Andrews, J.S., Turski, L., Stephens, D.N. (1989) Drug Dev. Res. 16, 247-256.
Ator, N.A, Griffiths, R.R. (1985) Fed. Proc. 44, 495.
Carlton, P.L. (1983) A primer of behavioural pharmacology. New York, Freeman and Company.
Colpaert, F.C., Balster, R.L. (1988) Transduction Mechanisms of Drug Stimuli. Springer-Verlag, Heidelberg.
Colpaert, F.C., Janssen, P.A.J. (1984) J. Pharmacol. Exp. Ther. 230, 193-199.
Colpaert, F.C., Rosecrans, J.A. (1978) Stimulus Properties of Drugs Ten Years of Progress. Amsterdam, Elsevier-North Holland.
Colpaert, F.C., Slangen, J.L. (1982) Drug Discrimination Applications in CNS Pharmacology. Amsterdam, Elsevier.
D'Mello, G.D., Stolerman, I.P. (1977) Br. J. Pharmacol. 59, 453-454.
De Vry, J., Slangen, J.L. (1985) Psychopharmacology, 85, 483-485.
De Vry, J., Slangen, J.L. (1986) Psychopharmacology, 89, S45.
Emmet-Oglesby, M.W., Mathis, D.A., Moon, R.T.Y., Lal, H. (1990) Psychopharmacology, 101, 292-309.
Herling, S., Shannon, H.E. (1982) Life Sci. 31, 2105-2112.
Ho, B.T., Richards, W.I.I.I., Chute, D.L. (1978) Drug Discrimination and State Dependent Learning. Academic Press, New York.
Janssen, P.A.J., Niemegeers, C.J.E., Awouters, F., Schellekens, K.H.L., Megens, A.A.H.P., Meert, T.F. (1988) J. Pharmacol. Exp. Ther. 244, 685-693.
Jarbe, T.U.C., Swedberg, M.D.B. (1982) A conceptualization of drug discrimination learning. Drug Discrimination Applications in CNS Pharmacology. Amsterdam, Elsevier.
Koek, W., Colpaert, F.C. (1989) In: Behavioural pharmacology of 5-HT (P.Bevan, A.R.Cools and T.Archer, Eds), Hillsdale, New Yersey, Lawrence Erlbaum Ass. 407-424.
Koek, W., Woods, J.H. (1989) Drug Dev. Res. 16, 169-181.
Lal, H. (1977) Discriminative Stimulus Properties of Drugs. Advances in Behavioural Biology. New York, Plenum.
Lal, H., Fielding, S. (1984) Drug Dev. Res. 4, 3-21.
Mathis, D.A., Emmett-Oglesby, M.W. (1990) J. Neurosci. Meth. 31, 23-33.
Mathis, D.A., Emmett-Oglesby, M.W., Harris, C.M., Lal, H. (1987) Psychol. Rec. 37, 161-165.
Meert, T.F., De Haes, P., Janssen, P.A.J. (1989) Psychopharmacology, 97, 206-212.
Meert, T.F., De Haes, P.L.A.J., Vermote, P.C.M., Janssen, P.A.J. (1990) Drug Dev. Res. 19, 353-373.
Overton, D.A. (1971) Discriminative control of behaviour by drug states. Stimulus Properties of Drugs. New York, Appleton-Century Crofts.
Overton, D.A., Batta, S.K. (1977) In: Predicting Dependence Liability of Stimulant and Depressant Drugs (T.Thompson, K.R.Unna, Eds), Baltimore, University Park Press, pp 125-135.

Animal Models in Psychopharmacology
Advances in Pharmacological Sciences

A DRUG DISCRIMINATIVE ANALYSIS OF R 79598, A NEW NEUROLEPTIC

T.F. Meert, F.H.L. Awouters, P.L.A.J. de Haes and P.C.M. Vermote

Janssen Research Foundation, B-2340 Beerse, Belgium

SUMMARY: R 79598, a new potent neuroleptic, was compared with haloperidol and risperidone in a drug discrimination test procedure in rats. Generalization and antagonism studies were performed in rats trained to discriminate either LSD, DOM, d-amphetamine, cocaine or xylazine from saline. None of the three compounds produced a stimulus generalization to any of the training drugs. In terms of DOM and LSD antagonism, the relative order of potency was R 79598 $\geq$ risperidone > haloperidol (inactive) and for d-amphetamine it was R 79598 > haloperidol > risperidone. The three drugs were only partial cocaine antagonists and none of the drugs antagonized xylazine. All three compounds reduced response rate. Despite marked effects on response rate, R 79598 showed the greatest differentiation between drug antagonism and sedation. These results confirm R 79598 to be a very potent serotonin 5-HT_2 and dopamine D_2 antagonist.

INTRODUCTION

All clinical active neuroleptics, such as haloperidol, have the ability to antagonize central dopamine (Niemegeers and Janssen, 1979). Clinically, these drugs are effective against the positive but not the negative symptoms of schizophrenia. They all have the risk to induce extrapyramidal side-effects (EPS) (Berger, 1978). Selective 5-HT_2 antagonists, such as ritanserin (Awouters et al., 1988), given in combination with neuroleptics, can overcome the negative symptoms of schizophrenia and reduce EPS (Gelders, 1989; Reyntjens et al., 1986). Blockade of serotonin 5-HT_2 receptors seems to add an additional therapeutic activity and to reduce the side-effects of dopamine D_2 receptor blockade. Risperidone is a potent serotonin 5-HT_2 and dopamine D_2 antagonist, with the serotonin antagonism being more pronounced than its antidopaminergic activity (Janssen et al., 1988). Clinically, risperidone was found active against both the positive and negative symptoms of schizophrenia and the drug had a very low incidence of EPS (Castelao et al., 1989; Meco et al., 1989; Mesotten et al., 1989).

Recently R 79598, a benzisoxazol derivative, was characterized as a new potent neuroleptic with an equipotent serotonin 5-HT_2 and dopamine D_2 antagonism. In rats, R 79598 was at least as potent as haloperidol and several times more potent than risperidone in inhibiting dopaminergic overactivity. In dogs, R 79598 was even more potent than haloperidol. In order to further characterize the pharmacological profile of R 79598, the drug was further compared with risperidone and haloperidol in a drug discrimination test procedure in rats. LSD and DOM were selected as training drugs because both drugs have a well defined serotonin 5-HT_2 component and because risperidone was demonstrated to be a long acting LSD antagonist (Meert et al., 1989, 1990). Cocaine and d-amphetamine have a dopaminergic involvement and both haloperidol and risperidone antagonize the discriminative stimulus properties of these training drugs (Meert et al., 1990).

MATERIALS AND METHODS

Wistar rats, weighing 240 $\pm$ 20 g at the beginning of the experiment, were used. The animals were housed individually and allowed to feed freely for 1 h daily.

Tests were performed in six test cages with a house light, two levers and a food pellet receptacle. The drug discrimination test procedure itself has been described in detail elsewhere (Meert et al., 1989, 1990). Daily discrimination training started after habituation and initial shaping to lever press for food on a fixed ratio 10 (FR = 10) schedule during a 15 min test session. At a fixed time before being placed in the test cage, the rats were injected with either the training drug or physiological saline. Depending on whether they were injected with drug or saline, the animals obtained food pellets by pressing either the drug lever or the saline lever, respectively. Responses on the incorrect lever had no consequences. The training drugs were: 0.16 mg/kg LSD (IP, T=-15 min), 0.63 mg/kg DOM (IP, T=-15 min), 1.25 mg/kg d-amphetamine (SC, T=-30 min), 10.0 mg/kg cocaine (IP, T=-15 min) and 2.5 mg/kg xylazine (IP, T=-30 min). After training, animals were tested on Fridays while the normal training procedure was continued on the remaining days.

To test for stimulus generalization, either R 79598, risperidone, haloperidol or the vehicles were injected subcutaneously at T=-60 min before testing, followed by an injection with saline, given according to the treatment conditions of the training drug.

To test for an antagonism, the injections of the three test compounds or the vehicles were followed by a treatment with the training drug.

Table 1 Effects of R 79598, risperidone and haloperidol on various training drugs in terms of generalization and antagonism. Given are ED_{50}s (and 95 % confidence limits) or the maximal obtained activity at the highest dose tested. Doses are given in mg/kg.

Training drug test condition	R 79598	Risperidone	Haloperidol
DOM (0.63 mg/kg)			
generalization	> 0.16	> 0.63	> 0.63
antagonism	0.0027 (0.0018-0.0040)	0.024 (0.016-0.037)	> 0.63
LSD (0.16 mg/kg)			
generalization	> 0.16	> 0.63	> 0.63
antagonism	0.021 (0.011-0.043)	0.028 (0.014-0.056)	> 0.63
Cocaine (10.0 mg/kg)			
generalization	> 0.16	> 0.63	> 0.63
antagonism	60 % (at 0.63 mg/kg)	40 % (at 2.5 mg/kg)	40 % (at 0.63 mg/kg)
d-Amphetamine (1.25 mg/kg)			
generalization	> 0.16	> 0.63	> 0.63
antagonism	0.043 (0.026-0.069)	0.34 (0.21-0.54)	0.11 (0.091-0.14)
Xylazine (2.5 mg/kg)			
generalization	> 0.16	> 0.63	> 0.63
antagonism	> 0.16	> 0.63	> 0.63

RESULTS

Rats can be trained to discriminate either 0.63 mg/kg DOM, 0.16 mg/kg LSD, 10.0 mg/kg cocaine, 1.25 mg/kg d-amphetamine or 2.5 mg/kg xylazine from saline. The ED_{50}s (and 95 % confidence limits) of the training drug for stimulus generalization with the training condition at 60 min after treatment were: 0.17 (0.11-0.25), 0.26 (0.16-

0.41), 1.55 (1.03-2.31), 0.51 (0.34-0.76) and 0.77 (0.52-1.16) mg/kg for DOM, LSD, cocaine, d-amphetamine and xylazine, respectively.

The effects of the 3 neuroleptics on the different training drugs is given in Table 1. Neither R 79598, risperidone or haloperidol produced a stimulus generalization to any of the training drugs. Also with relation to xylazine antagonism, no activity was measured. R 79598 and risperidone completely antagonized DOM, with the lowest ED_{50}s being 0.0027 and 0.024 mg/kg, respectively. Haloperidol up to 0.63 mg/kg was inactive. Both 79598 and risperidone also completely antagonized LSD, with corresponding lowest ED_{50}s of 0.021 and 0.028 mg/kg. Again, haloperidol was inactive. For cocaine, a partial antagonism was observed with all three test compounds; the maximal antagonism ranged between 40 (2.5 mg/kg risperidone and 0.63 mg/kg haloperidol) and 60 % (0.63 mg/kg R 79598). All three test compounds antagonized d-amphetamine. The ED_{50}s were 0.043 mg/kg for R 79598 and 0.34 and 0.11 mg/kg for risperidone and haloperidol.

All three compounds reduced response rate. The rate reducing effects started in the different experiments at doses between 0.04 and 0.16 mg/kg for R 79598 and haloperidol and between 0.16 and 0.63 mg/kg for risperidone.

DISCUSSION

The results presented here indicate that R 79598 is a potent and complete DOM, LSD and d-amphetamine antagonist with lowest ED_{50}s of 0.0027, 0.021 and 0.043 mg/kg, respectively. In terms of DOM antagonism, R 79598 was 8.9 times more potent than risperidone whereas for LSD antagonism, both drugs were equipotent. Haloperidol was inactive against both DOM and LSD. R 79 598 was the most potent d-amphetamine antagonist; 2.6 to 7.9 fold higher doses of haloperidol and risperidone were needed. Neither R 79598, risperidone nor haloperidol were able to completely antagonize the discriminative stimulus properties of 10.0 mg/kg cocaine. A maximal reduction with 60 % was observed at 0.63 mg/kg R 79598. Because the discriminative stimulus properties of LSD and especially DOM are serotonin 5-HT_2 mediated and because those of d-amphetamine, and to some extent those of cocaine, are dopamine D_2 mediated (Meert et al., 1989, 1990), these results confirm R 79598 to be a potent serotonin 5-HT_2 and dopamine D_2 antagonist. The ratio between the ED_{50}s for DOM and d-amphetamine antagonism, or between 5-HT_2 and D_2 blockade, was relatively similar for R 79598 and risperidone. Haloperidol did not reveal a 5-HT_2 antagonism.

None of the tested neuroleptics affected the discriminative stimulus properties of xylazine, demonstrating the lack of an α_2-adrenoceptor interaction. Based on response rate, R 79598 was demonstrated to have sedative effects at doses similar to haloperidol; sedation was less pronounced with risperidone. However, because antagonism of the stimulus properties occurred at much lower doses with R 79598, higher differentiation ratio's can be obtained.

REFERENCES

Awouters, F., Niemegeers, C.J.E., Megens, A.A.H.P., Meert T.F., Janssen, P.A.J. (1988) Drug Dev. Res. 15: 61-73.

Berger P.A. (1978) Science 200, 974-981.

Castelao J.F., Ferreira, L., Gelders Y.G., Heylen, S.L.E. (1989) Schizophr. Res. 2: 411-415.

Gelders, Y.G. (1989) Br. J. Psychiat. 155 (suppl 5), 33-36.

Janssen, P.A.J., Niemegeers, C.J.E., Awouters, F., Schellekens, K.H.L., Megens, A.A.H.P., Meert, T.F. (1988) J. Pharmacol. Exp. Ther. 244: 685-693.

Meco G., Bedini, L., Bonifati, V., Sonsini, U. (1989) Curr. Ther. Res. 46, 876-883.

Meert, T.F., De Haes, P., Janssen, P.A.J. (1989) Psychopharmacology 97, 206-212.

Meert, T.F., De Haes, P.L.A.J., Vermote, C.M., Janssen, P.A.J. (1990) Drug Dev. Res. 19, 353-373.

Mesotten, F., Suy E., Pietguin, M., Burton, P., Heylen, S., Gelders, Y. (1989) Psychopharmacology 99, 445-449.

Niemegeers, C.J.E., Janssen, P.A.J. (1979) Life Sci. 24, 2201-2216.

Reyntjens, A., Gelders, Y.G., Hoppenbrouwers, M.L.J.A., Van Den Bussche, G. (1986) Drug Dev. Res. 8, 205-211.

Animal Models in Psychopharmacology
Advances in Pharmacological Sciences

HALOPERIDOL SUBSTITUTES PARTIALLY FOR BUSPIRONE IN RATS TRAINED TO DISCRIMINATE BUSPIRONE FROM SALINE

H.J. Rijnders and J.L. Slangen

Department of Psychopharmacology, Faculty of Pharmacy, University of Utrecht, Sorbonnelaan 16, 3584 CA, Utrecht, The Netherlands

SUMMARY: Using rats in a drug discrimination model, this study revealed a dopaminergic involvement in the discriminative stimulus effects of buspirone in addition to its serotonergic activity. It is suggested that buspirone may produce a compound discriminative stimulus complex.

In psychopharmacology, drug discrimination has proven to be a useful tool for assessing the pharmacological correlates of the behavioural effects of drugs. The pharmacological specificity of this animal model has made it a test procedure regularly incorporated in screening procedures of new drugs. Drug discrimination tests involving the new anxiolytic compounds buspirone, flesinoxan and 8-OH-DPAT have confirmed in vivo the findings obtained by in vitro research indicating that the pharmacological properties of these compounds are not related to those of the benzodiazepines.

In drug discrimination buspirone's mechanism of action is believed to be through the 5-HT_{1A} receptor where it acts as an agonist. Rats trained to discriminate 8-OH-DPAT from saline gave rise to 8-OH-DPAT-appropriate responding when administered buspirone (Tricklebank et al., 1987). In vitro receptor binding studies however, have shown that buspirone has affinity for both the 5-HT_{1A} and the dopamine-D_2 receptor sites (Eison and Temple, 1986). Until now, drug discrimination experiments with rats have not been able to demonstrate dopaminergic involvement in the discriminative stimulus properties of buspirone. Barrett and Olmstead (1989) however produced data suggesting that the D_2 activity of buspirone plays a role in drug discrimination using pigeons. Further, Kamien and Woolverton (1990) showed that in rhesus monkeys the discriminative stimulus properties of the dopamine agonist apomorphine could be blocked by buspirone, indicating D_2 antagonist activities of buspirone.

It was the aim of the present experiment to study the involvement of the dopaminergic D_2 receptor in the discriminative stimulus effect of buspirone in rats.

MATERIALS AND METHODS

Twelve rats (male, Wistar) were trained to discriminate 2.0 mg/kg buspirone HCl (i.p.) from saline in a standard two lever drug discrimination procedure (FR10, food reward). Different doses of the following compounds were tested for substitution: the 5-HT_{1A} agonists buspirone, 8-OH-DPAT and flesinoxan (Schipper et al., 1990); eltoprazine (5-$HT_{1A/1B}$ agonist; Schipper et al., 1990); spiroxatrine (5-HT_{1A}, DA_2 ligand; Barrett et al., 1989b) and haloperidol (DA_2-antagonist). Additionally, 0.05 and 0.1 mg/kg 8-OH-DPAT were combined with 0.5 and 1.0 mg/kg haloperidol. Test sessions were terminated after six reinforcements or after 15 minutes of testing whichever occurred first. During the test sessions both levers were reinforced. The percentage of buspirone appropriate responses out of the total number of responses during the test session served as independent variable. All drugs were injected intraperitoneally 15 minutes before testing. In reference to the report by Barrett and Olmstead (1989), the effect of the order in which compounds were tested for substitution was taken into consideration. After the buspirone generalization tests, 6 animals were tested first with 8-OH-DPAT and then with spiroxatrine and haloperidol. The other 6 animals were tested first with spiroxatrine and then with by 8-OH-DPAT and haloperidol. Next all rats were tested with flesinoxan, eltoprazine and the combinations of haloperidol and 8-OH-DPAT.

RESULTS

The discriminative control of buspirone was dose-dependent and reached a level of 96% drug lever responding at the training dose of 2.0 mg/kg (Table I). Spiroxatrine did not substitute for buspirone when it was tested before 8-OH-DPAT. However, the group that was tested first with 8-OH-DPAT, emited 45 % buspirone-appropriate responding when tested with 1.0 mg/kg spiroxatrine (table 1). There was no difference between the two groups for the remaining substitution tests and therefore the results of both groups, as presented in table 1 and figure 1, were combined.

bb=1>

Table 1 Drugs and doses with the correspondent percentages of buspirone appropriate responding. Due to a shortage of spiroxatrine, the group that was tested after 8-OH-DPAT could only be tested with 1.0 mg/kg spiroxatrine.

drug	dose (mg/kg)	% responding on buspirone lever
buspirone	0	0
	0.5	31.9
	1.0	39.8
	2.0	96.3
spiroxatrine tested before 8-OH-DPAT	0.25	1.0
	0.50	1.2
	1.0	1.0
	2.0	0
spiroxatrine tested after 8-OH-DPAT	1.0	45.0
flesinoxan	0.1	10.0
	0.3	16.1
	1.0	84.3
eltoprazine	0.5	0
	1.0	3.4
	2.0	6.2

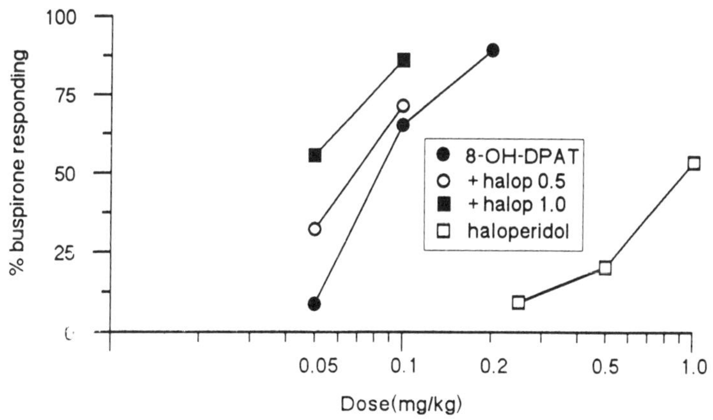

Fig. 1 Generalization curves of 8-OH-DPAT (filled circles), haloperidol (open squares) and combinations of these two compounds (open circles and filled squares), obtained in rats to discriminate between buspirone (2.0 mg/kg, i.p.) and saline.

As can be seen in figure 1, 8-OH-DPAT substituted completely and dose-dependently for buspirone. Haloperidol substituted partially for buspirone yielding an average maximum level of buspirone-appropriate responding of 53 percent at a dose of 1.0 mg/kg. The generalization to flesinoxan was complete and dose-dependent; at a dose of 1.0 mg/kg rats emitted 84.3 percent buspirone-appropriate responding (table 1). Eltoprazine did not substitute for buspirone at the doses tested (table 1).

The results of the 8-OH-DPAT-haloperidol combination tests are depicted in figure 1. The percentages of drug lever responding generated by 0.05 and 0.1 mg/kg 8-OH-DPAT were dose-dependently increased when 0.5 and 1.0 mg/kg haloperidol were co-administered with these doses of 8-OH-DPAT.

DISCUSSION

The results of the substitution tests with 8-OH-DPAT and flesinoxan in buspirone trained rats are consistent with other reports. Mansbach and Barrett (1987) reported a complete substitution by 8-OH-DPAT in pigeons trained to discriminate buspirone from saline, as did Nader et al. (1989). Tricklebank et al. (1987) found a generalization to buspirone in rats trained to discriminate 8-OH-DPAT from saline. In our own laboratory Ybema et al (personal communication) also observed a generalization to buspirone in 8-OH-DPAT trained rats. Further, a generalization to buspirone has been reported in flesinoxan trained rats (Ybema et al., 1989) and vice versa in pigeons (Barrett et al., 1989a). Together these findings provide ample evidence for 5-HT_{1A} receptor mediation of the discriminative stimulus effects of buspirone in pigeons as well as in rats. However, the partial generalization to the buspirone cue observed in the present study after haloperidol administration may reflect additional dopaminergic involvement in the buspirone cue. The dopaminergic involvement is consistent with the results obtained by the 8-OH-DPAT-haloperidol combination tests. Buspirone-appropriate responding at intermediate doses of 8-OH-DPAT was increased when haloperidol was administered in addition to 8-OH-DPAT. It has been suggested that haloperidol may disturb discriminative behaviour by disrupting the relation between response and reinforcement (Colpaert et al., 1977). However, as both levers were reinforced during the test sessions, a disruption of the discrimination due to the additional haloperidol administrations most likely would have resulted in nonsystematical changes of drug appropriate responding, which is clearly not the case (fig. 1).

The failure of the non-selective 5-HT_1 receptor agonist eltoprazine to substitute for buspirone suggests that the discriminative stimulus effect of buspirone is specifically related to activation of 5-HT_{1A} receptor sites. Additionally, Ybema et al. (personal communication) reported a lack of generalization when eltoprazine was tested in rats trained to discriminate either 8-OH-DPAT or flesinoxan from saline.

The failure of spiroxatrine to substitute for buspirone in the group that was tested with spiroxatrine immediately after completion of the generalization tests with buspirone is remarkable. It may reflect a difference in sensitivity to spiroxatrine between rat and pigeon, for a complete substitution has been reported in the latter animal (Nader et al., 1989). The tests with 8-OH-DPAT may have caused a higher sensitivity to spiroxatrine as shown by the partial generalization in rats that were tested with spiroxatrine after the 8-OH-DPAT tests.

Finally, this study has shown that the dopaminergic actions of buspirone do have some relevancy in the discriminative stimulus properties of buspirone. The ability of buspirone to block the apomorphine cue in monkeys, as reported by Kamien and Woolverton (1990), provides additional support for its dopaminergic actions, although this study did not directly evaluate the discriminative stimulus properties of buspirone. In contrast, our findings concerning haloperidol do not fit with the results of Mansbach and Barrett (1987) who tested haloperidol in pigeons trained to discriminate buspirone from saline and found that haloperidol did not substitute for buspirone. Again this may reflect a difference in sensitivity between rats and pigeons. One should bear in mind though that in the present study, the generalization of haloperidol to buspirone was partial and that the selective 5-HT_{1A} agonists 8-OH-DPAT and flesinoxan produced a complete generalization to buspirone. A possible explanation for these findings is that both dopamine and serotonin mediate the discriminative stimulus properties of buspirone, i.e. buspirone produces a compound discriminative stimulus complex and the serotonergic actions of buspirone overshadow the dopaminergic actions. Stolerman et al. (1987) and Garcha and Stolerman (1989) have shown that rats can be trained to discriminate a mixture of two drugs (nicotine-midazolam) from saline. Their results suggest that elements of compound interoceptive stimuli are perceived separately, rather than being merged into new, homogeneous stimuli. In addition they observed effects which could be interpreted in terms of overshadowing and it was speculated that overshadowing may be a key process in the acquisition of discriminations produced by single drugs with multiple effects.

Obviously, further research is required to disclose the mechanisms by which the discriminative effects of buspirone are mediated. The concept of discrimination of drug mixtures may provide a basis to extend our knowledge of these mechanisms.

REFERENCES

Barrett, J.E., Gleeson, S., Nader, M.A., Hoffman, S.M. (1989a) J. Psychopharmacol. 3, 64-69.

Barrett, J.E., Hoffmann, S.M., Olmstead, S.N., Foust, M.J., Harrod, C., Weissmann, B.A. (1989b) Psychopharmacology 97, 319-325.

Barrett, J.E., Olmstead, S.N. (1989) Drug Dev. Res. 16, 365-374.

Colpaert, F.C., Niemegeers, C.J.E., Janssen, P.A.J. (1977) Psychopharmacology 53, 169-173.

Eison, A.S., Temple, D.L. (1986) Am.J.Med. 80, 1-9.

Garcha, H.S., Stolerman, I.P. (1989) Behav. Pharmacol. 1, 25-31.

Kamien, J.B., Woolverton, W.L. (1990) Pharmacol. Biochem. Behav. 35, 117-120.

Mansbach, R.S., Barrett, J.E. (1987) J. Pharmacol. Exp. Ther. 240, 364-369.

Nader, M.A., Hoffmann, S.M., Gleeson, S., Barrett, J.E. (1989) Behav. Pharmacol. 1, 57-67.

Schipper, J., Tulp, M.Th.M., Sybesma, H. (1990) In: Reviews on: Drug metabolism and drug interactions (M.Raghoebar, B.Olivier, D.L.Rasmussen, and J.Mos, Eds), Vol 8 Nos. 1-2. Freund Publishing House LTD., London, pp 85-114.

Stolerman, I.P., Rauch, R.J., Norris, E.A. (1987) Psychopharmacology 93, 250-256.

Tricklebank, M.D., Neil, J., Kidd, E.J., Fozard, J.R. (1987) Eur. J. Pharmacol. 133, 47-56.

Ybema, C.E., Slangen, J.L., Olivier, B., Mos, J. (1990) Pharmacol. Biochem. Behav. 35, 781-784.

Animal Models in Psychopharmacology
Advances in Pharmacological Sciences

FURTHER CHARACTERIZATION OF THE DISCRIMINATIVE STIMULUS INDUCED BY FLESINOXAN

Caroline E. Ybema, Ingrid van den Broek, Jef L. Slangen, Berend Olivier* and Jan Mos*

Psychopharmacology, Faculty of Pharmacy, University of Utrecht, Sorbonnelaan 16, Utrecht, *CNS-Pharmacology, Duphar B.V., P.O.Box 900, 1380 DA Weesp, The Netherlands

SUMMARY: Using a two lever operant drug discrimination procedure, rats were trained to discriminate the 5-HT_{1A} agonist flesinoxan (0.5 mg/kg) from saline. Hereafter, several serotonergic compounds were tested in generalization and antagonism tests. The flesinoxan-stimulus generalized to the 5-HT_{1A} agonist gepirone, but not to the mixed 5-$HT_{1A/1B}$ agonist RU 24969, the 5-$HT_{1B/1C}$ agonist TFMPP, the 5-HT_2 agonist DOI and the 5-HT_2 / 5-HT_3 drug quipazine. The non-specific 5-HT antagonist methysergide, the 5-$HT_{1C/2}$ antagonist mesulergine, the 5-HT_2 antagonist ketanserin and the 5-HT_3 antagonist ICS 205-930 neither blocked nor mimicked the cue. The present results are in favour of a 5-HT_{1A} mechanism for the discriminative stimulus of flesinoxan and consistent with the results of previous drug discrimination studies with flesinoxan.

INTRODUCTION

The centrally acting anti-hypertensive drug flesinoxan shows a high affinity and selectivity for 5-HT_{1A} receptors in the brain (Wouters et al., 1987). Results of recent drug discrimination studies with flesinoxan are in agreement with a 5-HT_{1A} mechanism of action of flesinoxan. Barrett et al. (1989) found that flesinoxan dose-dependently substitutes for the 5-HT_{1A} drug buspirone in pigeons, whereas Ybema et al. (1990a) reported that buspirone produced nearly 80% generalization in flesinoxan-trained rats. In the latter study it was also found that cross-generalization occurs between the selective 5-HT_{1A} agonist 8-OH-DPAT (Hamon et al., 1984; Middlemiss and Fozard, 1983) and flesinoxan. Furthermore, flesinoxan generalized to the 5-HT_{1A} agonist ipsapirone (Peroutka, 1985; Traber et al., 1984) and both the flesinoxan-stimulus and the ipsapirone-induced stimulus generalization were dose-dependently blocked by pindolol, a 5-HT_{1A} and β-adrenergic antagonist (Middlemiss et al., 1977; Nahorski and Willcocks, 1983; Ybema et al., 1990b).

To explore further the mechanism of action of flesinoxan, rats were trained to discriminate flesinoxan from saline in a two-lever operant task and consequently tested with a range of serotonergic drugs having varying degrees of affinity for the different 5-HT_1 subtypes, 5-HT_2 or 5-HT_3 sites.

MATERIAL AND METHODS

Subjects: Twenty-six male Wistar rats, weighing approximately 200 g at arrival, were obtained from CPB-TNO (Zeist, The Netherlands). They were housed individually under a non-reversed 12h light-dark cycle and a room temperature of 20 - 22°C. Tap water was freely available. Subjects were maintained at approximately 85% of their expected free-feeding weight by providing them with a diet of 13 g food (Hope Farms) one hour after each daily session. Food was freely available from Friday afternoon until Sunday morning.

Apparatus: Seven ventilated operant chambers (Campden) equipped with two levers and housed in sound-insulated boxes were used. A pellet dispenser delivered 45 mg pellets (Noyes) in a tray placed between the levers. A Digital Equipment Corporation PDP-11 microcomputer with software (SKED) supplied by State Systems Incorporated (Kalamazoo, MI) programmed session events and recorded data.

Training: After initial lever-press training, the rats (N=26) were trained to lever-press according to a fixed ratio 10 (FR 10) schedule of reinforcement. Hereafter the rats were trained to discriminate flesinoxan from the solvent saline (0.9% NaCl). Depending on the injection-condition reinforcement could be obtained by pressing either the drug-appropriate (D) or the saline-appropriate (S) lever. The position of the D and the S lever was counterbalanced across rats. Fifteen minutes before the daily session animals were injected with either drug or saline, according to a two-weekly alternating schedule: (D-S-D-D-S, S-D-S-S-D). For half of the animals the discrimination training started with the first half of the sequence, for the other half training started with the second one. Drug-induced stimulus control was assumed to be present when the animal, in ten consecutive sessions (5 D and 5 S), accurately selected the appropriate lever. The lever on which the rat first emitted ten responses was scored as the selected lever. Accurate lever selection was defined as the accumulation of ten responses on the injection-appropriate lever with five or less responses on the inappropriate lever (i.e., number of responses made on both levers before the first reinforcement, between 10 and 15). Testing was started as soon as at least 6 animals had attained the criterion for drug-induced stimulus control.

Testing procedure: Generalization- or antagonism-tests were carried out on Wednesday and Friday. On the remaining days the training procedure was continued. Throughout the test-session, responding on the selected lever was rewarded according to a FR 10 schedule. Both the training and test sessions ended after fifteen minutes or sooner, i.e. when the animal had obtained fifty reinforcements. Stimulus generalization was said to occur when at least 80 % of the animals, after given a dose of the test compound, selected the drug-appropriate lever. The definition of stimulus antagonism was that at least 80% of the animals should select the saline-appropriate lever. The criterion for accurate drug- or saline-lever selection was defined as the accumulation of a maximum of 13 responses on the injection-condition-appropriate lever (i.e. less than four responses were made on the inappropriate lever) before the first reinforcement. The sequence in which different doses of a test-drug were given was counterbalanced across rats. The sequence in which drugs were tested was the same for all rats. Each dose was tested in at least 6 different animals and was ,in most cases, administered to the animals once. All drugs were administered fifteen minutes before the test, except for the antagonists which were injected one hour before the test (i.e. 45 minutes before either the standard dose of 0.5 mg/kg of flesinoxan or a control injection with saline). Response rates were analysed by means of one-way ANOVA with one within factor (i.e. dose) having between two and five levels. Only the data of rats which showed no missing values were used. If the overall effect was significant, planned comparisons with univariate F-tests were performed between the test- and saline baseline-conditions. For all effects a significance level of 5% was chosen.

Drugs: Flesinoxan (Duphar), TFMPP(1-(3-trifluoromethylphenyl)piperazine, Duphar), RU24969 (5-methoxy-3-(1,2,3,6-tetrahydropyridinyl) 1H indole, Roussel UCLAF), DOI (Research Biochemicals Incorporated), gepirone (Bristol-Myers), quipazine (Duphar), methysergide (Sandoz) and mesulergine (Duphar) were dissolved in 0.9% NaCl. Ketanserin (Janssen Chemica) and ICS 205-930 (Research Biochemicals Incorporated) were dissolved in 10% acetic acid (25μl/ml) and brought up to volume with 0.9% NaCl. All solutions were prepared fresh daily and administered via intraperitoneal injection in a volume of 2 ml/kg.

Table 1: Results of generalization studies using flesinoxan (0.5 mg/kg) as the training drug

Drug	Dose (mg/kg)	N°	% Rats selecting DL	Responses (± SEM) #
SALINE	-	26/26	· 92.3	1.17 (0.05)
FLESINOXAN	0.02	19/19	5.3	1.08 (0.08)
	0.1	21/21	19.0	1.11 (0.12)
	0.3	23/23	61.1	1.23 (0.07)
	0.5	26/26	88.5	1.07 (0.05)
	0.8	16/16	90	0.97 (0.15)
GEPIRONE	0.3	9/9	33.3	1.22 (0.11)
	1.0	9/9	88.8	1.17 (0.09)
	3.0	9/9	77.8	0.76 (0.07)*
	9.0	9/9	85.7	0.50 (0.16)*
RU24969	0.1	8/8	12.5	1.03 (0.18)
	0.3	8/8	25	0.69 (0.19)
	0.9	6/8	0	0.50 (0.18)*
TFMPP	0.3	9/9	22.2	1.03 (0.18)
	0.6	9/9	11.1	1.03 (0.16)
	1.2	7/9	0	0.42 (0.15)*
QUIPAZINE	0.5	9/9	0	0.94 (0.10)
	1.0	6/9	0	0.48 (0.15)*
	2.0	5/9	0	0.36 (0.17)*
MESULERGINE	0.25	7/7	0	1.39 (0.10)*
	0.5	7/7	14.3	0.76 (0.21)
	1.0	5/7	0	0.58 (0.20)*
	2.0	5/7	0	0.29 (0.14)*
DOI	0.1	8/8	0	0.93 (0.09)
	0.3	7/8	14.3	0.62 (0.16)
	0.9	3/8	0	0.50 (0.16)*
KETANSERIN	0.25	8/8	14.3	0.86 (0.12)
	0.5	8/8	0	0.80 (0.16)
	1.0	8/8	0	0.61 (0.18)
ICS 205-930	0.005	6/6	0	1.20 (0.18)
	0.01	6/6	0	1.40 (0.12)
	0.05	6/6	0	1.54 (0.12)*
	0.2	6/6	0	1.08 (0.20)
METHYSERGIDE	9.0	9/9	22.2	1.02 (0.21)

° Number of animals responding/number to receive drug
Average responses / s (± SEM) during session
* Significantly different from saline baseline-sessions ($p \leq 0.05$)

RESULTS

Learning to discriminate between flesinoxan (0.5 mg/kg) and saline required an average of 47 training sessions (range: 26-90 sessions).

In generalization-studies (table 1) gepirone substituted in a dose-related manner for the discriminative stimulus produced by 0.5 mg/kg of flesinoxan. TFMPP, RU24969, DOI and quipazine did not substitute for flesinoxan. Ketanserin, ICS 205-930, mesulergine and methysergide neither blocked nor mimicked the flesinoxan-cue (table 2). The results of the statistical analysis of the response rates showed that the following conditions had a main effect on response rate: Gepirone ($F(4,32)=11.94$, $p<0.01$), RU24969 ($F(3,21)=6.09$, $p<0.01$), TFMPP ($F(3,24)=7.03$, $p<0.01$), quipazine ($F(3,24)=6.93$), $p<0.01$), mesulergine ($F(4,24)=12.98$, $p<0.01$), DOI ($F(3,21)=7.54$, $p<0.01$), ICS 205-930 ($F(4,20)=3.20$, $p<0.01$), mesulergine + 0.5 mg/kg of flesinoxan ($F(2,18)=27.99$, $p<0.01$). Response rates were significantly decreased after either 3.0 or 9.0 mg/kg of gepirone ($F(1,8)=22.00$, $p<0.01$ and $F(1,8)=13.96$, $p<0.01$, respectively), 0.9 mg/kg of RU24969 ($F(1,7)=10.76$, $p<0.05$), 1.2 mg/kg of TFMPP ($F(3,24)=7.03$, $p<0.01$), either 1.0 or 2.0 mg/kg of mesulergine ($F(1,6)=9.35$, $p<0.05$ and $F(1,6)=41.92$, $p<0.01$, respectively), 0.9 mg/kg of DOI ($F(1,7)=21.18$, $p<0.01$) and either 1.0 or 2.0 mg/kg of mesulergine given in combination with 0.5 mg/kg of flesinoxan ($F(1,9)=22.91$, $p<0.01$ and $F(1,9)=30.14$, $p<0.01$, respectively). A significant increase in response rate was found after 0.25 mg/kg of mesulergine ($F(1,6)=13.84$, $p<0.05$) and 0.05 mg/kg of ICS 205-930 ($F(1,5)=9.03$, $p<0.05$).

DISCUSSION

The results of the present study suggest that effects of flesinoxan can serve as a discriminative stimulus and that stimulation of 5-HT_1 neuronal systems are important to these discriminative stimulus effects. Since it has been shown that the anti-hypertensive effects of flesinoxan do not occur at doses below 1.0 mg/kg it seems unlikely that the internal stimulus of flesinoxan is caused by a reduction of the blood-pressure (Calis et al., 1986; Dreteler et al., 1990).

The finding that the specific 5-HT_{1A} agonist gepirone (Peroutka, 1985) dose-dependently substitutes for flesinoxan is consistent with previous drug discrimination studies, showing a complete generalization of the flesinoxan-cue to the the 5-HT_{1A}

agonists 8-OH-DPAT and ipsapirone, almost 80% stimulus generalization with the somewhat less specific 5-HT_{1A} compound buspirone and a dose-dependent blocking of the flesinoxan-stimulus and the ipsapirone-induced stimulus generalization with the 5-HT_{1A} and β-adrenergic antagonist pindolol (Ybema et al., 1990a, 1990b). In contrast, the mixed 5-$HT_{1A/1B}$ agonist RU24969 (Hunt and Oberlander, 1981; Sills et al., 1984; Tricklebank et al., 1986), and the 5-$HT_{1B/1C}$ agonist TFMPP (Fuller et al., 1978; Sills et al., 1984), which apart from its affinity for 5-HT_{1B} and 5-HT_{1C} receptors also possesses some affinity for 5-HT_{1A} sites, did not substitute for flesinoxan. Similarly, methysergide (Sills et al., 1984), a 5-HT antagonist binding to both 5-H_{1A} and 5-HT_2 sites, neither blocked nor mimicked the cue. The inability of these non-specific 5-HT_{1A} compounds to mimick or block the flesinoxan-stimulus might be a reflection of the specificity of the flesinoxan-cue.

Table 2: Results of antagonism studies using flesinoxan (0.5 mg/kg) as the trainingdrug

Drug	Dose (mg/kg)	N°	% Rats selecting DL	Mean Responses (± SEM) #
FLES (0.5) + METHYSERGIDE	1.0	11/11	81.8	1.20 (0.11)
FLES (0.5) + METHYSERGIDE	3.0	11/11	66.7	1.12 (0.12)
FLES (0.5) + METHYSERGIDE	9.0	11/11	81.8	0.97 (0.16)
FLES (0.5) + ICS 205-930	0.005	8/8	87.5	0.89 (0.16)
FLES (0.5) + ICS 205-930	0.01	16/16	81.2	1.18 (0.15)
FLES (0.5) + ICS 205-930	0.05	8/8	87.5	0.94 (0.17)
FLES (0.5) + ICS 205-930	0.1	8/8	100	1.06 (0.20)
SALINE + ICS 205-930	0.05	7/7	14.3	1.33 (0.21)
FLES (0.5) + KETANSERIN	0.25	10/10	90	1.27 (0.11)
FLES (0.5) + KETANSERIN	0.5	20/20	80	1.17 (0.14)
FLES (0.5) + KETANSERIN	1.0	20/20	90	1.02 (0.11)
FLES (0.5) + KETANSERIN	2.0	10/10	80	1.00 (0.14)
SALINE + KETANSERIN	1.0	8/8	12.5	1.09 (0.09)
FLES + MESULERGINE	1.0	2/12	75	0.75 (0.10)*
FLES + MESULERGINE	2.0	11/11	72.7	0.47 (0.13)*

For explanation of °, . and *, see legend to table 1.

Additional evidence for specificity of the stimulus effect of flesinoxan can be derived from the finding that substances with no or less affinity for 5-HT_{1A} receptor sites are ineffective in both generalization- and antagonism-tests. Thus, the relatively selective 5-$HT_{1B/1C}$ agonist TFMPP was not able to mimick the flesinoxan-cue. Similarly, quipazine (Fuller et al., 1981) which binds to 5-HT_2 and 5-HT_3 sites did not substitute for flesinoxan, nor did the specific 5-HT_2 agonist DOI (Shannon et al., 1984) induce a flesinoxan-like stimulus. Furthermore, neither the 5-HT_2 antagonist ketanserin (Leysen et al., 1981), the 5-HT_3 antagonist ICS 205-930 (Richardson et al., 1985), nor the 5-$HT_{1C/2}$ antagonist mesulergine (Hoyer et al., 1985; Pazos et al., 1984) significantly affected flesinoxan-appropriate responding, nor did they substitute for flesinoxan.

In conclusion, the present results are in good agreement with both receptorbinding-studies and previous drug discrimination-studies and suggest that the discriminative stimulus effects of flesinoxan are specifically mediated via 5-HT_{1A} receptors in the brain.

REFERENCES

Barrett, J.E., Gleeson, S., Nader, M.A., Hoffmann, S.M. (1989) J. Psychopharmacol. 3 (2), 64-69.
Calis, J.I.M., Hartog, J., Wouters, W. (1986) Br. J. Pharmacol. 89, 469P.
Dreteler, G.H., Wouters, W., Saxena, P. (1990) Eur. J. Pharmacol. 180, 339-349.
Fuller, R.W., Snoddy, H.D., Mason, N.R., Molloy, B.B. (1978) Eur. J. Pharmacol. 52, 11-16.
Hamon, S., Bourgoin, S., Gozlan, H., Hall, M.D., Goetz, C., Artaud, F., Horn, A.S. (1984) Eur. J. Pharmacol. 100, 263-276.
Hoyer, D., Engel, G., Kalkman, H.O. (1985) Eur. J. Pharmacol. 118, 13-23.
Hunt, P., Oberlander, C (1981) In: Serotonin, Current aspects of neurochemistry and function (Haber et al., Eds), Plenum, New York, pp 547- 562.
Leysen, J.E., Wouters, F.A., Kennis, L., Laduron, P.M., Vandenberk J., Janssen, P.A.J. (1981) Life Sci. 28, 1015-1022.
Middlemiss, D.N., Blakeborough, L., Leather, S.R. (1977) Nature 267, 289-291.
Middlemiss, D.N., Fozard, J.R. (1983) Eur. J. Pharmacol. 90, 151-153.
Pazos, A., Hoyer, D., Palacios, J.M. (1984) Eur. J. Pharmacol. 106, 539-546.
Richardson, B.P., Engel, S., Donatsch, P., Stadler, P.A. (1985) Nature, 316, 126-131.
Nahorski, S.R., Willcocks, A.L. (1983) Br. J. Pharmacol. 78, 107P.
Peroutka, S.J. (1985) Biol. Psychiatry 20, 971-979.
Shannon, M., Battaglia, G., Glennon, R.A., Titeler, M. (1984) Eur. J. Pharmacol. 102, 23-29.
Sills, M.A., Wolfe, B.B., Frazer, A. (1984) J. Pharmacol. Exp. Ther. 231, 480-487.
Traber,J., Davies, M.A., Dompert, W.U., Glaser, T., Schuurman, T., Seidel, P.R. (1984) Brain Res. 12, 741-744.
Tricklebank, M.D., Middlemiss, D.N., Neill, J. (1986) Neuropharmacology, 25, 877-886.
Wouters, W., Tulp, M.Th.M., Bevan, P. (1987) Eur. J. Pharmacol. 133: 47-56.
Ybema, C.E., Slangen, J.L., Olivier, B., Mos, J. (1990a) Pharmacol. Biochem. Behav. 35, 781-784.
Ybema, C.E., Slangen, J.L., Olivier, B., Mos, J. (1990b) Eur. J. Pharmacol. 183, 1921-1922.

Animal Models in Psychopharmacology
Advances in Pharmacological Sciences

DISCRIMINATIVE STIMULUS AND RECEPTOR BINDING EFFECTS OF 5-HT$_{1A}$ AGONISTS IN PIGEONS

S. Gleeson, E.E. Gamble, M.A. Nader, and J.E. Barrett

Department of Psychiatry, Uniformed Services University of the Health Sciences, 4301 Jones Bridge Road, Bethesda, Maryland 20889-4799 U.S.A.

SUMMARY: Pigeons were trained to discriminate buspirone (1.0 mg/kg) from saline. Dose-dependent generalization occurred to BMY 14802, SM 3997, WY 47,846, and eltoprazine, but not to mCPP. The range of doses that occasioned 90% or greater generalization was similar for all four drugs. The drugs that substituted for buspirone were investigated for their ability to inhibit [^{3}H]8-OH-DPAT binding in pigeon cerebrum. The IC$_{50}$ for each was in the nanomolar range. The results suggest that BMY 14802, SM 3997, WY 47,846, and eltoprazine have substantial behavioural effects mediated by the 5-HT$_{1A}$ receptor in pigeons.

INTRODUCTION

The extension of operant stimulus control methodology to the study of the discriminative stimulus effects of drugs has proven to be a useful procedure in behavioural pharmacology. In conjunction with receptor binding studies, drug discrimination procedures can provide information about similarities in mechanisms underlying the behavioural effects of drugs. In some cases, drug discrimination procedures are more selective than are other methods of studying drug effects on operant behaviour, and therefore can clarify the results obtained in such studies. For example, benzodiazepines and 5-HT$_{1A}$ agonists both exhibit anti-conflict effects, but cross-generalization between these two pharmacological classes typically is not observed in drug discrimination studies (e.g., Ator & Griffiths, 1986; Mansbach & Barrett, 1987). This suggests a different mechanism underlies the discriminative stimulus and perhaps behavioural effects of the two classes, and has implications for understanding the different neurobiological bases of their similar anxiolytic effects.

The discriminative stimulus effects of 5-HT_{1A} agonists in pigeons is of particular interest because the anti-conflict effects of these drugs seem to be more pronounced in this species than in rats (Howard & Pollard, 1990) or monkeys (Gleeson & Barrett, 1990). In a recent report, Nader et al. (1989) showed that in pigeons trained to discriminate buspirone from saline, generalization occurred to other 5-HT_{1A} agonists, but not to drugs acting at other 5-HT receptors, or at dopaminergic or noradrenergic receptors. Thus the buspirone drug discrimination procedure in pigeons appears to be a useful technique for assessing the 5-HT_{1A} effects of novel compounds, including those with actions at other receptor sites.

In the present experiment, we examined the discriminative stimulus effects of SM 3997 and WY 47,846, both of which appear to be 5-HT_{1A} agonists; BMY 14802, an antagonist at the putative sigma receptor; eltoprazine, a mixed 5-$HT_{1A/1B}$ agonist; and mCPP, a mixed 5-$HT_{1B/1C}$ agonist. These drugs were tested in pigeons trained to discriminate buspirone from saline. Additionally, we investigated the ability of these drugs to inhibit [^{3}H]8-OH-DPAT binding in pigeon cerebrum.

MATERIALS AND METHODS

Behavioural studies

Subjects and Apparatus: Eight White Carneau pigeons obtained from Palmetto Pigeon Plant (Sumter, SC, USA) were maintained at approximately 85% of free-feeding weight. Each bird was housed individually with free access to water and crushed oyster shell grit in a colony room with a 14 h light cycle.

Behavioural studies were conducted daily Monday-Friday in a standard operant conditioning chamber equipped with two response keys illuminated by white lights. Reinforcers were 3-sec access to mixed grain provided in a food tray recessed behind the front panel of the chamber.

Procedure: Pigeons received intramuscular injections of either saline or buspirone (1.0 mg/kg) and were placed in the chamber. A 20-min timeout in the darkened chamber preceded each session, which terminated after 30 reinforcers or after 20 min had elapsed. Responses were reinforced, on the left key if saline had been injected or on the right key if buspirone had been injected, according to a fixed-ratio 30 (FR 30) schedule of reinforcement.

Substitution tests were conducted on Tuesdays and Fridays provided that the discrimination was stable on the preceding day. Stability was defined as at least 90%

responding on the injection-appropriate key over the course of the session, and at least 25 of the first 30 responses of the session on the injection-appropriate key. All drugs were dissolved in saline and injected in 1.0 ml/kg volume. Each drug was tested in 5 pigeons, except mCPP which was tested in 4 pigeons. During substitution test sessions, responses on either key were reinforced according to the FR 30 schedule.

RECEPTOR BINDING STUDIES

Experiments were carried out according to the general procedures described by Gleeson et al. (in press). Briefly, reaction tubes contained 100 μl Tris buffer (pH 7.4); 50 μl drug, 5-HT, or buffer; and 100 μl [^{3}H]8-OH-DPAT (approx. 1.4 nM). After addition of 250 μl tissue homogenate, tubes were incubated, cooled in ice water, and filtered using a Brandel M-24R cell harvester with GF/B glass fiber filters. Radioactivity was counted for 2 min in a Packard 2000CA TRI-CARB liquid scintillation analyzer. Tissue from at least 2 pigeons was used in each experiment.

Table 1 ED_{50} values, IC_{50} values, and Hill coefficients.

DRUG	ED_{50} (μmole/kg)	IC_{50} (nM)	nH
Buspirone	0.6	29 ± 9.8	0.67 (± 0.06)[a]
SM 3997	0.8	47 ± 5.3	0.89 (± 0.17)
WY 47,846	1.8	41 ± 3.6	0.98 (± 0.04)
BMY 14802	3.3	156 ± 10	0.99 (± 0.08)
Eltoprazine	5.1	34 ± 2.0	0.82 (± 0.03)
mCPP	-.	220 ± 34	0.99 (± 0.05)[b]

a. Witkin et al., 1987; b. Gleeson et al., in press

RESULTS

All of the drugs except mCPP substituted for buspirone at one or more doses. The degree of generalization was dose-dependent, and the range of doses that occasioned 90% or greater drug-key responding was similar for BMY 14802, SM 3997, WY 47,846, and eltoprazine. Response rates were not substantially altered at these doses. Each of the four drugs that substituted for buspirone also displaced [^{3}H]8-OH-DPAT with IC_{50} values in the nanomolar range. The rank order of the drugs in terms of IC_{50}

values was similar, but not identical, to the rank order in terms of ED_{50} for producing generalization as shown in Table 1.

DISCUSSION

The results provide evidence that BMY 14802, SM 3997, WY 47,846, and eltoprazine act as agonists or partial agonists at the 5-HT_{1A} receptor site. Eltoprazine has been described as having high affinity for the 5-HT_{1B} receptor, but in pigeons its effects appear to be mediated through its actions at the 5-HT_{1A} site. In this respect it resembles the mixed $5\text{-HT}_{1A/1B}$ agonist RU 24969, which also substitutes for buspirone (Nader et al., 1989) and increases the punished responding of pigeons (Gleeson et al., 1989). BMY 14802 has received recent experimental attention primarily in light of its effects as an antagonist at the putative sigma receptor site (Taylor & Dekleva, 1988), but the present results suggest that its 5-HT_{1A}-mediated effects are substantial. Finally, the results suggest that SM 3997, WY 47,846, BMY 14802, and eltoprazine should exhibit anti-conflict effects in pigeons, a particularly good model for potential novel anxiolytic drugs that act through the 5-HT_{1A} receptor.

Acknowledgements: This research was supported by PHS Grant DA-02873.

REFERENCES

Ator, N.A., Griffiths, R.R. (1986) J. Pharmacol. Exp. Ther. 237, 393-403.
Gleeson, S., Ahlers, S.T., Mansbach, R.S., Foust, J.M., Barrett, J.E. (1989) J. Pharmacol. Exp. Ther. 250, 809-817.
Gleeson, S., Weissman, B.A., Barrett, J.E. (1991) Eur. J. Pharmacol. (In press).
Howard, J.L., Pollard, G.T. (1990) Drug Dev. Res. 19, 37-49.
Mansbach, R.S., Barrett, J.E. (1987) J. Pharmacol. Exp. Ther. 240, 364-369.
Nader, M.A., Hoffmann, S., Gleeson, S., Barrett, J.E. (1989) Behav. Pharmacol. 1, 57-67.
Taylor, D.P., Dekleva, J. (1988) In: Sigma and Phencyclidine-like Compounds as Molecular Probes in Biology (E.F.Domino and J.M.Kamenka, Eds), NPP Books, Ann Arbor, MI. pp 345-355.
Witkin, J.M., Mansbach, R.S., Barrett, J.E., Bolger, G.T., Skolnick, P., Weissman, B. (1987) J. Pharmacol. Exp. Ther. 243, 970-977.

Animal Models in Psychopharmacology
Advances in Pharmacological Sciences

IS THE CONDITIONED TASTE AVERSION PROCEDURE A USEFUL TOOL IN DRUG DISCRIMINATION RESEARCH?

A. van Hest, J.L. Slangen and B. Olivier

Duphar B.V., Department of Pharmacology, P.O.Box 900, 1380 DA Weesp, The Netherlands. Psychopharmacology Laboratory, Faculty of Pharmacy, University of Utrecht, Sorbonnelaan 16, 3584 CA Utrecht, The Netherlands

SUMMARY: Rats were trained to discriminate CDP or 8-OH-DPAT from saline by means of a conditioned taste aversion (CTA) procedure in order to further evaluate the use of CTA in drug discrimination research. CTA was found to be more time-consuming, less accurate, and more bothersome to the animals than traditional drug discrimination procedures.

INTRODUCTION

The conditioned taste aversion (CTA) paradigm has recently been described as an alternative approach to two-lever choice procedures in drug discrimination learning (Lucki, 1988). In CTA thirsty rats are given daily access to sweetened (saccharin) water. Drug injections are given prior to saccharin-LiCl pairings. Vehicle treatment precedes saccharin-NaCl pairings. Alternating drug and vehicle injections are given until rats avoid saccharin consumption following drug administration. Mastropaolo and co-workers (1989) showed that rats will also learn to avoid saccharin following vehicle if vehicle has been given prior to saccharin-LiCl pairings. The aversion can thus be established independently of whether the presence or the absence of the drug signals LiCl toxicosis.

It has been argued that conditioning to the CTA paradigm occurs more rapidly and requires less frequent exposure to drugs than two-lever choice procedures. The differential effects on saccharin consumption were observed after only 2-3 drug-LiCl pairings. In addition, it has been claimed that CTA provides a more sensitive discrimination index than traditional procedures. Kautz and co-workers (1989) reported that rats were able to discriminate low (0.3 mg/kg) doses of naloxone, an opiate antagonist which otherwise fails to support drug discrimination training. Riley et al. (1989) reported that subjects are able to discriminate pentobarbital at lower doses using CTA than when two-lever operant techniques are employed. The CTA procedure does also not require expensive equipment. Given these advantages, the CTA seems to be a useful tool in drug discrimination learning. However, the use of CTA in these kinds of studies is relatively new, and its efficacy and potency must be further assessed.

The present experiments were designed to investigate the stimulus properties of the benzodiazepine chlordiazepoxide (CDP) and the 5-HT_{1A} agonist 8-OH-DPAT. LiCl was given in doses up to 1.2 mEq (48.0 mg/kg), the precise doses depending upon the amount of saccharin consumed. This manipulation was installed for two reasons. Firstly, from a theoretical point of view, it is undesirable to impose the same aversive contingencies following two opposing responses (i.e. drinking and avoiding the saccharin solution). Secondly, lowering the total amount of LiCl diminishes possible hazardous effects of prolonged LiCl treatment.

MATERIALS AND METHODS

Male Wistar rats (233.2 ± 2.3 g) were individually housed under a reversed light-dark cycle. They were adapted to a restricted drinking schedule (30 min access to a 0.1% w/v saccharin solution). Subjects drank an average of 21.9 (± 2.5) ml saccharin water. They were matched on saccharin consumption during adaptation and were assigned to different treatment conditions (n=8 per group). Subjects were injected with drug (groups E_d and C_d) or vehicle (groups E_v and C_v) 15 minutes prior to saccharin exposure. Immediately following the drinking period, subjects were treated with LiCl (groups E_d and E_v) or NaCl (groups C_d and C_v). On other days, subjects in groups E_d and C_d received vehicle injections prior to saccharin exposure, while subjects in groups E_v and C_v received drug. All groups received NaCl injections following saccharin exposure on these days. Drug and vehicle sessions were given according to a two-weekly alternating schedule.

During dose-substitution sessions, subjects were treated with different doses of the training drug. Saccharin exposure during these sessions was never followed by LiCl injections.

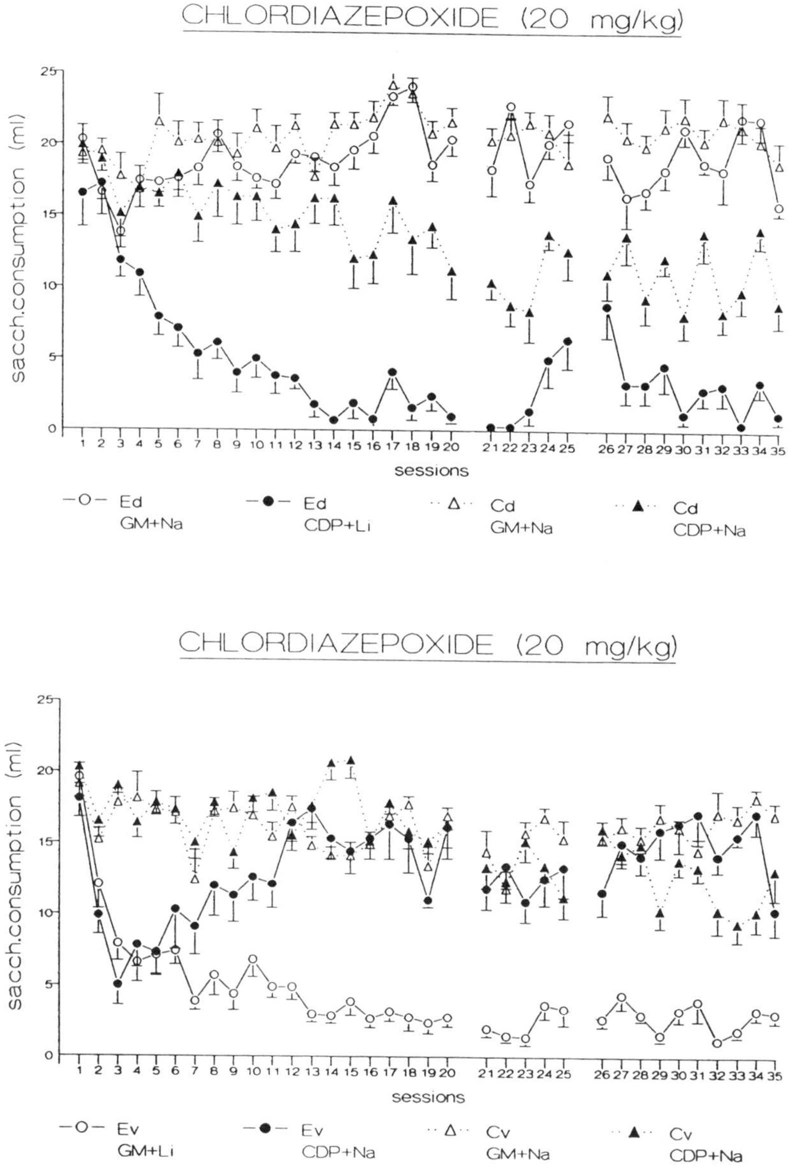

Fig. 1 Mean saccharin consumption for rats trained to discriminate CDP from saline. LiCl sessions (S) 1-20: baseline; S 21-25: dose-substitution; S 26-35: re-exposure to baseline. GM=vehicle (gelatin-mannitol); Li=Lithium chloride; CDP=chlordiazepoxide; Na=NaCl.

Thirty-two subjects were used to assess the discriminative stimulus properties of 20 mg/kg CDP. Another 32 subjects were trained to discriminate 0.4 mg/kg 8-OH-DPAT. All drugs were given i.p. at a volume of 2 ml/kg. LiCl was dissolved in distilled water and injected in a volume of 8 ml/kg. Data were subjected to analysis of variance (ANOVA). Significance level was set at $p<0.01$.

RESULTS

Chlordiazepoxide: Analysis showed that CDP alone, irrespective of treatment condition, accounted for a 2 ml reduction of saccharin intake. Saccharin consumption following CDP decreased for subjects in group E_d, but a similar decrease was observed for subjects in group C_d. However, saccharin intake was more suppressed in experimental subjects than in controls when such non-specific drug effects were taken into account (Fig. 1, upper panel). The results for subjects in groups E_v and C_v mirrored the results of groups E_d and C_d, suggesting that the vehicle came to serve as a discriminative stimulus for subjects in group E_v (Fig. 1, lower panel).

Orderly dose-response curves (Fig. 2) were obtained when substitution tests were interspersed between training sessions. However, experimental subjects consumed an increasing amount of saccharin water during LiCl sessions 21-25. All subjects were thus re-exposed to the training procedure (Fig. 1).

8-OH-DPAT: Subjects in groups E_d and E_v still consumed more than 5 ml saccharin water at the completion of 20 LiCl sessions (Fig.3). It was decided to give subjects free access to tap water at the end of each session to see whether the failure to obtain good discrimination performance was due to dehydration. Saccharin intake during LiCl sessions decreased, but also during NaCl sessions. The original contingencies were then re-installed until consumption returned to baseline levels. Finally all experimental subjects were injected with 1.2 mEq LiCl during LiCl sessions. Subjects in groups E_d and E_v drank less than 5 ml saccharin water preceding LiCl treatment at the completion of this acquisition phase.

ANOVA on bodyweights revealed that subjects that received LiCl treatment gained significantly less weight in the course of training than controls.

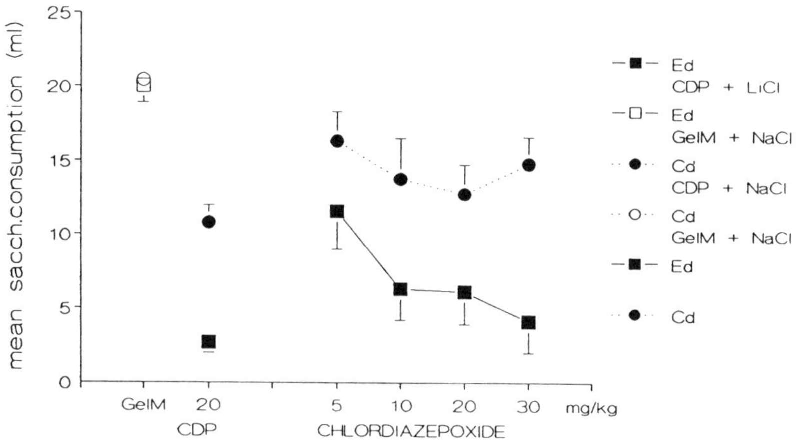

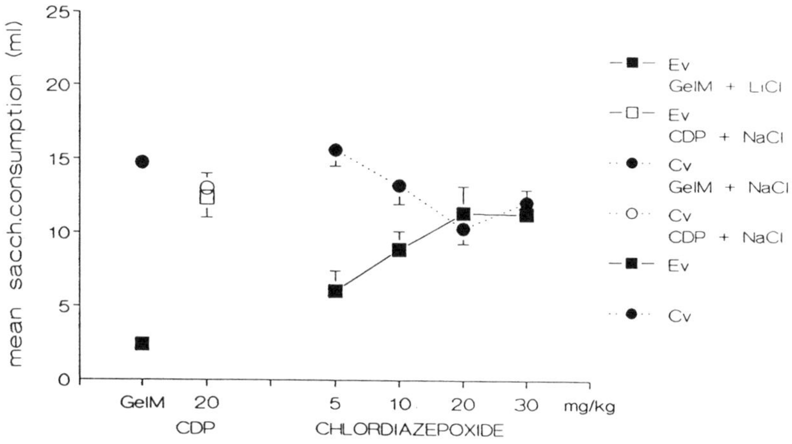

Fig. 2 Mean saccharin consumption after subjects were treated with increasing doses CDP. For further legends see fig. 1.

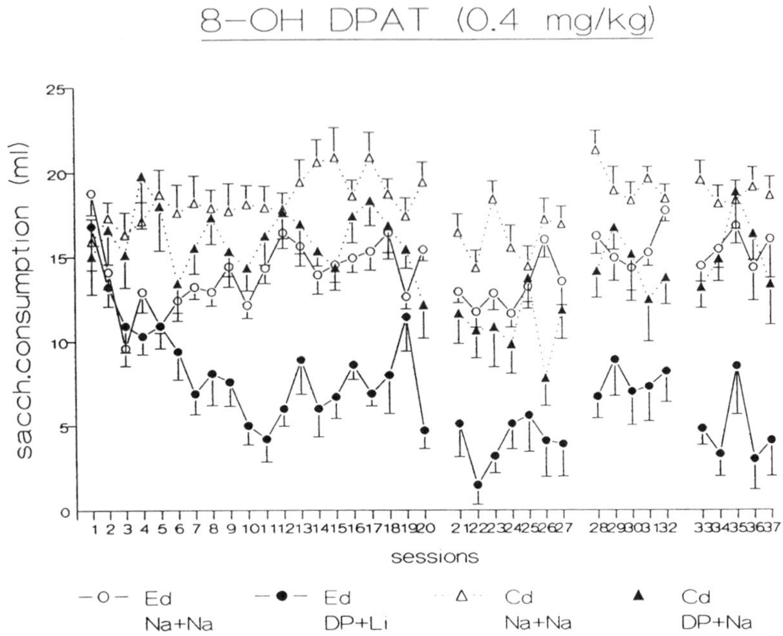

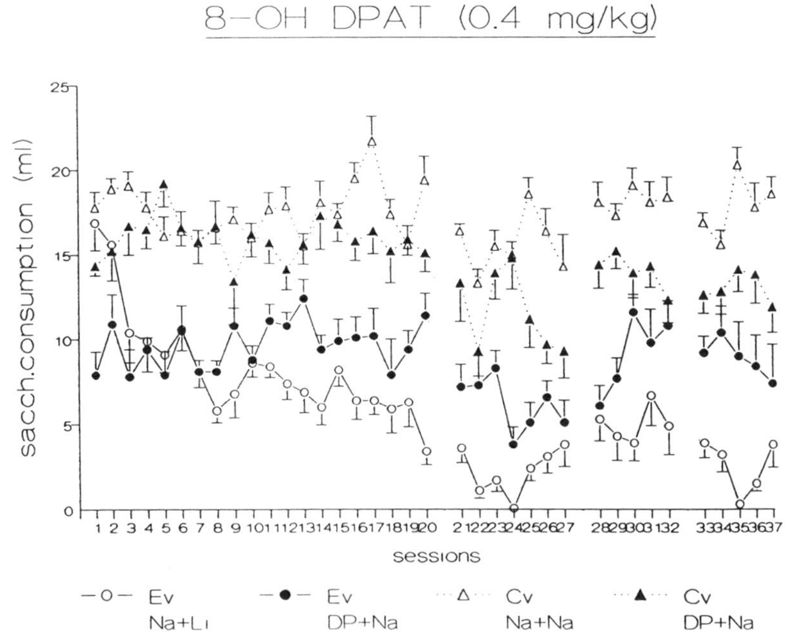

Fig. 3 Mean saccharin consumption for rats trained to discriminate 8-OH-DPAT from saline. LiCl sessions (S) 1-20: baseline; S 21-27: free access to tap water at the end of each session; S 28-32: return to baseline; S 33-37: fixed amount of LiCl. For further legends see fig. 1.

DISCUSSION

The present data show that all experimental subjects eventually learned the discrimination, drinking less than 5 ml saccharin water when consumption was followed by LiCl treatment. However, stable baseline performance for subjects trained to discriminate CDP was achieved only after subjects had been exposed to the experimental contingencies for a total of 40 sessions. Behaviour deteriorated when a dose-response relationship was established, probably due to the omission of LiCl treatment during dose-substitution sessions. Additional training (20 sessions) was necessary. 8-OH-DPAT discrimination training took even longer (74 sessions), and was observed only at the highest doses of LiCl. These findings, taken together with the observed large day-to-day fluctuations in behaviour, suggest that discrimination training under the present conditions may not yield results as robust as the results obtained with other procedures.

Non-specific drug effects on saccharin intake were also observed in the present set-up. Both CDP and 8-OH-DPAT unconditionally suppressed drinking, thereby attenuating discrimination performance. Separate groups are needed to control for such non-specific drug effects, thus doubling the number of rats needed. It is very likely that other manipulations that interfere with drinking (e.g. deprivation, saccharin concentration, rebound effects) also interfere with discrimination acquisition. These variables need to be carefully examined.

The daily testing routine (i.e. preparing drug and/or vehicle solutions, weighing saccharin, calibrating and reading off the drinking tubes) was very laborious. The test is also very bothersome to the laboratory animal. Rats are injected twice daily, and are regularly subjected to sickening effects of LiCl treatment. Relatively high doses of LiCl are necessary to establish and maintain the discrimination (see also Jaeger and Mucha, 1990). Experimental subjects gained considerable less weight over the experiment than controls. This observation may be indicative of possible hazardous effects of chronic LiCl treatment.

REFERENCES

Jaeger, T.V., Mucha, R.F. (1990) Psychopharmacology, 100; 145-150.
Kautz, M.A., Geter, B., McBride, S.A., Mastropaolo, J.P., Riley, A.L. (1989) Drug Devel. Res. 16, 317-327.
Lucki, I. (1988) J. Pharm. Exp. Therap. 247, 1120-1127.
Mastropaolo, J.P., Moskowitz, K.H., Dacanay, R.J., Riley, A.L. (1989) Pharmacol. Biochem. Behav. 32, 1-8.
Riley, A.L., Jeffreys, R.D., Pournaghash, S., Titley, T.L., Kufera, A.M. (1989) Drug Devel. Res. 16, 229-236.

part 6:

MISCELLANEOUS

Animal Models in Psychopharmacology
Advances in Pharmacological Sciences

ETHOLOGY AND NEUROPHARMACOLOGY: RODENT ULTRASOUNDS

Klaus A. Miczek, Walter Tornatzky and Jeffrey Vivian

Tufts University, Research Building, Tufts University, 490 Boston Ave, Medford MA 02155, USA

INTRODUCTION

One of the most fascinating and instructive illustrations of how neural networks mediate complex communication comes from the neuroethological studies with zebra finches and canaries that sing to court mates and defend their territories (e.g., Nottebohm, 1985, 1981). The results from oscine songbirds inform importantly on the plasticity of the neural network, its seasonal androgen-dependent expression, and the behavioural significance in reproductive and agonistic contexts. However, less is known about the neurochemical characteristics of these communication-specific networks and how potential neural receptors in these networks may be targeted by substances with known effects of affective expression in mammals. Nonetheless these investigations point to multisynaptic networks involving aminergic and steroid-sensitive cells that mediate the initiation and production of important communicative signals.

Anatomical studies on the neural control of mammalian vocalizations have delineated a hierarchical neural organization that includes cingulate forebrain, diencephalic, mesencephalic, pontine and medullary structures in cats and in squirrel monkeys (Bandler, 1988; Jürgens, 1988; Ploog, 1988; Siegel and Pott, 1988). Most significantly for the purposes of the present discussion are the limbic structures that modulate the central grey area from which in turn efferent pathways originate to the lower brain stem nuclei controlling laryngeal innervation and respiration. At present, the neurotransmitters and their receptors in this limbic-brain stem network for primate vocalizations remain to be specified. At these cingulate, septal, amygdaloid, thalamic, and hypothalamic and central grey sites, psychotropic substances are expected to exert significant effects on vocalizations that express pleasurable and aversive states.

In the past half dozen years systematic neuropharmacological and neurochemical studies in rodents have begun to identify cortical and subcortical receptors for opioid peptides, excitatory amino acids, benzodiazepine-GABA, and catecholamines as particularly significant in the mediation of ultrasonic vocalizations (e.g., Insel and Winslow, this volume; Panksepp, this volume).

ULTRASONIC VOCALIZATIONS IN RATS

In the context of maternal care, copulation, agonistic interactions, defense against predators and other stressful and aversive situations, several murine species emit vocalizations over a range of frequencies that extend far beyond those audible to humans, coined ultrasounds (e.g., Anderson, 1954; Barfield and Thomas, 1986; Lore and Flannelly, 1977; Sales, 1972). The application of high-capacitance condenser microphones allowed the detection of these ultrasonic vocalizations (Sales and Pye, 1974; Sewell, 1967), and in recent times microprocessor systems enable accurate and rapid spectrographic analysis of different types of ultrasounds, their biophysical characteristics and temporal organization (Kaltwasser, 1990a, Van der Poel and Miczek, in prep.). A whistle mechanism in the larynx is responsible for ultrasounds in the 20-30 kHz as well as 40-70 kHz range (Roberts, 1975). Sharp frequency breaks in the ultrasonic emission may be due to changes in the action of thyro-arytenoid folds of the larynx. Careful and systematic studies of ultrasounds have delineated the ontogenetic development as well as the social contexts for these types of sounds (e.g., Kaltwasser, 1990b; Sales, 1972).

1. Pup Ultrasounds

When rat pups are separated from their littermates and their dam, and particularly when they are exposed to an unfamiliar, cool environment or rough handling, they emit ultrasounds at a high rate (Allin and Banks, 1971; Hård et al., 1985; Hofer and Shair, 1978; Insel and Winslow, 1991; Kehoe and Blass, 1986; Noirot, 1966, 1968; Sewell, 1970). Pups as young as 2-days old emit ultrasounds, and peak levels are produced between 6-10 day after birth; older pups decline in their rate of ultrasounds even if exposed to low ambient temperature (Allin and Banks, 1971). Typically, pup ultrasounds are emitted as brief pulses during exhalation between 30-60 kHz (Fig. 1). Ultrasounding pups are quickly located and retrieved by the dam.

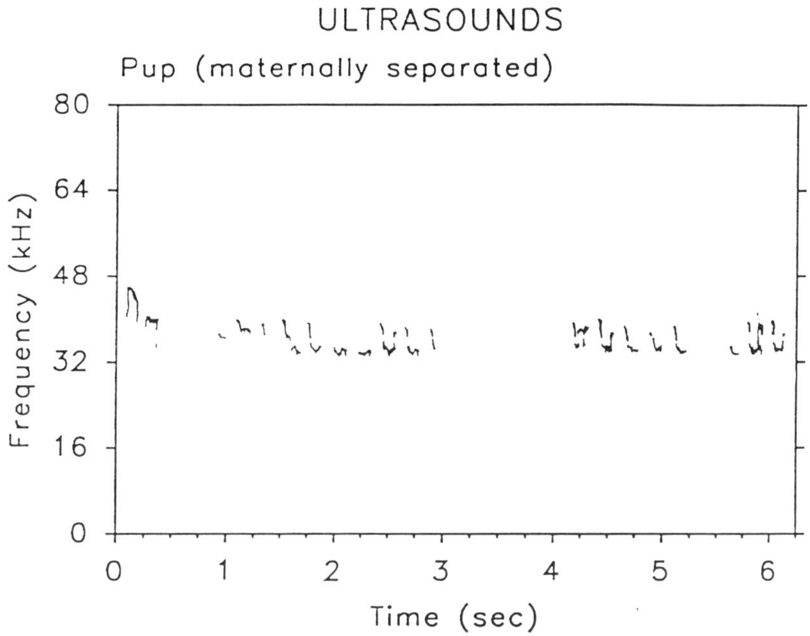

Fig. 1 Representative spectrogram of ultrasonic vocalizations by a maternally separated rat pup.

2. Sexual Ultrasounds

The ultrasonic postejaculatory song of the male rat (Barfield and Geyer, 1972) represents one of the most thoroughly studied ultrasonic vocalizations, particularly in terms of its hormonal mediation. The male rat emits vocalizations, centered around 22 kHz, during most of the post-ejaculatory refractory period, while being sexually and motorically inactive (Barfield and Geyer, 1972, 1975). These calls are frequency-modulated, most notably the medial and terminal portions of the calls with large up- and downswings (>5 kHz) at the beginning and ending portion of the calls (Van der Poel and Miczek, in prep.).

Male and female rats emit ultrasounds around 50 kHz during mating (Barfield and Thomas, 1986). These "higher" calls are more brief and modulated in frequency than the male's postejaculatory 22 kHz ultrasounds. The mating calls of the male and the female appear to facilitate the partner's pre-copulatory and proceptive behaviour (Thomas et al., 1981, White and Barfield, 1987).

electric shocks and to the lower frequency sounds associated with defense and submission.

6. Ultrasounds During Startle Noises

Sudden, loud noises evoke a startle reflex, and in 40-70% of laboratory rats this response is accompanied by ultrasound emission (e.g., Kaltwasser, 1990a; Farrell, Vivian and Miczek, unpublished observations). Bursts of 105-115 dB brief tones evoke bouts of 12-20 pulses of ultrasonic vocalizations, each 50 to 1200 msec in duration. The characteristics of these ultrasounds are most similar to the lower-frequency ultrasounds of submissive rats and of pain-exposed rats (Farrell, Vivian and Miczek, unpublished observations). It has been proposed that the startle-associated ultrasounds may express an affective state of fear (Kaltwasser, 1990a).

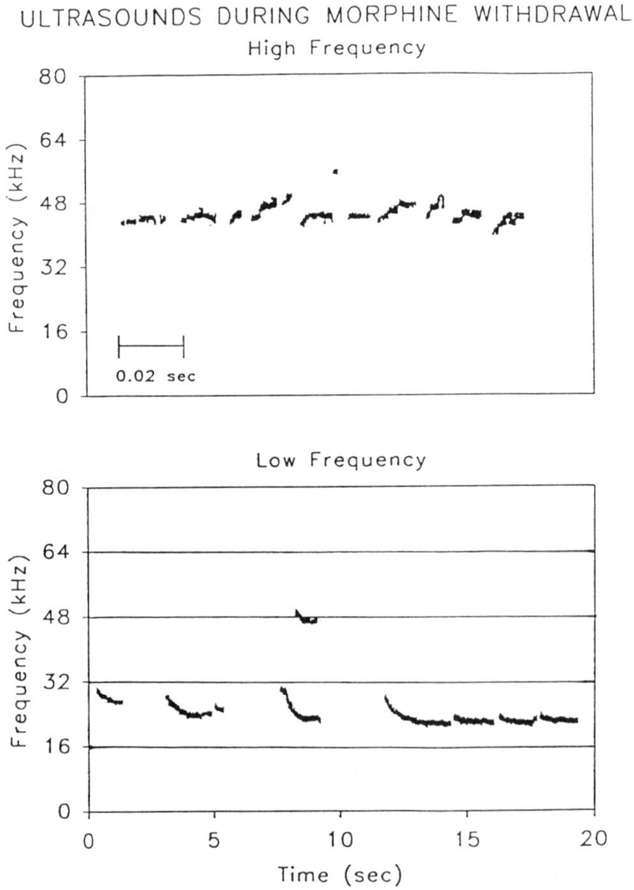

Fig. 3 Spectrograms of high (32-65 kHz) and low (20-32 kHz) frequency vocalizations for a defeat experienced morphine withdrawn rat. Note that the high frequency calls are representative samples and the time base between calls varies; low frequency calls are depicted in real time.

7. Ultrasounds during Opiate Withdrawal

Recently, ultrasonic vocalizations have been documented in pairs of rats that are wirhdrawn from morphine; these ultrasounds may provide an index of the affective components of the withdrawal syndrome (Vivian and Miczek, in press; Fig. 3). The high rate of ultrasound emission during the initial hours after removing morphine declines within the next four days (Fig. 4). Similar to the ultrasounds by the submissive rat, a long pulse (1-2 sec), lower frequency ultrasounds (21-25 kHz) as well as a shorter pulse (0.02-0.2 sec) higher frequency ultrasound (44-52 kHz) are emitted during morphine withdrawal. Prior experience in agonistic and copulatory encounters further enhances the rate of ultrasound emission in morphine-withdrawn rats.

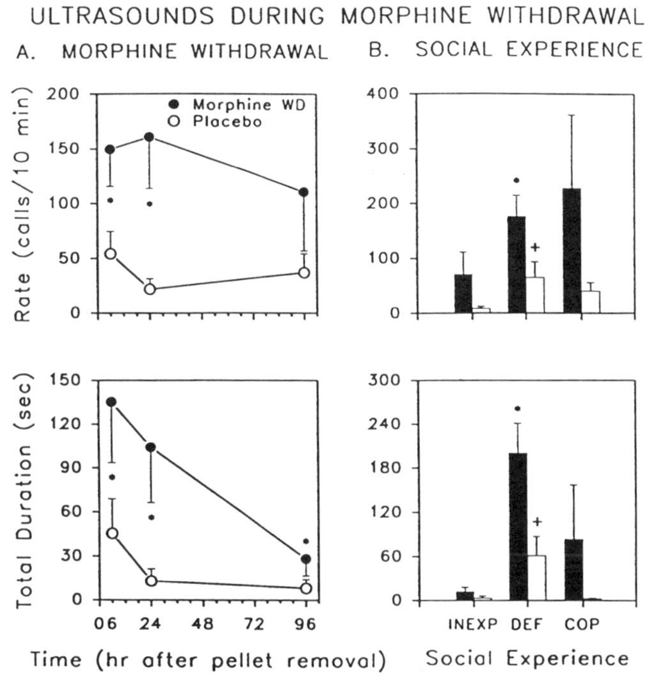

Fig. 4 A. Mean rate (calls/10 min) and total duration (sec) of ultrasonic vocalizations throughout 96 hr withdrawal for morphine and placebo implanted rats. Error bars denote SEM, asterisks denote significant differences ($p<0.05$) from placebo controls. Each point is the combined vocalizations of socially inexperienced, defeat and copulatory experienced rats.
B. Mean rate (calls/10 min) and total duration (sec) of ultrasonic vocalizations for socially inexperienced, defeat and copulatory experienced rats. Error bars denote SEM, asteriks and crosses denote significant differences ($p<0.05$) from socially inexperienced comparison group. Each bar is the combined vocalizations across the 96 hr withdrawal period.

8. Anti-predatory Ultrasounds

It has been proposed that ultrasonic vocalizations by colonially living rats when exposed to a cat or the cat's odors may alarm other colony members that also begin to ultrasound (Blanchard et al., 1990). Initial recordings indicate that these cat-induced ultrasounds are emitted in the 20-25 kHz range at a high rate for several hours, even after the cat has disappeared. In contrast to the deterring audible vocalizations in direct confrontations with the predator, the ultrasounds may serve a communicatory function within the burrow system of the colony.

9. Overview

Rats like other rodents emit different kinds of ultrasounds in the context of highly significant behavioural events throughout the life span. Many of these events share the feature of being aversive in nature. Are ultrasonic vocalizations epiphenomena or adjuncts to behaviour patterns comprising reproduction, agonism and defense? Is there a common message by the long 20-30 kHz ultrasounds? There is no lack of proposals as to the potential functions of ultrasonic vocalizations. It is feasible that pups who are separated from the litter call in order to be retrieved, that males and females call in order to synchronize and facilitate their copulatory sequence, that males signal their absolute refractory period after ejaculation, that males and females who are threatened and attacked signal their submissive mode of behaviour, that sentinels who encounter a potential predator alarm members of their colony, and that noxious and painful events in the internal and external "milieu" of the rat are accompanied by ultrasonic expressions of affect.

NEUROPHARMACOLOGY OF RODENT ULTRASONIC VOCALIZATIONS

Rodent ultrasounds may represent expressions of affect, and their neurobiological mechanisms may be modified by drugs that have anxiolytic and antidepressant effects in humans. Important neuropharmacological studies on opioid peptides and their receptors with regard to their role on "distress" vocalizations in young chicks, guinea pigs and dogs were initiated by Panksepp and his associates shortly after the discovery of their existence in brain in the late 1970s (see Panksepp et al., 1978 and this volume). In comparison to the effects of drugs that preferentially act upon catecholaminergic, serotonergic or cholinergic neurons, opiates and their antagonists are exquisitely potent and selective in their modulation of the audible "distress" vocalizations in puppies,

Table 1

Behavioural Context	Modal Frequency (and Range)/ Duration of calls (and Range)	Frequency Modulation and Temporal Pattern	References
Pup Calls	(36- 48) kHz/100 ms 39(33-50) kHz	see Fig 1.	Cagiano et al., 1986 Insel et al., 1986
Pre-ejaculatory	50 kHz (30-80) kHz/(10-300) msec. 22 kHz		Peters et al., 1988 White et al., 1990 Cagiano et al., 1989
Post-ejaculatory Calls	22 kHz 20-30 kHz/960 msec.	50% of the calls contain a 5 kHz up or downsweep. During medial part of the calls: 2.5 kHz modulations. Average bout duration: 2.73 sec.	Peters et al., 1988 Van der Poel & Miczek (submitted)
Calls during defense and submission "Low"	22 kHz/ < 3400 msec. (21-36) kHz/(300-920) msec.	Most of the constant frequency calls (bandwith 1-6 kHz) contain initial downwards drifts and short upwards steps at the end	Sales, 1972 Sales, 1979
	(20-29) kHz/(800-3400) msec (22-28) kHz/(200-1700) msec. 20-30 kHz / 960 msec.	in bouts 50% of the calls contain a 5 kHz up or downsweep. During medial part of the calls: minimal modulations. Average bout duration: 2.73 sec.	Corrigan & Flannelly, 1979 Kaltwasser, 1985 Van der Poel & Miczek (submitted)
	22 (21-29) kHz/650 (40-3600) msec.	see Fig. 2B. Single or in bouts.	Tornatzky & Miczek (in preparation)
"High"	50 kHz/3-65 msec. (20-116) kHz/53 (7-380) msec.	in bouts, frequency modulated with 2-82 kHz bandwith	Sales, 1972 Sales, 1979
	(45-70) kHz, 25 kHz bandwith / (30-70) msec.		Kaltwasser, 1985
	(45-70) kHz / (3-56) msec.	Corrigan and Flannelly 1979	
	50 (30-70) kHz / 50(20-200) msec.	see Fig 2B. wavelike or inverted U shape / single or in bouts	Tornatzky & Miczek (in prep.)
Antipredator calls	22 kHz		Blanchard et al., 1990
Pain-related calls	(20-30) kHz/(1000-4000) msec	Sounds with little frequency modulation, 50% of the calls contain downward drifts (2.5 kHz) and upwards sweeps. Calls occur as single pulses or in short bouts (2.73 sec.)	Van der Poel et al., 1989 Van der Poel & Miczek (in prep.)
	(22-28) kHz/(300-1200) msec. 23 (20-32) kHz / 1600(375-3000) msec.	33% of the rats vocalized 77% of the rats vocalized	Tonoue et al., 1986 Cuomo et al., 1988
Brain stimulation evoked calls	21 kHz 30 kHz		Yajima et al., 1980 Yajima et al., 1981
Opiate Withdrawal Calls "Low"	(22-25) kHz/(1000-8000) msec.		Vivian & Miczek (in press)
"High"	(44-52) kHz / (20-100) msec.		Vivian & Miczek (in press)
Startle related calls	(20-30) kHz / (50-1200) msec.	40% - 70% of the rats vocalized. Calls emitted in bouts (12-20 calls)	Kaltwasser, 1990b
	(20-30) kHz / 0.77 (90-2500) msec.	40% of the rats vocalized	Farrell et al., (in prep.)

chickens, guinea pigs and dogs (Panksepp et al., 1978). In the past decade, these studies of vocalizations have expanded at the neuropharmacological and the biophysical level. Importantly, studies on the neuropharmacology of vocalizations in primates and in common laboratory rodents have begun (e.g., Cuomo et al., 1988; Newman, 1988).

1. Opiates

The potent effects of opiate receptor agonists and antagonists as well as opioid peptides on ultrasonic vocalizations in infant and adult rodents in several behavioural situations have led to several investigations into the role of opioid systems in affective expressions.

μ-Receptor agonists potently and specifically decrease rat pup ultrasounds in a naltrexone-reversible fashion (Carden and Hofer, 1990a,b; Carden et al., 1990; Cuomo et al., 1988; Gardner, 1985, Robinson et al., 1985; Winslow and Insel, 1991). This effect is seen both in male and female pups of different strains, ranging in age from 4-12 days. Naltrexone in itself may actually increase the rate of pup ultrasounding (e.g., Carden and Hofer, 1990a; Kehoe, 1989; Robinson et al., 1985). Effects similar to those after systemic morphine are seen with DAGO that selectively acts on μ-receptors and with DPDE for δ-receptors; both substances decrease ultrasounds of isolated pups (Carden et al., 1990). By contrast, κ-specific agonists either have no effect on pup ultrasounds or actually increase these vocalizations (Carden et al., 1990; Kehoe et al., 1990). This latter effect may be due to the aversive effects of κ-receptor activation.

In adult male and female rats, ultrasounds are readily emitted during confrontations with an attacking and threatening opponent (Fig. 2A, B). Ultrasounds by submissive male and female rats are potently suppressed by morphine in a naltrexone-reversible manner (Vivian and Miczek, 1990, Haney and Miczek, in preparation). This conclusion is based on ongoing studies in which an intruder rat is exposed to: (1) attacks and threats by a resident opponent of the same sex and, subsequently, (2) to the threat of attack by a resident opponent (Miczek et al., 1991). The attack and threat situation consists of brief agonistic interactions until the intruder rat displays submissive postures (i.e. crouch, supine) for five seconds; subsequently, the intruder is exposed for 25-75 minutes to threats by the opponent while protected from physical contact by a wire mesh. Male and female intruder rats receive ca. 10 bites in 60 seconds, prompting submissive postures and frequent ultrasounds. Under these conditions, pharmacological manipulations of opioid receptors result in marked attenuation of ultrasounds emitted by threatened and defeated rats.

In male rats, morphine at doses as low as 3 mg/kg SC decreased the rate, but not the duration of 20-29 kHz ultrasounds by approximately 40%. Morphine had much more potent effects on the 30-70 kHz ultrasounds: 3.0 mg/kg suppressed the rate by approximately 66% and 6 mg/kg abolished these types of vocalizations. Similar suppressive effects on the rate of ultrasounds were found in submissive rats that were administered ethylketocyclazocine (0.1-1.0 mg/kg SC). Opioid specificity was demonstrated as 0.1 mg/kg IP naltrexone produced a 4-6 fold rightward shift in morphine's dose-effect curve for its suppressive effects on ultrasounds.

Female rats that were attacked and threatened by a lactating opponent required 10 mg/kg of morphine to suppress the rate of their 20-29 kHz ultrasounds. Similar to males, the rate of 30-70 kHz ultrasounds by females was decreased by approximately 66% with 3.0 mg/kg and further suppressed (94%) with 6 mg/kg morphine. These observations point to the capacity of both males and females to emit both types of ultrasounds and to the similarity in how opiates modulate these vocalizations in both sexes.

The decrease in the rate of 30-70 kHz ultrasonic vocalizations in male and female submissive rats after opiate treatment is seen in a dose range that produces analgesia as assessed in thermal pain tests. It is possible that some of the ultrasounds represent an affective component of this highly aversive situation and that opiates not only block the motoric reaction to pain, but also the affect that is associated with pain. However, at present it is unclear why stress-activated endogenous opioid mechanisms do not fully inhibit in a submissive rat potentially detrimental pain and fear responses. If the stress of being threatened by a potential attacker prompts acts, postures and ultrasonic vocal signals of a defensive and submissive nature, and if brain opioid receptors are concurrently activated by endogenous peptides, then the opiate suppression of ultrasonic "distress" signals is counterintuitive. Alternatively, if the physiological and behavioural demands of the social stress are of such high intensity, then many neurochemical systems, among them also endogenous opioid peptides and their receptors, are at their limits, and opoid-modulated ultrasonic calls may represent one of the coping responses in rats.

Noxious physical stimuli (e.g., footshock, pinch) evoke vocalizations in rats that are audible to the human ear as well as ultrasonic vocalizations. These latter ultrasounds precede the impending pain stimulus as well as follow it. The opiate peptides β-endorphin, dynorphin, met- and leu-enkephalin strongly suppressed 22-kHz ultrasounds in reaction to electric foot shock, but left the audible squeaks unaltered (Tonoue et al., 1986). Morphine (3, 6 mg/kg IP) decreased the amount of time that a rat emitted 20-30 kHz ultrasounds during the 5 min interval preceding volleys of brief electrical pulses to the tail (Van der Poel et al., 1989). Naloxone, while ineffective when given by itself, blocked the suppressive effects of morphine on ultrasonic vocalizations preceeding pain stimuli. Van der Poel et al., (1989) also analyzed the temporal structure of pain-related ultrasounds, and found that morphine obliterated the characteristic bout structure of these vocalizations. When given morphine, rats emitted ultrasounds mainly as single pulses. These observations are consistent with the interpretation that ultrasounds preceding the pain stimuli may represent affective expressions that are very sensitive to opoid modulation.

Withdrawal from opiates is highly aversive in humans as well as in laboratory rodents. While the explosive motor symptoms and signs of autonomic "distress" of opiate withdrawal are well known (e.g., Bläsig et al., 1973), it has been difficult to study the affective changes in experimental preparations with laboratory rodents. Recently, we have demonstrated that rats emit 21-25 kHz as well as 44-52 kHz ultrasounds for at least 96 h after their exposure to morphine was terminated (Vivian and Miczek, in press). Particularly in rats with agonistic and reproductive experience, ultrasounds are emitted at high rates during morphine withdrawal suggesting an opioid involvement in their generation.

The presently available pharmacological data implicate μ and δ opiate receptors as important sites for modulation of ultrasonic vocalizations in pups that are separated from dam and littermates as well as in adult male and female rats that are exposed to the social threat and attack. This modulatory role of opiates appears particularly prominent for higher frequency ultrasounds (i.e. 30-60 kHz). The large increase in ultrasounds during opiate withdrawal corroborates the important role of opiate receptors in this potential index of affective expression.

2. Benzodiazepine-type Anxiolytics

Ultrasonic vocalizations in rats have been proposed to serve as potential indices of "anxiety", "fear" or "distress" (e.g., Hofer and Shair, 1978; Tonoue et al., 1986). This interpretation has prompted a series of pharmacological investigations, particularly large-scale efforts to establish this relatively simple measure as screen for drugs with anxiolytic properties (e.g., Gardner, 1985; Mos and Olivier, 1989). The early results were indeed promising, since not only opiates, but also low doses of benzodiazepine-type anxiolytics reduced rat pup ultrasounds (e.g., Gardner and Budhram, 1987, Insel et al., 1986). However, in the meantime, several substances such as MDMA or cocaine that have no clinical anxiolytic action were found to decrease pup ultrasounds, and conversely, some agents such as clonidine with anxiolytic effect actually increase pup ultrasounds (Winslow and Insel, 1990b; Hård et al., 1988). Moreover, it has become apparent that a drug's effect on ultrasounds may be secondary to its primary action on respiratory and thermoregulatory activity.

Rat pups that are separated from dam and littermates and that were handled in a rough manner or exposed to cold ambient temperature, emit a high rate of ultrasounds that are effectively suppressed by low doses of agonists at benzodiazepine receptors (Carden and Hofer, 1990a; Gardner, 1985; Gardner and Budhram, 1987; Insel et al., 1986). Derivatives of benzodiazepines, quinolines and β-carbolines that act as agonists at benzodiazepine receptors decrease the high rate of rat pup ultrasonic vocalizations at low, non-sedative doses (e.g., Gardner and Budhram, 1987; Gardner et al., 1987; Mos and Olivier, 1989). The suppressive effects of benzodiazepines on rat pup ultrasounds are dose-dependently attenuated by antagonists at the benzodiazepine receptor such as flumazenil (e.g., Carden and Hofer, 1990a; Gardner and Budhram, 1987; Insel et al., 1986).

Anxiogenic drugs increase rat pup ultrasounds under appropriate conditions. For example, Insel et al., (1986) showed that the analeptic pentylenetetrazol increased the rate and the power of ultrasonic calls by isolated pups that were exposed to a warm ambient temperature where they normally vocalize very little. Inverse agonists at the benzodiazepine receptors such as FG 7142 increase ultrasounds by pups induced by rough handling (Gardner and Budhram, 1987). It is evident that agonists, inverse agonists and antagonists at the benzodiazepine recepror site modulate ultrasonic calls by separated rat pups in a behaviourally specific, systematic, dose-dependent manner.

Ultrasonic vocalizations by adult rats that are exposed to painful and startle stimuli are attenuated by benzodiazepines. For example, low, non-sedative doses of diazepam decreased ultrasounds in male rats that were presented with electric foot shock (Cuomo et al., 1988; Tonoue et al., 1986; Traber et al., 1988), but not with tail shock (Van der Poel et al., 1989).

Ongoing experiments with acoustic startle stimuli show that 40-50% of Long-Evans rats, either socially naive or previously defeated in a social confrontation, emit ultrasonic vocalizations (Farrell, Vivian and Miczek, unpublished observations). The rats are presented every 30 sec with acoustic startle stimuli, 18 at 105 dB, 18 at 115 dB, on a 60 dB continuous background noise. Diazepam (0.3, 1 and 3 mg/kg IP) decreased ultrasounds both in naive and socially defeated rats. In rats with a social defeat experience, diazepam (0.1-3.0 mg/kg IP) increased the amplitude of the startle response to 115 dB tones, while concurrently suppressing ultrasonic vocalizations.

These findings as well as those with electric foot shock stimuli demonstrate that even in the absence of an audience, rats emit ultrasounds in the 20-30 kHz range and that these ultrasounds can be attenuated by benzodiazepines at doses that do not compromise thermoregulatory and motoric activity. While reflexive responses to pain and startle stimuli remain unaltered or are even amplified, the concurrent benzodiazepine attenuation of ultrasonic vocalizations may reflect an action on affective responses to these noxious stimuli.

Ultrasonic vocalizations in different frequency ranges, some monotonic, some modulated, as brief pulses or in bursts, accompany agonistic behaviour in adult rats, particularly in the submissive rat (Van der Poel and Miczek, in prep.). Our ongoing work begins to identify which types of ultrasonic vocalizations are most readily modulated by benzodiazepines as compared to other substances. Of particular interest is a newly developed experimental preparation that permits the study of behavioural and physiological events before, during and after a significant social challenge (Miczek et al., 1991). Behavioural and physiological measurements are obtained via cardiovascular telemetry, permanent intravenous catheters as well as video- and audio-recordings from an intruder rat that is threatened by potential attacks from a resident but protected from injuries. In a 10-min period before confronting the resident, the intruder starts to emit ultrasonic vocalizations. A very high rate of short-pulse 30-60 kHz ultrasounds is emitted by a rat immediately after it has been physically attacked and while it is being actively threatened and shielded from physical injury. Diazepam (1, 3 mg/kg IP) decreases the rate of 30-60 kHz-ultrasonic calls before the confrontation, but not those calls emitted in the presence of the potential attacker. It is most remarkable that diazepam even at muscle-relaxant doses (6, 10 mg/kg) does not reduce 20-29 kHz ultrasounds when the intruder is actively threatened (Tornatzky and Miczek, 1990). The lack of suppression of 20-29 and 30-60 kHz ultrasounds by diazepam during the threat of attack has been confirmed in rats that are exposed to this intense social challenge either for the first time or on repeated occasions.

The differential effects of diazepam on 20-29 kHz and 30-60 kHz ultrasounds during the phase preceding the actual social challenge vs. the phase of actual intense threat of attack may be relevant to the clinical differentiation between general, anticipatory anxiety and panic attack. Diazepam is very effective in the treatment of general anticipatory anxiety, but considerably less so in panic attacks (Rickels and Schweitzer, 1987). Further pharmacological validation of the mechanisms for different ultrasonic vocalizations may reveal these sounds to represent the evolutionary precursors to human expressions of anxiety.

3. 5-HT_{1A}-type Anxiolytics

Buspirone, ipsapirone and gepirone represent prototypes of a group of anxiolytic substances for which traditional preclinical tests for detecting anxiolytic activity have been largely disappointing (see chapters by Barrett and by Sanger, this volume). Not only do these substances differ from benzodiazepine-type anxiolytics in terms of their effects on animal and human behaviour, but also in terms of their effects on GABA-A vs. 5-HT_{1A} receptors. Ethological approaches have begun in delineating the behavioural actions of 5-HT_{1A}-type anxiolytics (see Blanchard et al., this volume; Mos and Olivier, this volume). Ultrasonic vocalizations by maternally separated rat pups and by environmentally and socially stressed adult rats appear to be exquisitively sensitive to the effects of pharmacological manipulations at the 5-HT_{1A} receptors.

The selective 5-HT_{1A} receptor agonist 8-OH-DPAT, a substance void of anxiolytic activity, decreases ultrasounds in rat pups (Hård and Engel, 1988; Mos and Olivier, 1989). Very potent suppression of pup ultrasonic calls was achieved by 5-HT_{1A} agonists with anxiolytic effects such as buspirone, ipsapirone and flesinoxan (Mos and Olivier, 1989). These calls were virtually eliminated by 5-HT uptake inhibitors such as paroxetine, citalopram and clomipramine (Winslow and Insel, 1990a). 5-HT_1 agonists and 5-HT uptake inhibitors suppress pup ultrasounds in a behaviourally specific manner, i.e. in the absence of any changes in motor activity.

In preliminary studies in adult male rats that were exposed to the threat of attack, very low doses of gepirone (0.3, 0.6 mg/kg IP) produced potent and long-lasting suppression of the rate and duration of 20-29 kHz ultrasonic calls. By contrast, the rate of short "high"-frequency ultrasonic calls remained unaffected even after 10 times higher gepirone doses. The sensitivity of 20-29 kHz-ultrasounds to the suppressive effects of very low doses of gepirone was confirmed in separate series of experiments with rats that experienced social defeat for the first time as well as with those that were repeatedly exposed to this social stress (Vivian and Miczek, in prep.; Tornatzky and Miczek, in prep.).

These very recent, mostly unpublished findings with anxiolytics that act as relatively selective agonists at 5-HT_{1A} point to ultrasonic vocalizations by isolated rat pups and socially stressed adult rats as detecting possibly the anxiolytic activity of these drugs. Moreover, these early results differentiate the effects of benzodiazepine- and 5-HT_{1A}-type on discrete types of rat ultrasounds which may correspond to their differential clinical profiles.

4. Noradrenergic anxiolytics

Severe symptoms of withdrawal from opiates may be alleviated by clonidine treatment in animals and humans (e.g., Fielding et al., 1978; Gold et al., 1981; Redmond and Krystal, 1984). Clonidine's effects on ultrasonic vocalizations during opiate withdrawal in rats have not been investigated so far. However, there have been several demonstrations of clonidine's unusual effects on ultrasonic calls by isolated pups. The original observation by Hård et al., (1985) that clonidine greatly increases pup ultrasounds has been confirmed repeatedly (e.g. Hard et al., 1988, Kehoe and Harris, 1989; Van der Poel et al., 1990). This clonidine effect on pup ultrasounds is antagonized by α_2 antagonists idazoxan or yohimbine (Hard et al., 1988; Kehoe and Harris, 1989) or the 5-HT anxiolytic flesinoxan (Van der Poel et al., 1990, this volume).

In adult rats, clonidine actually prolongs the ultrasounds in anticipation of and in response to a social stress in a newly developed experimental situation (Miczek et al., 1991). Ongoing experiments show that after clonidine (0.03-0.1 mg/kg IP) an intruder rat emits longer 20-29 kHz ultrasonic calls before and immediately during the threats by an attacking opponent (Tornatzky and Miczek, 1990). This duration-increasing effect on ultrasonic vocalizations after clonidine contrasts with the concurrent attenuation of the tachycardia and the hyperthermia and with the minor changes in behavioural defensive responses to the threats and attacks. These clonidine effects differ markedly from those with diazepam and gepirone suggesting differential contributions to the mechanisms mediating discrete ultrasounds by benzodiazepine-, 5-HT- and noradrenergic-type anxiolytics.

D. Summary

Ultrasonic vocalizations in rats occur in several important life situations that span from the maternally separated pup to the phases before and after copulation as well as confrontations with an attacking and threatening opponent and with a predator. Aversive environmental and pharmacological stimuli may engender ultrasonic calls, even in a non-social context. Each type of ultrasonic call occurs in discrete frequency ranges, either as very short pulses or bursts of calls, monotonous or frequency-modulated. In spite of the poorly understood functionality of many of these calls, it is likely that some ultrasonic vocalizations may signal distress, represent affective expressions, or serve as warning calls. The neurobiological mechanisms need to be traced from the pontine nuclei innervating the larynx to periaqueductal grey area, diencephalic and forebrain nuclei. Neuropharmacological investigations point to important roles of opioid, benzodiazepine-GABA-A, 5-HT and noradrenergic receptors in the mechanisms underlying different types of ultrasonic calls in rats. Drugs with antidepressant, anxiolytic and anti-stress effects appear to be particularly effective in modulating ultrasonic calls by infant and adult rats.

REFERENCES

Allin, J.T., Banks, E.M. (1971) Devl. Psychobiol. 4, 149-156.
Anderson, J.W. (1954) Science 119, 808-809.
Bandler, R. (1988) In: Progress in Psychobiology and Physiological Psychology (A.N.Epstein, A.R.Morrison, Eds), Academic Press, New York, pp 67-154.
Barfield, R.J., Geyer, L.A. (1972) Science 176, 1349-1350.
Barfield, R.J., Geyer, L.A. (1975) J. Comp. Physiol. Psychol. 88, 723-734.
Barfield, R.J., Thomas, D.A. (1986) In: Reproduction: A Behavioural and Neuroendocrine Perspective (B.R.Komisaruk, H.I.Siegel, H.H.Feder, Eds), Annals of the New York Academy of Sciences, Vol 475. New York, pp 33-43.
Blanchard, R.C., Blanchard, D.C., Weiss, S., Agullana, R. (1990) Soc. Neurosci. Abstr. 16, 598.
Bläsig, J., Herz, A., Reinhold, K., Zieglgansberger, S. (1973) Psychopharmacologia 33, 19-38.
Cagiano, R., Barfield, R.J., White, N.R., Pleim, E.T., Cuomo, V. (1989) Pharmacol. Biochem. Behav. 34, 53-58.
Cagiano, R., Sales, G.D., Renna, G., Racagni, G., Cuomo, V. (1986) Life Sci. 38, 1417-1423.
Carden, S., Barr, G., Hofer, M. (1990) Soc. Neurosci. Abstr. 16, 1192.
Carden, S.E., Hofer, M.A. (1990a) Behav. Neurosci. 104, 160-166.
Carden, S.E., Hofer, M.A. (1990b) Behav. Neurosci. 104, 457-463.
Corrigan, J.G., Flannelly, K.J. (1979) J. Comp. Physiol. Psychol. 93, 105-115.
Cuomo, V., Cagiano, R., Desalvia, M.A., Restani, P., Galimberti, R., Racagni, G., Galli, C.L. (1988) Neurotoxicol. Teratol. 10, 465-469.
Fielding, S., Wikler, J., Hynes, M., Szewczak, M., Novick, W.J., Lal, H. (1978) J. Pharmacol. Exp. Ther. 207, 899-905.
Gardner, C.R. (1985) J. Pharmacol. Meth. 14, 181-187.
Gardner, C.R., Budhram, P. (1987) Eur. J. Pharmacol. 134, 275-283.
Gardner, C.R., Deacon, R., James, V., Parker, F., Budhram, P. (1987) Eur. J. Pharmacol. 142, 285-295.
Gold, M.S., Pottash, A.L.C., Extein, I., Stoll, A. (1981) In: Problems of Drug Dependence, 1980 (L.S.Harris, Ed), (NIDA Research Monograph Series, Vol. 34). National Institite on Drug Abuse, Rockville, MD, pp 95-100.
Hård, E., Engel, J. (1988) Neuropharmacology 27, 981-986.
Hård, E., Engel, J., Larsson, K., Liljequist, S., Musi, B. (1985) Acta Pharmacol. Toxicol. 56, 354-363.
Hård, E., Engel, J., Lindh, A.S. (1988) J. Neural Transm. 73, 217-237.
Hess, W.R. (1948) Das Zwischenhirn. Syndrome, Lokalisationen, Funktionen. Benno Schwabe and Co., Basel.
Hofer, M.A., Shair, H. (1978) Devl. Psychobiol. 11, 495-504.
Insel, T.R., Hill, J.L., Mayor, R.B. (1986) Pharmacol. Biochem. Behav. 24, 1263-1267.
Jürgens, U. (1988) In: Primate Vocal Communication (D.Todt, P.Goedeking, D.Symmes, Eds), Springer-Verlag, Berlin, pp 163-167.
Kaltwasser, M.T. (1985) Ph.D. dissertation. University of Tübingen, Tübingen.

Kaltwasser, M.T. (1990a) Psychopharmacology 101, S28.
Kaltwasser, M.T. (1990b) Physiol. Behav. 48, 13-17.
Kehoe, P. (1989) In: Molecular Biology of Stress (S.Breznitz, O.Zinder, Eds), Alan R. Liss, New York, pp 307-317.
Kehoe, P., Blass, E.M. (1986) Behav. Neurosci. 100, 624-630.
Kehoe, P., Boylan, C., Shoemaker, W. (1990) Soc. Neurosci. Abstr. 16, 211.
Kehoe, P., Harris, J.C. (1989) Behav. Neurosci. 103, 1099-1107.
Levine, J.D., Feldmesser, M., Tecott, L., Gordon, N.C., Izdebski, K. (1984) Brain Res. 296, 121-127.
Lore, R., Flannelly, K. (1977) Sci. Am. 236, 106-116.
Miczek, K.A., Thompson, M.L., Tornatzky, W. (1991) In: Stress: Neurobiology and Neuroendocrinology (M.Dekker, Ed), New York, pp 323-357.
Mos, J., Olivier, B. (1989) In: Behavioural Pharmacology of 5-HT (P.Bevan, A.R.Cools, T.Archer, Eds), Lawrence Erlbaum Associates, Hillsdale, NJ.
Newman, J.D. (1988) In: Primate Vocal Communication (D.Todt, P.Goedeking, D.Symmes, Eds), Springer-Verlag, Berlin, pp 146-153.
Noirot, E. (1966) Anim. Behav. 14, 459-462.
Noirot, E. (1968) Anim. Behav. 16, 129-134.
Nottebohm, F. (1981) Science 214, 1368-1370.
Nottebohm, F. (1985) Hope For a New Neurology. Annals of the New York Academy of Sciences, Vol 457. New York.
Panksepp, J., Herman, B., Conner, R., Bishop, P., Scott, J.P. (1978) Biol. Psychiatry 13, 607-618.
Peters, R.H., Koch, P.C., Blythe, B.L., Sufka, K.J. (1988) Physiol. Behav. 44, 749-751.
Ploog, D. (1988) In: Primate Vocal Communication (D.Todt, P.Goedeking, D.Symmes, Eds), Springer-Verlag, Berlin, pp 195-212.
Redmond, D.E., Krystal, J.H. (1984) Ann. Rev. Neurosci. 7, 443-478.
Rickels, K., Schweitzer, E.E. (1987) In: Psychopharmacology: The Third Generation of Progress (H.Y.Meltzer, Ed), Raven Press, New York, pp 1193-1201.
Roberts, L.H. (1975) J. Zool. (London) 175, 243-257.
Robinson, T.E., Becker, J.B., Moore, C.J., Castaneda, E., Mittleman, G. (1985) Brain Res. 343, 374-377.
Sales, G.D. (1979) Am. Zool. 19, 513-527.
Sales, G.D. (1972) Anim. Behav. 20, 88-100.
Sewell, G.D. (1967) Nature 215, 512.
Sewell, G.D. (1970) Nature 217, 682-683.
Siegel, A., Pott, C.B. (1988) Progr. Neurobiol. 31, 261-283.
Takahashi, L.K., Thomas, D.A., Barfield, R.J. (1983) J. Comp. Psychol. 97, 207-212.
Thomas, D.A., Takahashi, L.K., Barfield, R.J. (1983) J. Comp. Biol. 97, 201-206.
Thomas, D.A., Talalas, L., Barfield, R.J. (1981) J. Comp. Physiol. Psychol. 95, 630-637.
Tonoue, T., Ashida, Y., Makino, H., Hata, H. (1986) Psychoneuroendocrinology 11, 177-184.
Tornatzky, W., Miczek, K.A. (1990) Soc. Neurosci. Abstr. 16, 432.
Traber, J., Schuurman, T., Benz, U. (1988) Soc. Neurosci. Abstr. 14, 556.
Van der Poel, A.M., Molewijk, E., Mos, J., Olivier, B. (1991) In: Animal Models in Psychopharmacology (B.Olivier, J.Mos, J.L.Slangen, Eds), Birkhäuser Verlag, Basel, present volume.
Van der Poel, A.M., Mos, J., Molewijk, E., Olivier, B. (1990) Eur. J. Pharmacol. 183, 1980-1981.
Van der Poel, A.M., Noach, E.J.K., Miczek, K.A. (1989) Psychopharmacology 97, 147-148.
Vivian, J.A., Miczek, K.A. (1990) Soc. Neurosci. Abstr. 16, 443.

Vivian, J.A., Miczek, K.A. (In press) Ultrasounds during morphine withdrawal in rats. Psychopharmacology.
White, N.R., Barfield, R.J. (1987) J. Comp. Psychol. 101, 73-81.
White, N.R., Cagiano, R., Moises, A.U., Barfield, R.J. (1990) J. Comp. Psychol. 104, 255-262.
Winslow, J.T., Insel, T.R. (1990a) J. Pharmacol. Exp. Ther. 254, 212-220.
Winslow, J.T., Insel, T.R. (1990b) Neuropsychopharmacology 3, 51-59.
Winslow, J.T., Insel, T.R. (1991) Behav. Neurosci. 105, 31-41.
Yajima, Y., Hayashi, Y., Yoshi, N. (1981) Brain Res. 229, 353-362.
Yajima, Y., Hayashi, Y., Yoshi, N. (1980) Brain Res. 198, 446-452.

Animal Models in Psychopharmacology
Advances in Pharmacological Sciences

MINIMALIZATION OF ANIMAL USE IN MODELS OF PSYCHOPHARMACOLOGY: A MODIFIED "UP&DOWN" METHOD FOR ED_{50} ESTIMATION

H.E. Molewijk, A.M. van der Poel, J. Tolboom[#], B. Olivier

DUPHAR B.V., Dept. of Pharmacology, [#]Dept. of Statistics, P.O.Box 900, 1380 DA Weesp, The Netherlands

SUMMARY: Here we propose an experimental design to rapidly and accurately quantify potencies of large numbers of drugs, by means of the median effective dose (ED_{50}), with minimal use of laboratory animals. This modification of the "up&down" method (Dixon, 1965) attempts to more precisely define an ED_{50} with confidence intervals and consists of additional requirements to establish 1. an upper and lower limit and 2. at least two intermediate doses. Compared to "classical" procedures, comprising a pilot and a multi-dose study, we could reduce animal use with 50 to 75%.

INTRODUCTION

Traditionally in *in vivo* pharmacology, regardless the exact nature of the response, the estimation of an ED_{50} of an unknown drug demands two experimental phases. First, a pilot study on a small sample is required to select a dose-range which includes both the minimal and the maximal response. Second, a multi-dose study is performed, resulting either in a log dose-response curve or a log dose-percent curve, depending on whether the response is graded or quantal (absent/present) respectively. In the log dose-response curve the mean intensity of the response is plotted against the log dose. The log dose-percent curve however, is the result of plotting the percentage of individuals that exhibit the specified effect as a function of the log dose. Both these curves can assume various mathematical relationships, but are commonly sigmoidal. Therefore, besides a non-effective and a maximal effective dose, at least two intermediate-effective doses are needed to define the slope of the sigmoidal curve. In case of the graded response the particular dose that would produce a half maximal effect is often referred to as "ED_{50}". The meaning of the "ED_{50}" in this sense is derived from the "K_d" in ligand-receptor binding interactions. Unlike the *in vitro* pharmacology however, the *in vivo* responses are composed of several effects of the drug and thus need not directly be related to receptor occupancy. Originally the ED_{50} was defined as the "median effective dose", being the dose of drug required to produce a specified intensity of effect in 50% of the individuals.

Both these methods however, require at least 6 animals per drug dose. The number of animals to be used would approximately be: in the pilot-study 10 animals, in the dose-response experiment 4x6 animals, plus 6 vehicle treated control animals, which adds up to 40 animals per ED_{50} estimation.

METHODS

In this study the response variable was a behavioural phenomenon, called "lower lip retraction", which is an "in vivo" measure of $5\text{-}HT_{1A}$ agonism in the rat (Berendsen et al., 1989; Molewijk et al., 1989). Although this is a -more or less- graded response, it was transformed into a quantal response by setting a certain criterion. Rats were ip injected with drugs and the lower lip retraction response was assessed 10, 20 and 30 minutes after drug administration. The response was scored on an arbitrary 0-2 scale and scores for the three observations were summed. A particular rat was designated as a positive responder when total scores were greater or equal to 2.

The first step towards minimalization of animal use is the integration of the pilot and the multi-dose study. This can be achieved by using the classical "up&down" method for ED_{50} estimation (Dixon, 1965), which implies a sequential (multi-trial) design where each positive response is followed by testing a lower dose and each negative response is followed by testing a higher dose in the next trial. Thus, range-finding is automatically performed during the ED_{50} estimation procedure. Furthermore, the accuracy of the ED_{50} estimation is ascertained since most of the information is collected near the "induction threshold" (ED_{50}) of the quantal response. However, due to this last property, the method is less suited to determine the standard error in the ED_{50} estimate. In addition, depending on the spacing between the dose levels, a situation could occur in which a dose with just negative responses is followed by a dose with just positive responses. In that case the slope of the curve, and consequently the ED_{50} estimate and its variance, can not properly be determined.

The modified "up&down" method proposed here attempts to more precisely define the ED_{50} estimate with confidence intervals and consists of additional requirements to establish 1. an upper and a lower limit dose, defined by 3 just positive or negative responses and 2. at least two intermediate doses provoking positive as well as negative responses. For the selection of the dose level at the next trial the "up&down" method is used as a guide. When these criteria are met, the log dose-percent curve can be constructed. Hence the ED_{50} is estimated, using a logistic regression method (GENSTAT) which (rightly) assumes a binomial distribution of the response. The 95% confidence interval for the ED_{50} is determined by means of the so-called delta-method (GENSTAT).

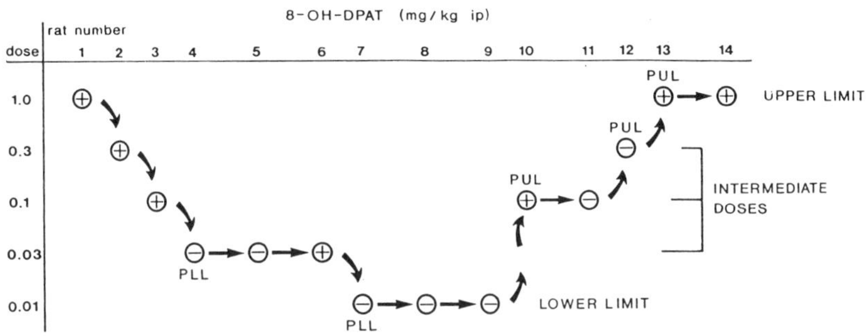

Fig. 1 Example of application of the "modified up&down" method in assessing the ED_{50} for the induction of the lower lip retraction response by 8-OH-DPAT after ip administration in the male Wistar rat. P.U.L. = potential upper limit , P.L.L. = potential lower limit.

RESULTS

In Fig. 1 a random example of an experiment using the "modified up&down" method is presented. The train of sessions starts with a fixed dose and fixed dose level-steps of $^{1}/_{2}$ log unit. When the lower lip retraction response is present at this dose, lower doses are applied until a negative response emerges. It is than investigated whether this dose represents the "lower limit". If positive responses should occur at such a dose, this dose apparantly is not a lower limit, but an "intermediate" dose. This procedure is repeated until three just negative responses at the same dose level have been obtained. Dosages are then raised in order to find the "upper limit". In the example-experiment using 8-OH-DPAT on ip administration, two potential upper limits turned out to be intermediate doses. Eventually the lower limit in this example was 0.01 mg/kg ip, the intermediate doses were 0.03, 0.1 and 0.3 mg/kg ip and the upper limit was 1.0 mg/kg ip. Plotting the log dose-percent curve and consulting the logistic regression method, the ED_{50} was estimated at 0.1 mg/kg ip. The 95% confidence interval for this estimate was 0.03-0.5 mg/kg ip. For this ED_{50} and confidence interval we used 14 rats.

In cases were less than two intermediate doses could be obtained using $^1/_2$ log dose steps, additional steps of $^1/_4$ or $^1/_6$ log units were inserted.

Table 1 ED_{50} estimates with 95% confidence intervals and number of animals used, for the induction of the lower lip retraction response by several 5-HT_{1A} (partial) agonists.

drug	ED_{50} (mg/kg ip)	confidence-interval	number of animals
8-OH-DPAT	0.1	0.03 - 0.5	14
flesinoxan	0.3	0.2 - 0.4	15
buspirone	1.0	0.4 - 2.7	13
ipsapirone	1.0	0.5 - 1.9	12
gepirone	1.7	0.6 - 5.2	10
BMY 7378	1.0	0.5 - 2.1	19

In table 1 more examples of the application of the modified "up&down" method are depicted, using well known 5-HT_{1A} (partial) agonist drugs. For our goals these ED_{50} values and confidence intervals were acceptable. And more important, the number of rats to be used was only 10 to 20 per ED_{50}, which means a reduction in animal use of 50 to 75% compared to classical ED_{50} estimation procedures.

DISCUSSION

Using sequential data acquisition, every animal is tested based on the information already provided by its predecessor(s), which enables concentration of observations at doses that are most informative. The ED_{50} estimates are now more precise than when subjects had been distributed in equal numbers to four doses. Obviously this efficient allocation of subjects reduces animal use and sometimes also testing time and amounts of drug.

The additional advantage of the modification of the "up&down" method proposed here is that more attention is paid to the definition of the eventual log dose-percent curve. The requirement to determine the upper and lower limit and intermediate doses promotes the accuracy of the ED_{50} estimate, because the slope of the sigmoidal curve can be better described. Depending on the specific question to be answered by the experiment, these criteria can of course be changed.

We have not performed direct comparisons with "classical" ED_{50} determinations, using identical sampling-times and criteria to define the presence/absence of this response, or replications of the ED_{50}s presented in Table 1. We can however refer to some extent to previous work with sc administration of 5-HT_{1A} agonists, in which the lower lip retraction response was measured in a 4 treatment study using 7 rats per dose on a slightly different time schedule (Molewijk et al., 1990). 8-OH-DPAT was found to be about 3 to 10 times more potent than flesinoxan and buspirone, and about 10 to 30 times more potent than ipsapirone. This largely parallels the results obtained using the "modified up&down" method.

It should be noted that in a sequential design treatments can only be given in series, not parallel, and the experimenter has to be prepared to change dose levels at every new session. If these factors do not trouble the set-up of experiments, the "modified up&down" method can be extremely efficient in terms of animal use in high capacity screens, where many ED_{50}s have to be estimated.

REFERENCES

Berendsen, H.H.G., Jenck, F., Broekkamp, C.L.E. (1989) Pharmacol. Biochem. Behav. 33, 821-827.

Dixon, W.J. (1965) J. Am. Statis. Ass. 60, 967-978.

Molewijk, H.E., van der Heyden, J.A.M., Olivier, B. (1989) Eur. J. Neurosci. suppl. 2, 64.23.

Molewijk, H.E., van der Heyden, J.A.M., Olivier, B. (1990) Psychopharmacology, 101 (suppl.): 141.

Animal Models in Psychopharmacology
Advances in Pharmacological Sciences

SEROTONIN AND PAIN; EFFECT OF FLUVOXAMINE

J. Gibert-Rahola, J. Casas, M.C. Gómez-Cama, J. Elorza, and J.A. Micó

Department of Neurosciences, School of Medicine, Cádiz, Spain

INTRODUCTION

Considerable evidence exists to implicate a role for the neurotransmitter 5-HT in the modulation of nociceptive transmission. Activation of the descending, serotoninergic bulbospinal system produces inhibition of behavioural and dorsal horn neuronal responses to noxious stimuli. For instance, increasing postsynaptic 5-HT receptor activity by administration of selective 5-HT uptake inhibitors has been reported. Antinociception as well as no effect in the hot plate and tail flick tests have been demonstrated in rats.

Moreover, in mice some selective 5-HT uptake inhibitors produced hyperalgesia in the formalin test. Antidepressants with 5-HT uptake inhibiting properties have been clinically used in the treatment of chronic pain. On the other hand, sex differences in the sensitivity of mice to serotonin agonists have been reported, females being more susceptible than males. The following experiments were undertaken to characterize the effects of fluvoxamine, a selective inhibitor of 5-HT uptake on several pain tests in male and female mice.

MATERIAL AND METHODS

Animals: Mice of the OF1 strain provided by the central service breeding facilities of the university of Cádiz, in the weight range of 20-30 gr, were fed on a PANLAB A04

diet. Food and water were "ad libitum" available. The temperature was constantly held at 21°C. The animals were placed in the experimental room 24 hours before the tests. A light-darkness cycle for 12 hours (8 a.m.-8p.m.) was kept.

Acetic acid method: (Koster et al., 1959). Mice are injected with an i.p. 0.8% acetic acid solution and a 6 minutes period is established before recording the numbers of contortions during 2 minutes.

Hot plate method: (Woolfe & McDonald 1944). Mice are exposed to a hot plate (Digital Ds37 socrel model) of 57°C. The criterion for pain is forelpaw licking. If this does not occur within 30 seconds exposure ("cut off"), the test is considered as concluded in order to avoid damage to the animal.

Tail flick method: (d'Amour & Smith 1941). Mice are exposed to a lamp (100 W) overheaded by a LI 7106 tail flick model with automatic arrest by photoelectric cell and a digital time counter. The tail of the mouse is exposed to this lightbeam on three ocassions after one minute each. Determinations are done at three differents points of the tail; their mean being taken as the pain latency. If the animals would not respond after 10 seconds exposure, the test is cut off in order to avoid damage to the animal.

Formalin method: (Dubuisson & Dennis 1977). 20 microlitres of 1% formol-saline was injected subcutaneously into the dorsal surface of the mouse's hindpaw. Inmediately thereafter an observer recorded the amount of time the injected paw was licked during 2 minutes.

Electrical tail stimulation test: (Nilsen, 1961). Two hypodermic needles were implanted at 1 cm from the base of the tail as electrodes. The two electrodes were connected to a stimulator (HUGO SACHS, type 215/II). The voltage consisting of 10 msec rectangular monophasic pulsed was increased by one unit every second until the threshold causing the first squeak was reached. The maximum voltage was 70 V.

Drugs: fluvoxamine (FVX) provided by Duphar was i.p. injected 30 minutes before starting the tests at the doses of 2.5; 5; 10; 20 and 40 mg/kg. The volume of injection was 0.25 ml/25 gr of weight.

Statistics: Results obtained are expressed as the mean ± S.E.M. and for the statistical comparison the non parametric Mann-Whitney's U test is employed. $p < 0.05$ was considered as the significant threshold.

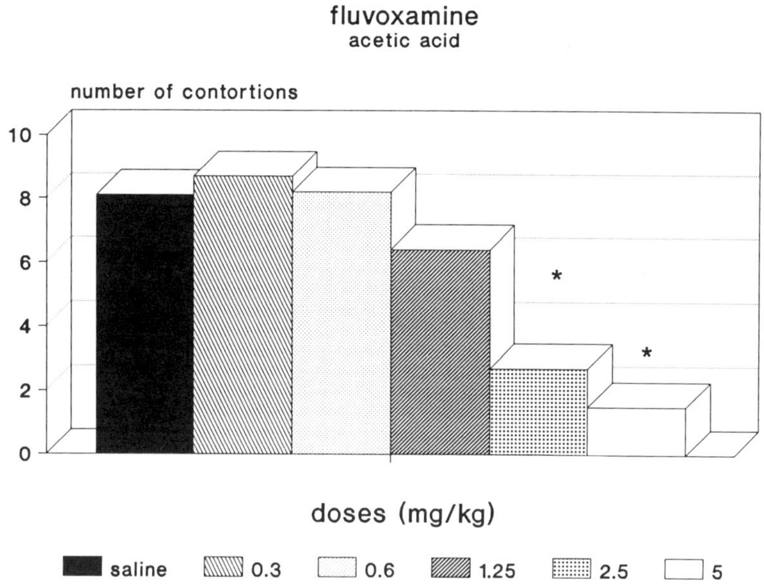

Fig. 1 Effect of fluvoxamine on female mice in acetic acid test. *$p < 0.01$ vs saline.

Fig. 2 Effect of fluvoxamine on male mice in acetic acid test. *$p < 0.01$ vs saline.

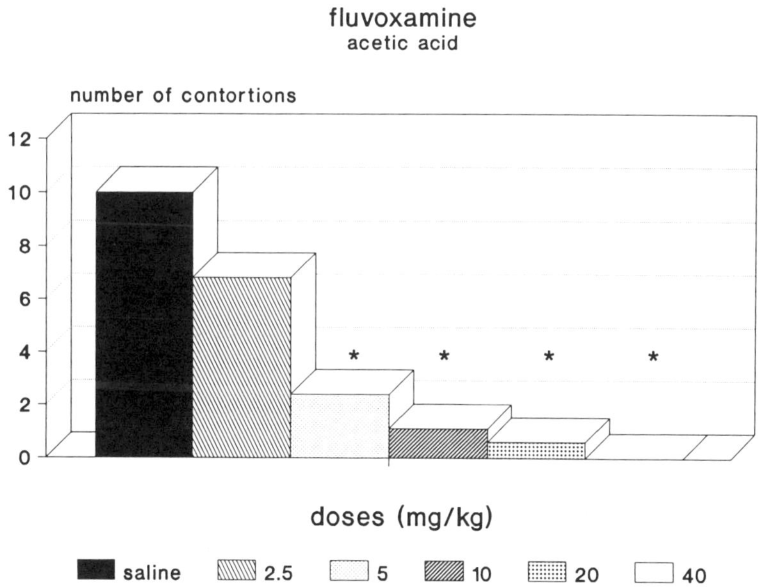

RESULTS

Acetic acid method: When FVX was injected to female mice (fig. 1) the number of contortions was reduced after the dose of 2.5 mg/kg (U=O; $p < 0.01$). In male mice (fig. 2) it was effective with 5 mg/kg (U=2; $p < 0.01$); with the highest dose employed (40 mg/kg) contortions were completely inhibited.

Hot plate method: FVX exerts analgesic effect at the dose of 20 mg/kg (U=16; $p < 0.05$) in male mice (fig 3). In females (fig 4) the analgesic effect was observed with the dose of 2.5 mg/kg, (U=20; $p < 0.05$) and an apparent dose effect was established.

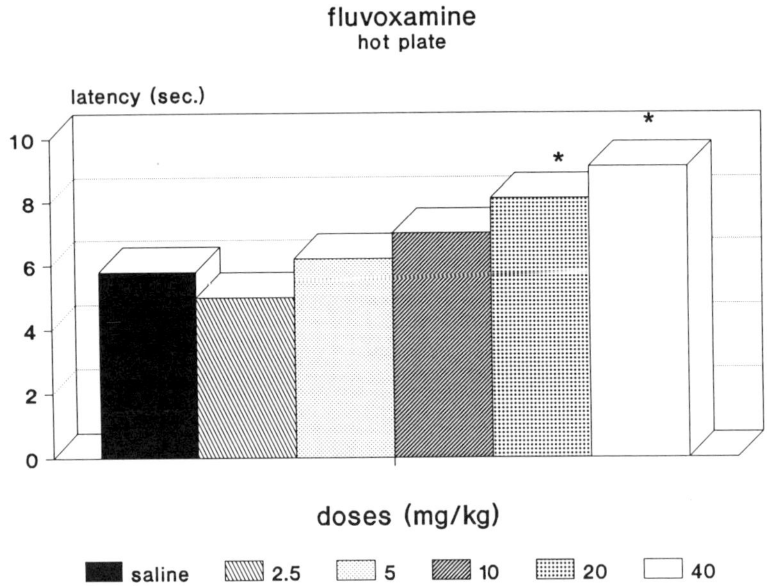

Fig. 3 Effect of fluvoxamine on male mice in hot plate test. *$p < 0.01$ vs saline.

Tail flick method: In this test, when FVX is injected to male mice (fig 5) no analgesic statistically significant effect is shown. At intermediate doses (5, 10 and 20 mg/kg), a moderate increase of the pain threshold was observed, but without statistical significance (U=34.5, U=30 and U=34.5, respectively). Nevertheless in female mice (fig 6) analgesia was enhanced with the doses of 20 mg/kg (U=17; $p < 0.01$).

Formalin method: In this test FVX exerts analgesic effects from the dose of 0.3 mg/kg in female mice (U=4; $p < 0.01$), (fig 7) and from the dose of 2.5 mg/kg on in male mice (U=21; $p < 0.05$) (fig 8).

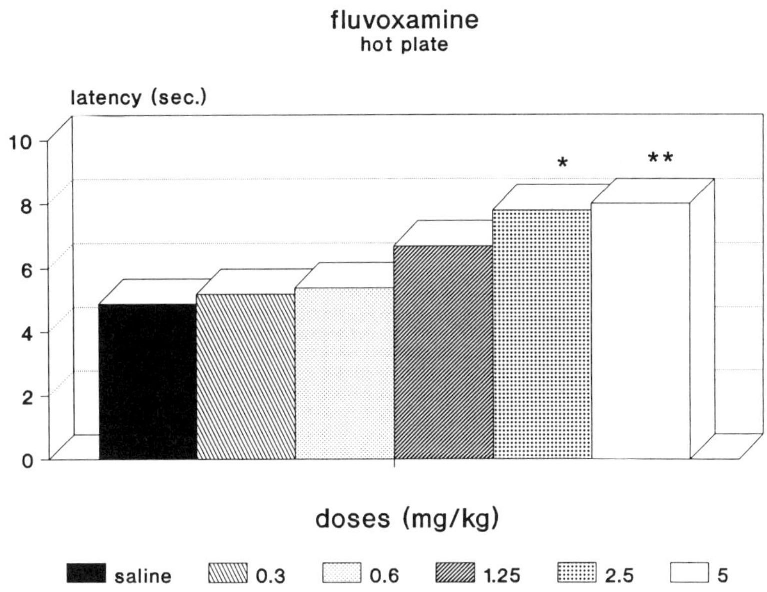

Fig. 4 Effect of fluvoxamine on female mice in hot plate test. *p<0.05 & **p<0.01 vs saline.

Fig. 5 Effect of fluvoxamine on male mice in tail flick test.

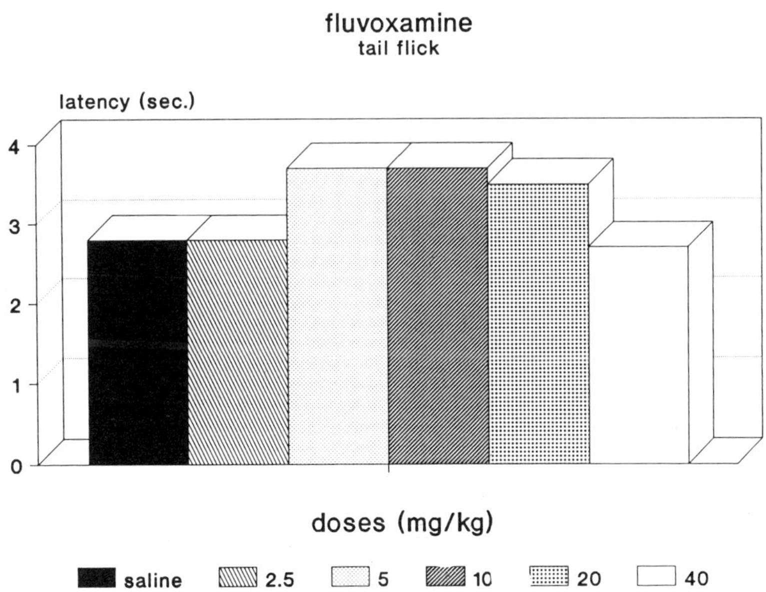

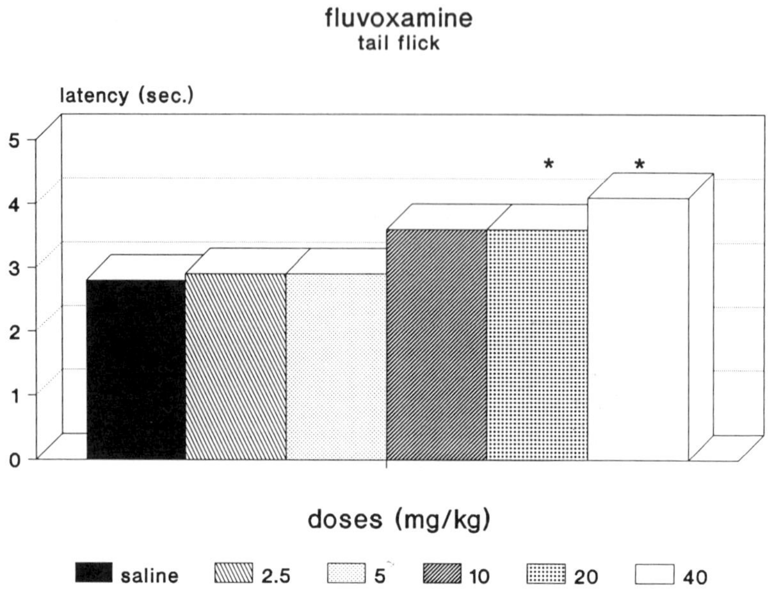

Fig. 6 Effect of fluvoxamine on female mice in tail flick test. *p <0.01 vs saline.
Fig. 7 Effect of fluvoxamine on female mice in formalin test. *p <0.01 vs saline.

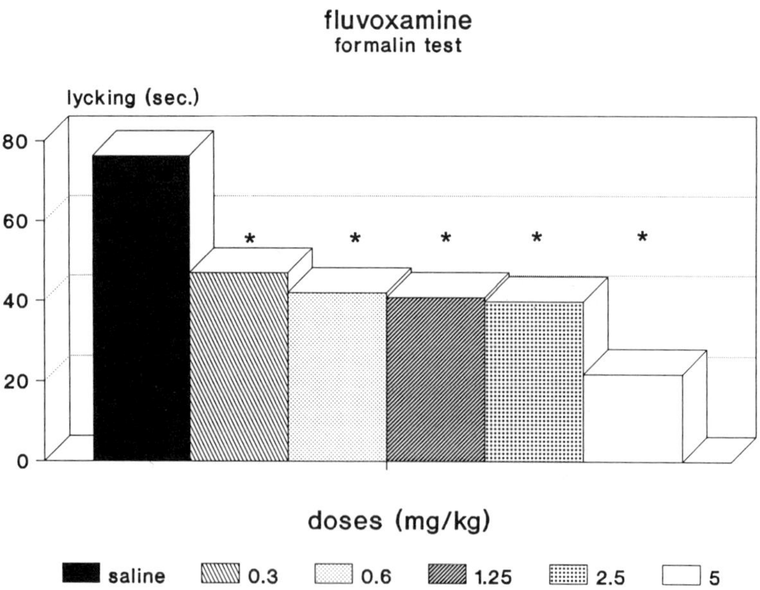

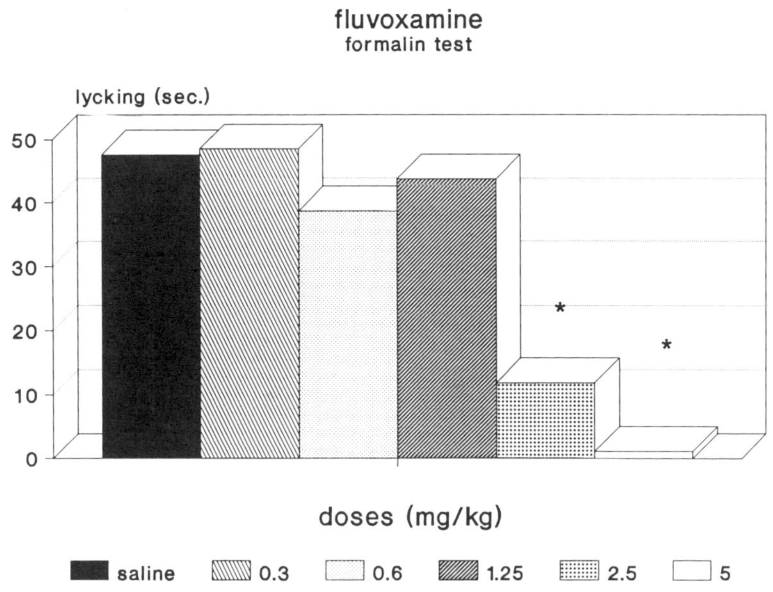

Fig. 8 Effect of fluvoxamine on male mice in formalin test. *p <0.01 vs saline.
Fig. 9 Effect of fluvoxamine on male mice in electrical tail stimulation test.

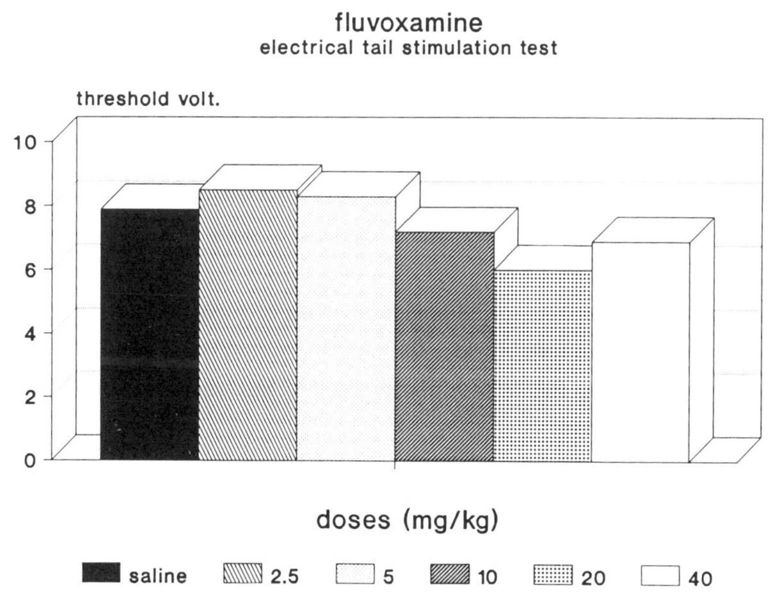

INDEX

D

T